Workshop and International Seminar on Science of Complex Natural Systems

IOP Conference Series: Earth and Environmental Science
Volume 31

Bogor, Indonesia
9 – 10 October 2015

Editor:

Hendradi Hardhienata

ISBN: 978-1-5108-2226-9
ISSN: 1755-1307

Printed from e-media with permission by:

Curran Associates, Inc.
57 Morehouse Lane
Red Hook, NY 12571

Printed by Curran Associates, Inc. (2016)

For permission requests, please contact the Institute of Physics at the address below.

Institute of Physics
Dirac House, Temple Back
Bristol BS1 6BE UK

Phone: 44 1 17 929 7481
Fax: 44 1 17 920 0979

techtracking@iop.org

Additional copies of this publication are available from:

Curran Associates, Inc.
57 Morehouse Lane
Red Hook, NY 12571 USA
Phone: 845-758-0400
Fax: 845-758-2633
Email: curran@proceedings.com
Web: www.proceedings.com

Table of contents

Volume 31

Workshop and International Seminar on Science of Complex Natural Systems
9–10 October 2015, Bogor, Indonesia

Accepted papers received: 6 January 2016
Published online: 22 February 2016

Preface

Papers

IOP Conf. Series: Earth and Environmental Science **31** (2016) 011001 doi:10.1088/1755-1315/31/1/011001

ISS-CNS **Workshop and International Seminar on Science of Complex Natural Systems**

"Understanding Complexity for Sustainable Agriculture and Green Development"

October 09-10, 2015, IPB Science Park, Taman Kencana IPB Campuss, Bogor, Indonesia

Preface

The International Seminar on Science of Complex Natural Systems 2015 (ISS-CNS 2015 was organized by the Faculty of Mathematics and Natural Sciences, Bogor Agricultural University. This event aims at sharing knowledge and expertise, as well as building network and collaborations among scientists from various institutions at national and international level. ISS-CNS 2015 was organized as a scientific seminar, which covers a keynote speech, some invited talks, and some contributed presentations. The parallel presentations came from 67 papers, which were presented in the form of oral as well as poster presentations. These papers are contributed from scientist from various research and higher education institutions in Indonesia as well as some neighboring countries. The talks were categorized into three categories namely Mathematical and Computational Modeling and Big Data (MCMB), Agricultural, Biological and Environmental System (ABES), and Socio-Economics Network Systems (SENS).

In this occasion, we were honored to have our distinguished keynotespeaker Prof. Peter M. A. Sloot, Director of Complexity Institute (Nanyang Technological University, Singapore) and invited speakers Prof. Stéphane Douady (Université Paris Diderot, France), Dr. Reed Beaman (University of Florida, USA), Dr. Ahmad Faqih (Bogor Agricultural University, Indonesia), and Dr. Roby Muhamad (University of Indoensia, Indonesia)

We also thank all participants for their contributions especially those who submit their full papers to the editorial board. In total, 44 papers are published and we are honored to have authors coming from various institutions around the world namely Netherland, Austria, Mexico, France, and Indonesia. Finally, we also thank our reviewers for maintaining a high quality standard of the accepted manuscripts for publication.

Bogor, December 2015

Dr. Husin Alatas
ISS-CNS 2015 Chairperson
Bogor Agricultural University (IPB)

Dr. Hendradi Hardhienata
ISS-CNS 2015 Proceeding Editor
Bogor Agricultural University (IPB)

IOP Conf. Series: Earth and Environmental Science **31** (2016) 011002 doi:10.1088/1755-1315/31/1/011002

Peer review statement

All papers published in this volume of *IOP Conference Series: Earth and Environmental Science* have been peer reviewed through processes administered by the proceedings Editors. Reviews were conducted by expert referees to the professional and scientific standards expected of a proceedings journal published by IOP Publishing.

IOP Conf. Series: Earth and Environmental Science **31** (2016) 012001 doi:10.1088/1755-1315/31/1/012001

Influence of Contactopy on Two-Dimensional Brazil-Nut Effect

Teja Kesuma*[1], Dimas Praja Purwa Aji[2], Sparisoma Viridi[3], Suprijadi[4]
[1, 2] Master Program in Computational Science, Institut Teknologi Bandung, Indonesia
[3, 4] Department of Physics, Institut Teknologi Bandung, Indonesia

E-mail: [1] teja.kesuma@gmail.com, [2] dmspraja2105@gmail.com,
[3] dudung@fi.itb.ac.id, [4] supri@fi.itb.ac.id

Abstract. Using hard sphere collision scheme and molecular dynamics method simulation, two-dimension Brazil-Nut Effect (BNE) is conducted. BNE inital configurations are artificially created to obtain number of contacts between the grains, called as contactopy. It is observed that from initial condition until the end of BNE observation, center-of-mass in vertical direction tends to decrease, while contactopy tends to increase, which are similar to the reported results in a *Gedankenexperiment* (Viridi et al., 2014). Initial configurations which are also dependent on contactopy influence the rise time since it can override influence of density ratio that should affect rise time (Möbius et al., 2001). We investigate on how the intruder rise time influenced by contactopy, as well as density.

1. Introduction

Brazil-Nut Effect (BNE) is a granular material phenomena, where larger grains (usually known as intruder) could rise, reaching the surface of the granular system when vibrated. Some results of experiments have shown the that intruder is able to rise due to the system normalized acceleration factor, Γ, which is a function of the vibrational frequency and amplitude [1]. It is possible to design BNE observation environment under two- [2,7] or three-dimensional system [3]. When BNE phenomenon occured, it has been investigated that the intruder rise time, t_{rise}, is influenced by the density ratio between the intruder and the granular bed [5].

Computer simulation via molecular dynamics (MD) has been applied to predict a variety of BNE behaviour types [4]. Influence of BNE properties, such as grain and vibrational parameters, can be measured comprehensively [6]. Furthermore, initial configurations which are dependent on position of the grains can be artificially designed to observe the trajectory as well as its rise time. This approach implies a wide opportunity to explore related mechanism, describing the effect of system parameters of the phenomena.

In this paper, we report the results of our simulation investigating the intruder rise time influenced by initial number of contacts between the grains (called as contactopy), for certain ratio density on two-dimensional granular bed. The simulation method and the initial configuration setup is presented in §2. We show the evolution of intruder position, contactopy, and average rise time in §3. Based on our trend analysis on the results, we describe future works and conclusions in §4.

* To whom any correspondence should be addressed.

2. Simulation Method

A single intruder two-dimensional BNE system is modeled using MD simulation of hard spheres collision scheme. The total number of grains (intruder and granular bed) is 157. The grains are placed inside a two-dimensional rectangle container. The width of the container is fixed, while the container base fluctuates following the sinusoidal motion when vibrated. Diameter ratio between the intruder and the granular bed is $d / d_m = 7.5$ and the density ratio ρ / ρ_m is varied from 0.1 to 1.8. The restitution coefficient is 0.99 for the grains and 0.26 for the container. There is no friction force acting between the grains. The coefficient of friction for the container is 0.98.

The simulation is initialized by arranging in order the granular bed above the intruder. The container is not vibrated at this stage as the granular bed move downwards and condensed under gravity. Snapshots of the situation for this initial configuration is shown in figure 1.

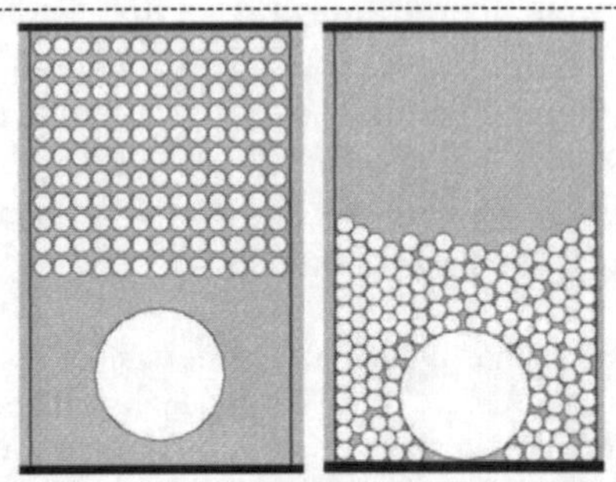

Figure 1. Configuration of BNE system. (left) granular bed initially in order on top. (right) Condensation state.

The contactopy, C, can be determined from its geometry as

$$C = \sum_{i=0}^{N}\sum_{j>i}^{N}\left[\frac{\max(0, R_i + R_j - |r_i - r_j|)}{R_i + R_j}\right] \quad (1)$$

which describe the number of contacts between the grains, where R is the grain radius and r is the distance from a reference point to its center of mass [8]. We could generate different contactopy from the initial configuration by setting up certain arrangement of the granular bed (table 1).

Table 1. Grains configuration and its relationship with C.

Configuration	Contactopy
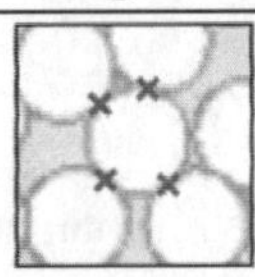	4
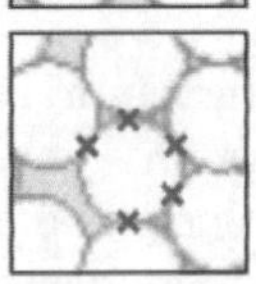	5
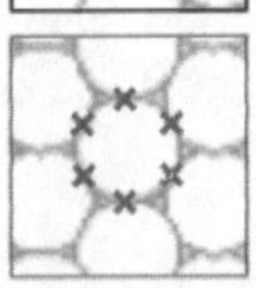	6

From this condensation state, the initial configuration is recorded. As the container vibrated, the intruder rise time counter is started. The system normalized acceleration is specified by using constant vibrational frequency $f = 13$ Hz and the amplitude A as

$$\Gamma = \frac{A(2\pi f)^2}{g} \tag{2}$$

where g is gravitational acceleration. We consider the normalized acceleration to be relatively high, $\Gamma = 5$, to ensure that BNE has a high probability to occur. Thus, the intruder rise time can be determined by the time it reaches the surface of the granular bed.

3. Results and Discussion

We generated 82 random initial configuration for 20 different values of ratio density and obtained possible numbers of contactopy between 328 and 355. Some of the contactopy snapshots are shown in figure 2. When the vibration applied, we examined the system in correspondence with some published experiments of similar normalized acceleration, regardless the distinction of contactopy [4,5]. It has been confirmed that all 1640 initial configurations used in our simulation model exhibit BNE phenomenon.

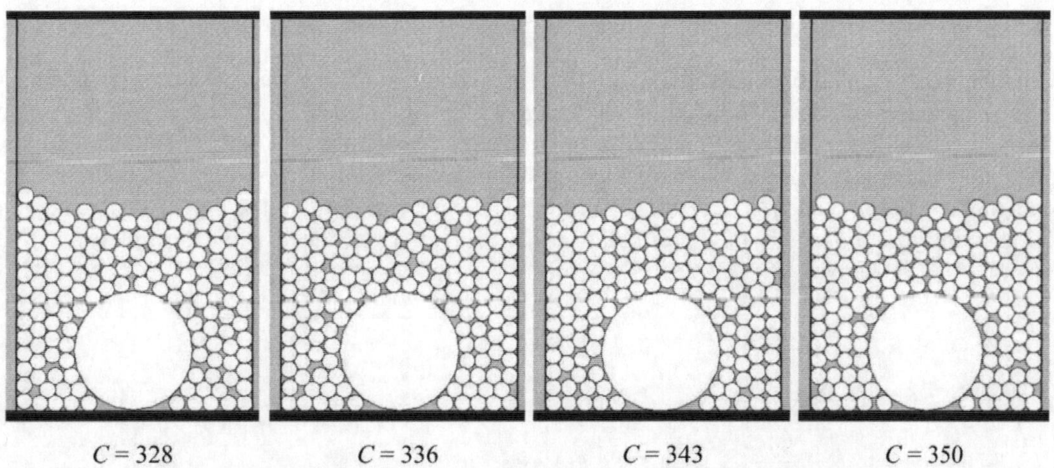

Figure 2. Samples of contactopy from 157 grains of BNE system.

It has been observed that from the initial condition until the end of BNE observation, the center-of-mass in vertical direction y_{CM} tends to decrease, while contactopy tends to increase as shown in figure 3 (left panel), which are similar to the reported results in a *Gedankenexperiment* (Viridi et al., 2014) [8]. Let C' denote the number of contactopy at $t = t_{rise}$ representing the highest number of contactopy from the system obtained in our simulation. As the system vibrates, grains configuration tends to be more compact toward C' and grains average speed in the vertical direction v_y^2 decreases as the vertical center-of-mass y_{CM} decreases. This relation is given in figure 3 (right panel). Since all grains in average do not moves in horizontal direction, value of v_x^2 is nearly zero.

To determine the intruder rise time in average as a function of density ratio, we classified the initial configurations into 0.1-1.8 range of density ratio. We did not, however, obtain the similar pattern as reported by Möbius et al. [5], that the density ratio should affect the rise time. One possible reason is that the simulation used two-dimensional system, which contains fewer degrees of freedom than the three-dimensional setup used by Möbius.

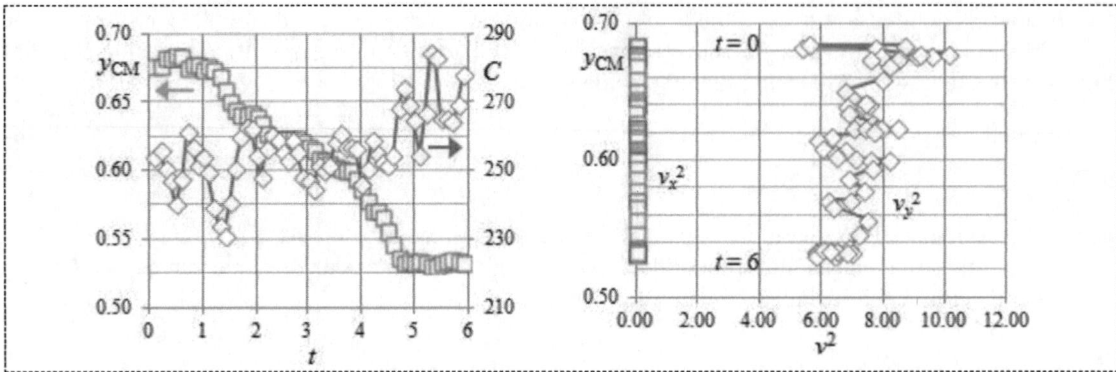

Figure 3. (left) Evolution of y_{CM} and initial configuration at $C = 328$, as a function of time. (right) Evolution of average speed in x and y direction at certain vertical direction.

In the next step, we plotted the intruder rise time as a function of contactopy. We found that the rise time minimum values vary in the range from 0.3 to 1.5 seconds as shown in figure 5 (left panel), while the maximum values have a wider range from 8.6 to 50.4 seconds. The associated density ratios for these values does not seem to have a unique relationship. This means that from all initial configurations, the minimum and maximum rise time cannot be distinguished from its contactopy, as well as density.

Interestingly, we found that the rise time in average tends to decrease as the number of contactopy increases as shown in figure 4. We suggest that this is due to the tendency of the vibrated system to become more compact before rising intruder out of the surface. Our simulation also indicates that there is a related pattern of compaction at the time it raises the intruder. Figure 3 presents that contactopy tends to increase toward C'. This may explain that the higher contactopy from the initial configurations used in the simulation, the nearer its distance from C' and result in the decrement of the rise time.

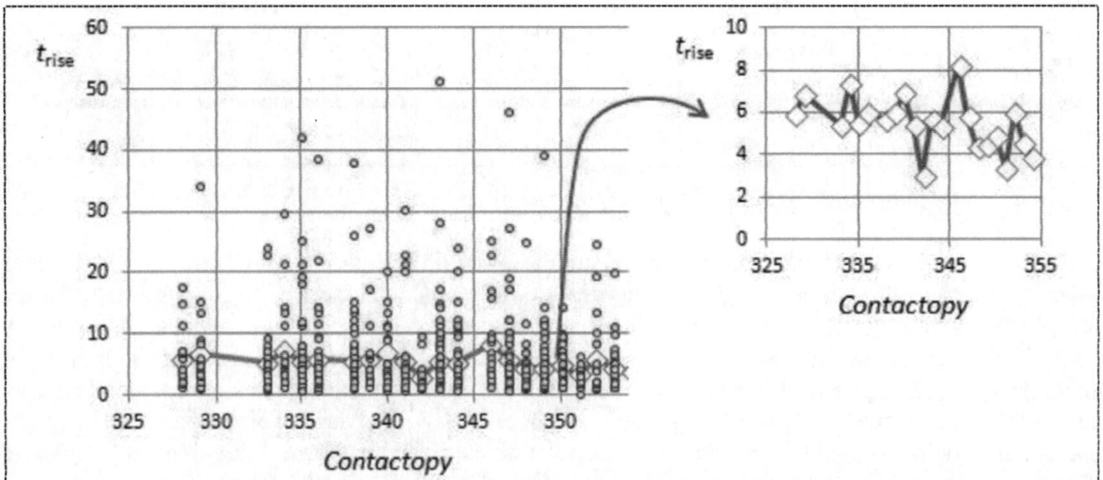

Figure 4. Contactopy does influence the average of rise time, $\langle t_{rise} \rangle$.

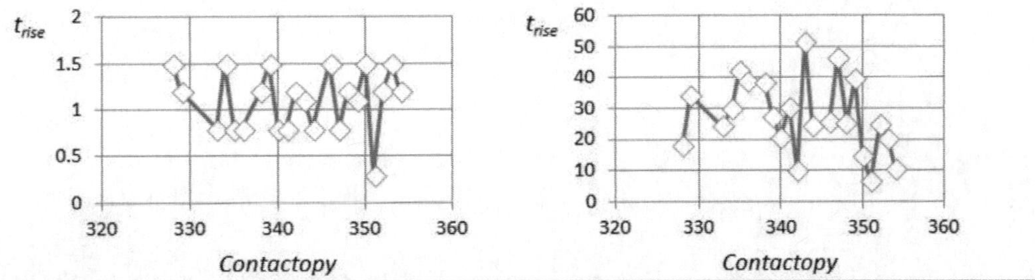

Figure 5. (left) Minimum and (right) maximum intruder rise time, t_{rise}, obseved as a function of contactopy from the initial configuration.

4. Conclusions

Contactopy can be generated into a certain range of number from the initial BNE system configurations of two-dimensional MD simulation. 157 grains in our simulation generates numbers of contactopy between 328 and 355. As the system vibrated at $\Gamma = 5$, the intruder rise time observed as a function of contactopy from the initial configurations. It shows that the rise time in average tends to decrease as the number of contactopy increases.

The future work should be focused on compaction detail during the intruder rising process. We hope to examine our assumption that the dynamic contactopy tends to a certain compaction in order to raise the intruder.

5. Acknowledgement

This work is funded by a Penelitian Unggulan Perguruan Tinggi - Desentralisasi DIKTI Tahun 2015 with project number 310i/I1.C01/PL/2015.

References

[1] A. P. J. Breu, H. -M. Ensner, C. A. Kruelle, and I. Rehberg, "*Reversing the Brazil-nut effect: Competition between percolation and condensation*", Physical Review Letters, vol. 90, no. 1, pp. 014302 (3 pages), January 2003.

[2] K. Liffman, K. Muniandy, M. Rhodes, D. Gutteridge, G. Metcalfe, "*A segregation mechanism in a vertically shaken bed*", Granular Matter, vol. 3, pp. 205–214, December 2001.

[3] M. E. Möbius, B. E. Lauderdale, S. R. Nagel, and H. M. Jaeger, "*Size separation of granular particles*", Nature, vol. 414, no. 6861, p. 270, November 2001.

[4] Hong, D. C., Quinn, P. V., dan Luding, S. "*Reverse Brazil Nut Problem: Competition between Percolation and Condensation*". Physical Review Letters, vol. 86, no. 15, pp. 3423-3426, April 2001.

[5] M. E. Möbius, X. Cheng, P. Eshuis, G. S. Karczmar, S. R. Nagel, and H. M. Jaeger, "*The effect of air on granular size separation in a vibrated granular bed*", Physical Review E, vol. 72, no. 1, pp. 011304 (13 pages), July 2005.

[6] M. P. Ciamarra, M. D. De Vizia, A. Fierro, M. Tarzia, A. Coniglio, and M. Nicodemi, "*Granular species segregation under vertical tapping: Effects of size, density, friction, and shaking amplitude*", Physical Review Letters, vol. 96, pp. 058001 (4 pages), February 2006.

[7] J. Ellenberger, C. O. Vandu, and R. Krishna, "*Vibration-induced granular segregation in a pseudo-2D column: The (reverse) Brazil nut effect*" Powder Technology, vol. 164, no. 3, pp. 168-173, June 2006.

[8] S. Viridi, Novitrian, Siti Nurul Khotimah, and Widayani, "*Evolusi energi potensial dan jumlah titik kontak material butiran dua dimensi dalam kasus Efek Kacang Brasil*", Prosiding Seminar Nasional Fisika 2014, pp. 149-153, ISSN 2088-4176.

IOP Conf. Series: Earth and Environmental Science **31** (2016) 012002 doi:10.1088/1755-1315/31/1/012002

Leaf Shape Recognition using Centroid Contour Distance

Abdurrasyid Hasim[1], Yeni Herdiyeni[1], Stephane Douady[2]
[1]Department of Computer Science, Faculty of Mathematics and Natural Science, Bogor Agricultural University, West Java, Indonesia
[2]Laboratoire : Matiere et system complex, Université Paris Diderot, France

E-mail: abdurrasyid.hasim@gmail.com, yeniherdiyeni@gmail.com

Abstract. This research recognizes the leaf shape using Centroid Contour Distance (CCD) as shape descriptor. CCD is an algorithm of shape representation contour-based approach which only exploits boundary information. CCD calculates the distance between the midpoint and the points on the edge corresponding to interval angle. Leaf shapes that included in this study are ellips, cordate, ovate, and lanceolate. We analyzed 200 leaf images of tropical plant. Each class consists of 50 images. The best accuracy is obtained by 96.67%. We used Probabilistic Neural Network to classify the leaf shape. Experimental results demonstrated the effectiveness of the proposed approach for shape recognition with high accuracy.

1. Introduction

Plant conservation becomes an urgent necessity for preservation of ecosystems. The growing of human population and climate change increasingly the environmental destruction. Extinction of existing plants may happen in next few years therefore identification of plant is important as conservation needs. Plant identification with traditional approach still be a difficult thing and can only be performed by trained botanists. This has lead to an increasing interest in automating the process of species identification and related tasks. The development and ubiquity of relevant technologies, such as digital cameras and portable computers has bought these ideas closer to reality [1].

In the field of image processing, the leaves are the most appropriate parts of the plant are used for identification as relatively easy to be photographed and analyzed [2]. Leaf is part of plant that is the basic feature in the identification of plants. Types of leaf can be described in terms of leaf shape, leaf margin, leaf venation, leaf texture, leaf vein, leaf color, leaf tip and leaf base. Based on leaf shape, the leaves are grouped into 20 classes [3]. There are two techniques to represent the shape of an object. Contour-based approach that only exploits the information contained in boundary and region-based which involving all parts of an object [4]. Centroid Contour Distance (CCD) is belonging to contour-based approach shape descriptor. Research has been conducted to compare CCD and CCG (Centroid Contour Gradient) in leaf recognition based on leaf tip and leaf base. CCG method is derived from CCD approach. The difference between these two approaches are CCD is used to find out the distance between centroid point and the boundary points while CCG is used to calculate the gradient between two points on the leaf's contour corresponding to the interval angle. The study included four types of leaf tip and leaf base (acuminate, cuspidate, obcus, and acute). The average accuracy for recognizing the leaf was 96.6% for CCG and 74.4% for CCD [5].

This research will identify leaf shape. CCD is used to extract the features of the leaf boundary. In classification process, we used Probabilistic Neural Network (PNN) to classify and measures accuracy. PNN is adopted for it has many advantages. Its training speed is many times faster than a BP network. PNN can approach a Bayes optimal result under certain easily met conditions. Additionally, it is robust to noise examples [6].

2. Leaf Shape

Botanist observes so many plant features using dichotomous key just to identify a species. Leaf shape is one of features to determinite plant species. Based on leaf shape, leaves are grouped to 20 classes.

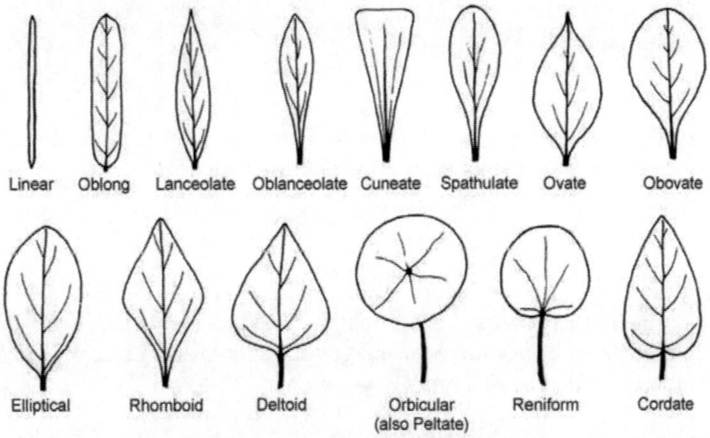

Fig. 1. Examples of leaf shapes

In general, leaf shape characteristics are determined by ratio of leaf height and leaf width and position of the widest part of the leaf [3]. We do not have leaf images of each class therefore we only used the most leaf shape of our collection: elliptical, cordate, ovate, and lanceolate.

Tracing an object contour can be considered as circling around its centroid. Point P on the boundary is determined by the midpoint of C, the distance between the point P and the center point C, and the angle α.

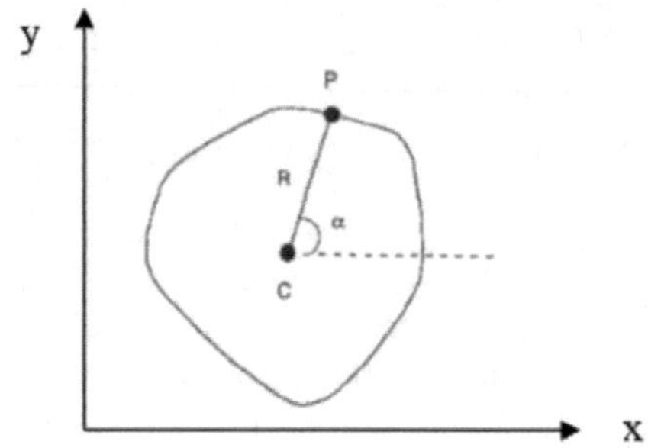

Fig. 2. Illustration of CCD

Centroid-distance function R(t) is the distance between the boundary points (x(t), y(t)) from the midpoint of the formula (1) [7].

$$R(t) = \sqrt{(x(t) - Xc)^2 + (y(t) - Yc)^2} \quad (1)$$

4. Method

Our method is to classify the leaf shape with CCD as the feature. We focus on how effective CCD in representing the leaf shape. In this part we explain how to extract the leaf boundary points and use them to determinate the class.

4.1 Experimental Data

We used 200 leaf images of tropical plant in Indonesia. We only used four leaf shapes, i.e., elliptical, cordate, ovate, lanceolate. Each class consists of 50 images. The images were taken from Arboretrum of Biopharmaca IPB. The representative's samples of images are shown in Figure 3. The leaf images were not separated from its background and have a random position.

Fig. 3. Samples of taken images from Arboretrum of Biopharmaca IPB for each class. From top left, clockwise consecutively, there are ellips, cordate, lanceolate, and ovate.

4.2 Preprocessing

A series of image preprocessing were done to ensure accurate edge and noise-free. In this research the leaf should has a white background and has standing position perpendicular to the horizontal line. At this manual preprocessing, since the leaf images were not separated from its background and have random position, at first we manually separated it from the background and rotated it so the leaf position is perpendicular to the horizontal line. If the leaf has big holes, the holes closed first. The petiol was also

IOP Conf. Series: Earth and Environmental Science **31** (2016) 012002 doi:10.1088/1755-1315/31/1/012002

eliminated in order to simplify the process of finding the midpoint.

The images are uniform with length and width of 700 pixels and a resolution of 72 dpi. To change the image to binary Threshold operation is done to change the image into a binary. The leaf image turned to black and this will simplify edge detection process because there will no wrong detection venation as an edge. To ensure there are no slightest noises outside the leaf object, dilation operation was performed. Erosion operation was also performed to close small holes in the leaf object.

4.3 Edge Detection

To perform edge detection, Canny Edge algorithm was used. The image that has been detected edges leaving only the outline of the leaf contour. These images are ready to be processed at the stage of leaf representation.

4.4 Feature Extraction

CCD calculates the distance between each boundary point and the midpoint. Midpoint (C_x, C_y) obtained from a simple calculation, C_x is half of the leaf width and C_y half of the leaf height. Assuming that the shape of the leaf is always symmetrical, CCD was only done on the right side of the leaf.

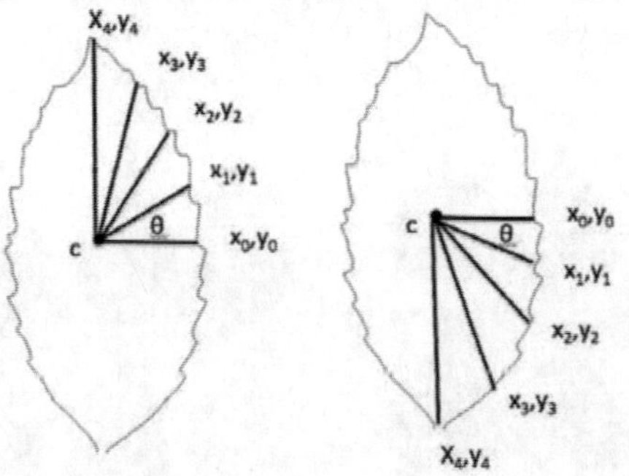

Fig. 4. Illustration of boundary points detection

Point selection was done in two stages, to the top and bottom of leaves. Edge point marked with (X_i, Y_i) and (i = 1,2,3, .., *n*-1, *n*) selected only that satisfy equation (2) for the upper leaves and equation (3) for the lower leaf.

$$Y_i = [tan(\theta) \times (X_i - C_x)] + C_y \quad (2)$$

$$Y_i = [tan(-\theta) \times (X_i - C_x)] + C_y \quad (3)$$

For the upper leaves, here n indicates the number of intervals with n = (90 / θ)+1. For example, if the value of θ = 15° then the boundary points are selected from the angles θ = {0, 15, 30, 45, 60, 75, 90}. As for the bottom leaves, n = (-90/-θ) +1. For angles θ = 15 ° then the selected boundary points are the angles θ = {0, -15, -30, -45, -60, -75, -90}. Coordinates (C_x, C_y) shows the central point of the leaf. In this study, we used 10° and 5° for the interval angle. After getting the boundary points, distance was calculated between the midpoint of the distance and the boundary points using the CCD formula.

IOP Conf. Series: Earth and Environmental Science **31** (2016) 012002 doi:10.1088/1755-1315/31/1/012002

5. Experimental Result

CCD performed with interval angle of θ=10° and θ=5°. Figure 5 shows with θ=10° obtained 19 points, and the θ=5° obtained 37 points. The distance between midpoint and boundary points R(t) calculates consecutively with t=1 at the tip of leaf.

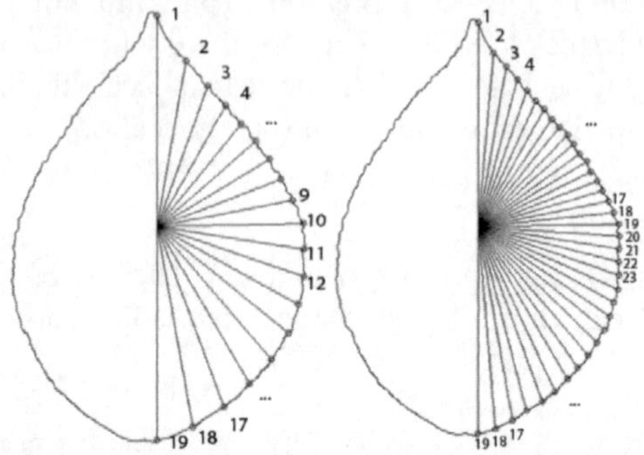

Fig. 5. The number of boundary points depends on interval angle

Figure 6 shows the samples of manually preprocessed leaf images that were used. The images have separated from their background, similar size, and upright position.

Fig. 6. Preprocessed sample of 4 classes' plant leaves

Graphical representation of the CCD results for four classes leaves with θ=10° is shown in Figure 7. Graph shows consecutively distance between midpoint and the boundary points R(t) with point t=1 starting from a point located at the leaf tip.

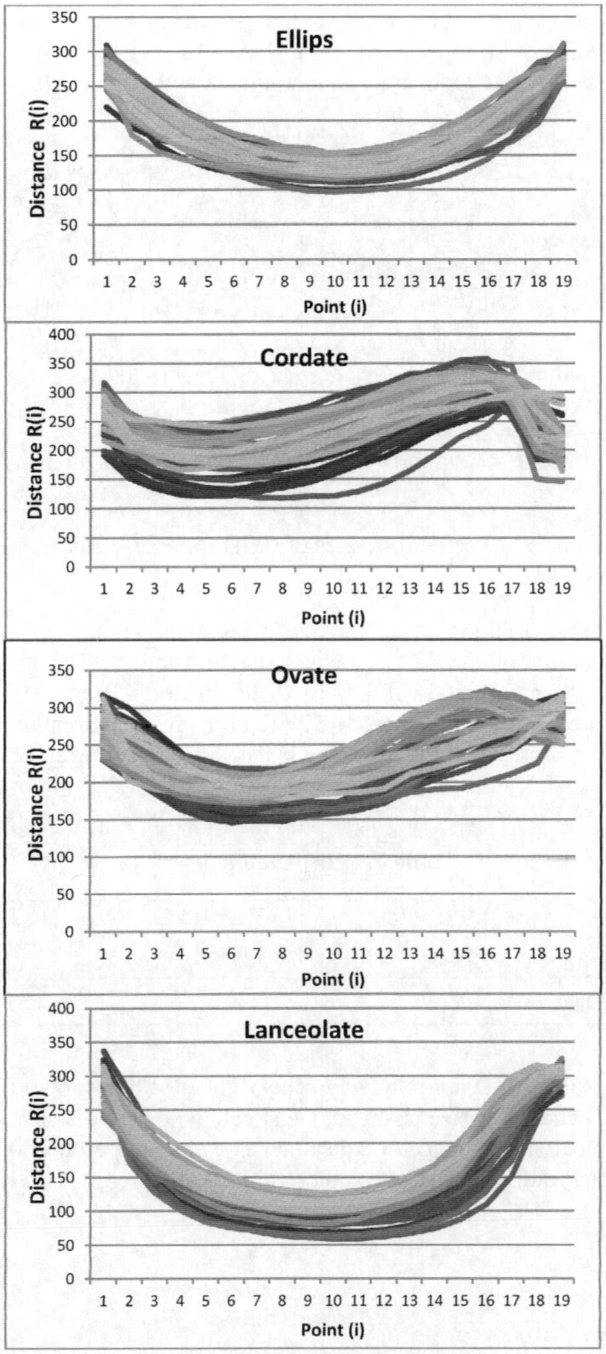

Fig. 7. CCD graphs for each class

In Figure 6, graphics of CCD has a distinctive pattern for each class. The shape of the graph class of ellips and lanceolate look similar, but lanceolate graph's tend to more concave downward because its shape is thin. The average value of R(t) for each class shown in Figure 8. Four leaf shape classes that have different graphs show that the CCD feature can distinguish leaf shape well.

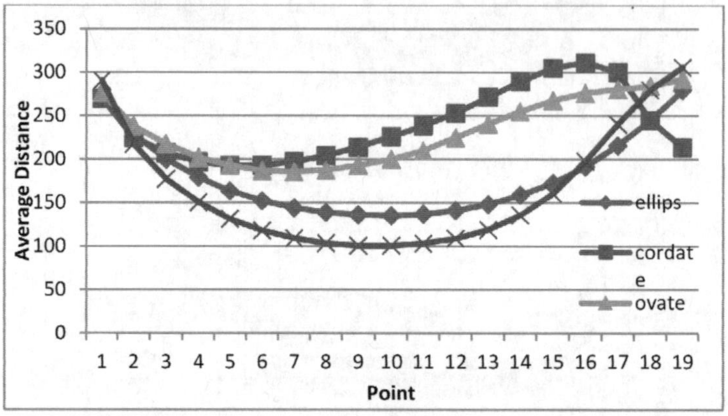

Fig. 8. Average distances of CCD for each class

5.1 Classification

Experiment were conducted 15 training data for each class and 140 testing data, the results of the classification of PNN to features of the CCD can be seen in Table 1. The number of CCD features depends on interval angles. Interval θ=10° will obtain 19 features and θ=5° get 37 features. In order to increase the number of features, those 19 features and 37 features from both interval were merged. In this case, the more features does not increase the accuracy because CCD achieved best accuracy by 96.67% with θ=10°.

Table 1. PNN Accuracy

Features	19 features (θ=10°)	37 features (θ=5°)	19+37 features
Accuracy	96.67%	95.00%	93.33%

According to experiment results, CCD are relatively robust under scale change. Figure 9 shows that the distance values of CCD will change if size changed. CCD calculates the distance between midpoint and boundary points. When the leaf size changed, the distance also changed proportionally. Both graph have the relatively similar pattern altough the distance values are different.

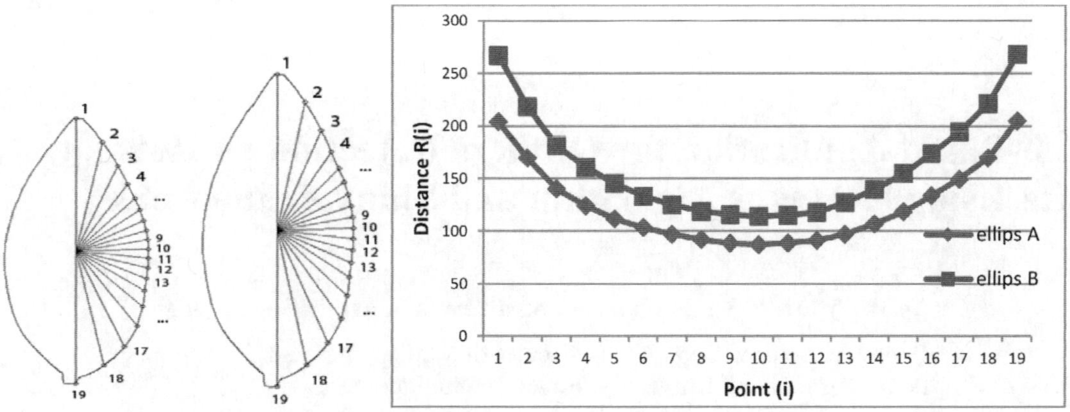

Fig. 9. Ellips A (left) and Ellips B (right) are the same leaf but different in size

6. Conclusion

This study successfully implements CCD to represent leaf shape. CCD is good to represent leaf shape because it can distinguish the characteristics of each class. The best accuracy obtained using the CCD 96.67% with interval angle θ=10° (19 features). In this experiment, the less interval angle which is the more points extracted does not mean the better accuracy obtained. CCD also relatively robust to the scale change so it could be applicable against leaves with various size.

7. Acknowledgment

We specially acknowledge to Prof. Stéphane Douady from Laboratoire: Matiere et System Complex, Université Paris Diderot for research collaboration. He support is gratefully acknowledged. Also, we are most grateful to a grant from DIKTI (Republic of Indonesia) who made this research possible through International Research Collaboration and Scientific Publication (2014-2015).

8. References

[1] Benson L 1957 *Plant Classification* (Boston : D.C. Heath and Company)
[2] Bong MF, Sulong GB, and Rahim M S MM 2013 *International Journal of Computer Science Issues* p 477
[3] Cerutti G, Coquin D, Mille J, Tougne L and Vacavant A 2011 *Conference on Multilangual and Multimodal Information Access Evaluation* p 1
[4] Cope J S, Cormey D, Clark J Y, Remagnino P, and Wilkin P. 2012 *Expert Systems with Applications* vol 39 p **7562** 7573
[5] Lu G and Zhang D 2004 *Pattern Recognition* 37 p **1** 19
[6] Pahalawatta K 2008 *Plant species biometric using feature hierarchies* (Department of Computer Science and Software Engineering University of Canterbury : Thesis)
[7] Wu S G, Bao F S, Xu E Y, Wang Y, Chang and Xiang Q 2007 *Signal Processing and Information Technology* IEEE International Symposium

IOP Conf. Series: Earth and Environmental Science **31** (2016) 012003 doi:10.1088/1755-1315/31/1/012003

Web-Based Application for Outliers Detection on Hotspot Data Using K-Means Algorithm and Shiny Framework

Agisha Mutiara Yoga Asmarani Suci[1], **Imas Sukaesih Sitanggang**[2]

Computer Science Department, Faculty of Mathematics and Science
Bogor Agricultural University, Bogor 16680, Indonesia

agishamyas@gmail.com[1], imas.sitanggang@ipb.c.id2

Abstract. Outliers analysis on hotspot data as an indicator of fire occurencesin Riau Provincebetween 2001 and 2012have been done, but it was less helpful in fire prevention efforts. This is because the results can only be used by certain people and can not be easily and quickly accessed by users. The purpose of this research is to create a web-based application to detect outliers on Hotspot data and to visualize the outliers based on the time and location. Outliers detection was done in the previous research using the k-means clustering method with global and collective outlier approach in Riau ProvinceHotspot data between 2001 and 2012. This work aims to develop a web-based application using the framework Shiny with the R programming language. This application provides several functionsincluding summary and visualizationof the selected data, clustering hotspot data using k-means algorithm, visualization of the clustering results and sum square error (SSE), and displaying global and collective outliers and visualization of outlier spread on Riau Province Map.

1. Introduction

In early March 2014, forest and peatland fires in Riau province soared to beyond the current crisis haze in Southeast Asia in June 2013. Since 20 February to March 11th, 2014, Global Forest Watch 3101 detects fires warning with a high confidence level in Sumatra using active hotspot data from NASA [1]. These numbers exceeds the 2643 total number of hotspots detected warnings at the peak of the crisis fires and haze on 13-30 June 2013 [1]. NASA active hotspot data obtained from satellite observations that use the sensor called the Moderate Resolution Imaging Spectroradiometer (MODIS).

Strategies that can be done as an effort to prevent the occurrence of such fires is the fire information system approach to system monitoring hotspots [2]. The hotspots data can be used as one indicator of forest and land fires events, so that the necessary analysis. Analysis can be done to the incident hotspots, namely the analysis of outliers using clustering algorithm [2].

Previous research about analysis outliers on hotspot data have been carried out [3] using the K-Means algorithm. However, in these studies only use the functions contained in R and not created a Web-based system. Results of the research that object is used as outliers is an object with a highest hotspots frequency. In other previous research, data mining technique has been applied on hotspot dataset in Riau from 2000 until 2014 using sequential pattern mining algorithm namely prefix span to find frequent sequences [4]. Clospan (Closed Sequential Pattern Mining) that available in the SPMF (Sequential Pattern Mining Framework) program to generate sequential patterns on hotspots data was used on hotspots, precipitation, and temperature data data that are grouped by year of events starting from 2001 to 2010 [5] . Another, spatial decision tree algorithm on data of forest fires has been done. The algorithm is the improvement of the conventional decision tree algorithm in which distance and topological relationships are included to grow up spatial decision trees [6]. Decision tree to predict hotspot occurences in Bengklis district, Riau Province using the spatial entropy-based decision tree algorithm was used. Data was used by that algorithm include city centre, river, road, income source, rain cover, population, precipitation, school, temperature, and wind speed [7]. Not only decision tree, extended spatial ID3 algorithm on the forest fires data in Bengkalis district, Riau Province Indonesia used too. The data include hotspots and non-hotspots,weather data, socio-economic data, and geographical characteristic. The result is a decision tree with the income source layer as the label of root node [8]. Association rule algorithm namely apriori to discovers the possible influence factors on the occurrence of fire events in the study area of Rokan Hilir Riau Province Indonesia was done. The Apriori algorithm was applied on a forest fire dataset which contained data on physical environment (land cover, river, road and city center), socio-economic (income source, population, and number of school), weather (precipitation, wind speed, and screen temperature), and peatlands [9]. ID3 and C4.5 that are decision tree algorithm was used in previous research on spatial data of forest fires for Rokan Hilir District in Riau Province in Indonesia. The result show that C4.5 algorithm has better perfomance than the ID3 algorithm in terms of accuracy and the number of generated rules [10]. Improvement of the ID3 decision tree algorithm such that it can be utilized on spatial data in order to develop a classification model for hotspots occurrence. As the ID3 algorithm that uses information gain in theattribute selection, the proposed algorithm uses spatial information gain to choose the best splitting layer from a set of explanatory layers. The new formula for spatial information gain is proposed using spatial measures for point, line and polygon features. The proposed algorithm has been applied on the forest fire dataset for Rokan Hilir district in Riau Province in Indonesia. The dataset contains physical data, socio-economic, weather data as well as hotspots and non-hotspots occurrence as target objects [11]. Classification module for hotspot occurrences in a GIS using OpenGeo Suite 3.0. The classification module was made using the decision tree method namely the C4.5 algorithm. The GIS enables users to determine whether an area is potential for hotspot incidence or not based on its characteristics. Potential hotspot occurrences can be used in decision making related to forest fire prevention [12].

Almost all the people of Indonesia need the internet. By using the internet makes it possible to perform long-distance communications quickly and cheaply. Results of the outlier detection of hotspots data need to be distributed quickly for stakeholders to be able to take a decision as early as possible in an effort to anticipate the fires and fire suppression.

Shiny is a package of R programming language that allows users to build web-based applications. Shiny provide queries and summary of the data to the user via a modern web browser easily. In addition, Shiny also equipped with a variety of widgets for building interactive user interface and can be easily integrated with HTML and CSS, JavaScript and JQuery can even be used to expand the scope of application Shiny [13].

2. K-Means Algorithm

K-Means algorithm is a technique based centroid and is one method of partitioning. $p \in C_i$ difference between an object and c_i, a representation of the cluster measured by $dist(p,c_i)$. C_i is the i-th cluster, while the cluster centroid c_i adalah to-i. $Dist(x,y)$ is the Euclidean distance between two objects or points x and y. The quality of a cluster C_i can be measured by Sum Square Error (SSE). SSE is defined as follows [14]:

$$\text{SSE} = \sum_{i=1}^{k} \sum_{p \in Ci} dist(p, c_i)^2 \sum_{i=1}^{k} \sum_{p \in Ci} dist(p, c_i)^2 \sum_{i=1}^{k} \sum_{p \in Ci} dist(p, c_i)^2 \qquad (1)$$

SSE is the total of the squared error for all objects in the data set, p is a point that represents the object, k as the number of clusters, and c_i is the focal point cluster C_i [6].

3. Outlier detection Clustering Based on Hotspot Data

This section describes fundamentals of outlier detection on hotspot data in Riau, including definition of outlier, detection outliers method, type of outliers, and degree of being an outliers. Outlier is a data object that significantly deviates from a collection of objects that are generated by different mechanisms [14].

Outliers detected with clustering method. Clustering is the process of grouping a set of data objects into groups or clusters so that objects within a cluster have high similarity, but very different from the other groups [14]. A cluster is a collection of records that are similar to one another and different from the other records in the group [15].

There are two various outliers will be detected in application that will be built.. That outliers are global outlier and collective outlier [14]. Global outlier is a data object that significantly deviates from hotspot data or not in accordance with the pattern in general. Collective outlier is part of the data objects are entirely deviate significantly from the overall hotspot data

In k-means clustering method, an objects came as global outlier outliers when the distance between the object to the center or centroid of cluster is large. Global outlier approach can be measured by the value of outlier score. Object has greater outlier score value is become a global outlier.

In addition, a collective approach to outliers in clustering method when the object is part of a small cluster member or cluster minority, then all objects in the cluster is an outlier [14]. Outlier detection also provide specific methods to calculating the degree of which data measurements deviate from the normal pattern of hotspot data. In hotspot data, outliers are measured in two scales, i.e., *scalar* and *outlier score* [16].

The scalar scale is a zero-one classification measure [16], which classifies each data measurement into outlier cluster. Thus, the output of techniques of scalar scale is provide a list of outlier that has a minimum member in clusters.

Techniques of scale value outliers set scores for each measurement data that depends on the degree of measurement are considered as outliers and provide a ranked list outliers. An analyst can choose to analyze the top n outliers that have the biggest outlier values or use a cut-off threshold to select outliers. This limit is often difficult to choose and are usually determined and set by the user [16].

Ratio to measure an object distance is defined as follows [14]:

$$\textit{Outlier Score} = \frac{\mathrm{dist}(o, c_o)}{l_{c_o}} \qquad (2)$$

o : object of data,
c_o : *centroid*,
$\mathrm{dist}(o, c_o)$: Euclid distance between object with the nearest centroid of object
l_{c_o} : mean $\mathrm{dist}(o, c_o)$.

4. Web-based Application for Outlier Detection

Application was built by using Shiny framework and R language for outlier detection on hotspot data between 2001 until 2012 in Riau Province with k-means clustering algorithm. This application consists of 5 menus, data menu, clustering menu, outlier detection menu, help menu, and about menu. 5 display menu available using NavbarPage, that is menus that are available in the application form of the navigation menu.

Users can choose which datasets to be used for the process of clustering on the data with a dropdown menu button. The dataset consists of datasets daily frequency of hot spots Riau Province between 2001 and 2012. The object of this dataset is made reactive to respond to any changes in user demand data. Data menu consists of two sub-menus, which is a data summary sub-menu and plot data sub-menu.

In the sub-menu of data summary, the data that has been selected will be displayed in the form of summaries minimum value, first quartile, median, average value, the third quartile, and the maximum value of the frequency. Display of data summary sub-menu can be seen in Figure 1.

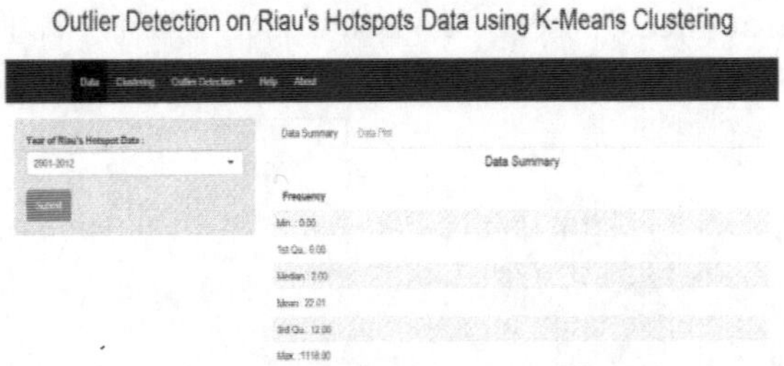

Figure 1. Data summary sub-menu

While in the data plot sub-menu is displayed line graph that showing the trends or distribution of the frequency of hotspots on the day index. The graph on the data plot sub-menu made reactive. Reactive means responsive to changes in inputs, so that the plot will vary according to the selected data input by the user. Display sub-menu of data plot can be seen in Figure 2.

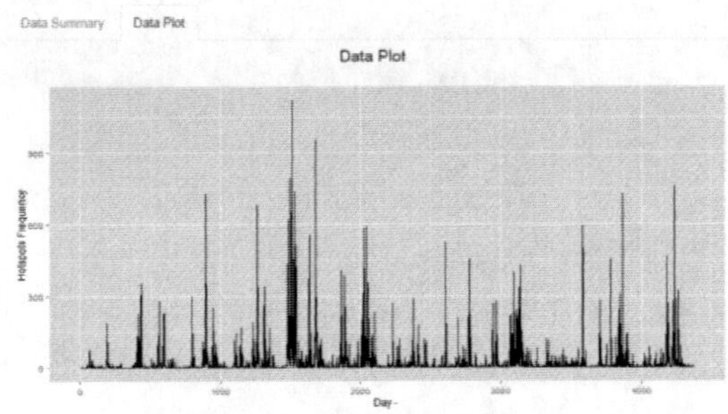

Figure 2. Data plot sub-menu

If users want to do clustering, then the user must select the menu clustering. K-means function on this application is reactive, so any changes in the value of the input will cause the redial function of the k-means and generate new cluster without the need to reload the browser page. Clustering menu consists of three sub-menus, namely clustering summary sub-menu, sum square error sub-menu, and k-means plot sub-menu.

Objects cluster, size of the cluster, the center of the cluster and the percent that is the result of clustering are combined to form a new frame of data to be displayed on the sub-menu clustering summary. Display of clustering summary sub-menu can be seen in Figure 3.

Object number of clusters, SSE, total withins and percent which is the clustering results are combined to form a new frame of data to be displayed on error square sum sub-menu. Display sub-menu sum square error can be seen in Figure 4.

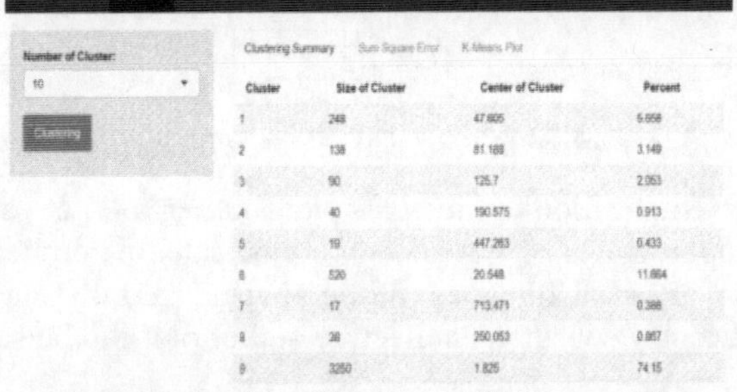

Figure 3. Clustering summary sub-menu

IOP Conf. Series: Earth and Environmental Science **31** (2016) 012003 doi:10.1088/1755-1315/31/1/012003

In the k-means plot sub-menu, displayed visualization of the clustering results. To visualize the clustering results are used ggplot2 library. Display sub-menu K-Means plot can be seen in Figure 5.

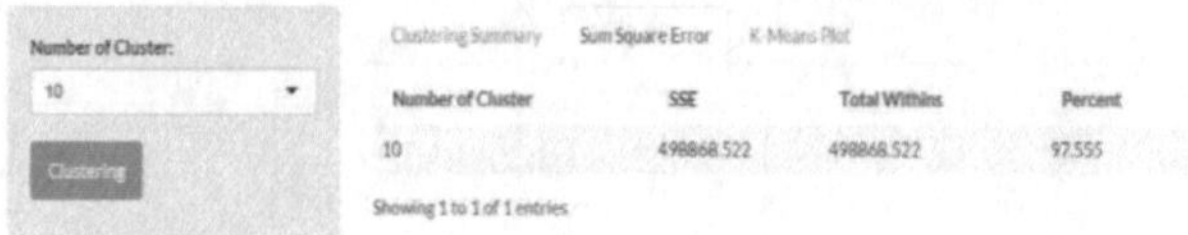

Figure 4. Sum square error sub-menu

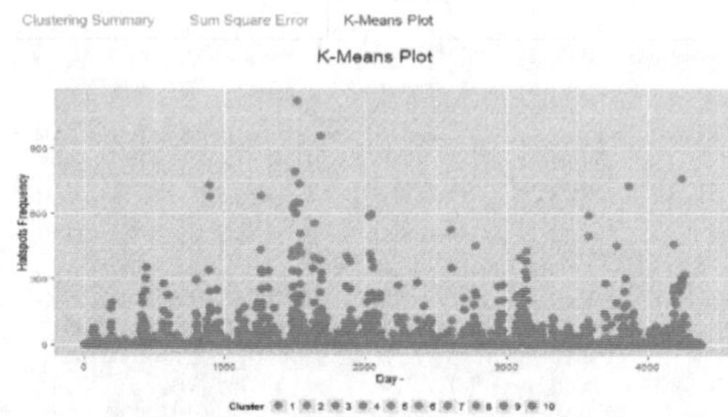

Figure 5. Kmeans plot sub-menu

Outlier detection menu is a menu that serves to detect outliers. The menu consists of two menus, that is collective and global outlier menus. If the user wants to detect collective outliers, then the user must select the collective outlier menu, whereas if you want to detect global outlier, then select a global menu.

On the menu collective outlier users can choose a minimum percentage of the number of members in a cluster by entering the percentage value in the column that has been provided. The menu consists of collective outlier summary sub-menu and collective outlier plot sub-menu.

Collective outlier summary sub-menu displays the results of the collective outlier detection in general. Detection results in the form of a table which consists of cluster objects and percent. If the user wants to see more clearly the objects which are included collective outliers, then the user can select or press the button detailed information available on the outlier collective summary sub-menu. Display collective outlier summary sub-menu can be seen in Figure 6.

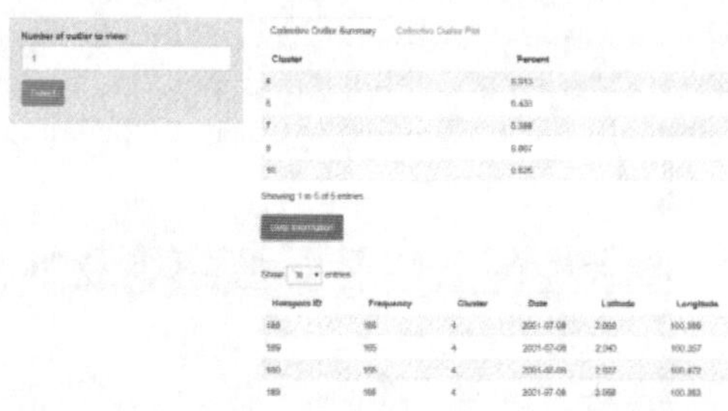

Figure 6. Collective outlier summary sub-menu

Sub-menu displays visualization collective plot outlier outlier detection results collectively. Latitude and longitude coordinates of hot spots were detected as outliers collective visualized into a map of the province of Riau. Collective outliers visualized by month for hotspots which are distinguished by color. Display sub-menu collective outlier plot can be seen in Figure 7.

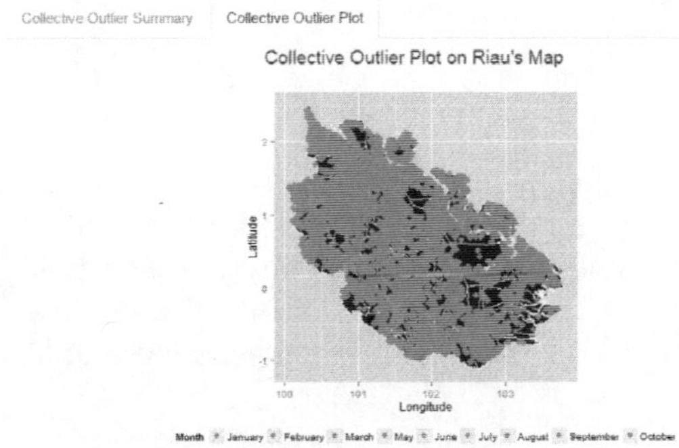

Figure 7. Collective plot sub-menu

On the global outlier menu, user can select the desired amount based on the value outlier best scores in the column that has been provided. Outlier global menu consists of two sub-menus, they are global outlier summary sub-menu and global outlier plot sub-menu. Global outlier summary sub-menu displays the outlier detection results. Detection results in the form of a table which consists of no object, index, frequency, score, date, and clusters. Display sub-global menu outlier summary can be seen in Figure 8.

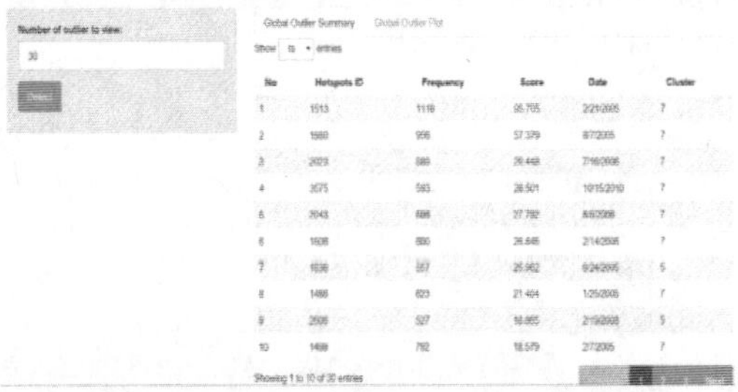

Figure 8. Global outlier summary sub-

Global outlier plot sub-menu display the visualization of the results of a global outlier detection. Latitude and longitude coordinates of hotspots were detected as a global outlier visualized into maps Riau Province. Global outliers visualized by month for hotspots which are distinguished by color. Display sub-global menu outlier plot can be seen in Figure 9.

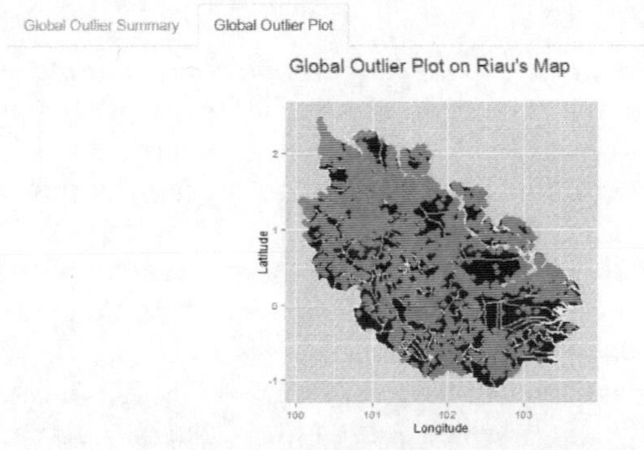

Figure 9. Global outlier plot sub-menu

Available main functions testing on the application done using blackbox testing. After the tests blackbox, the available main function in the application is running according to its function. That functions are data summary, data plot, clustering summary, SSE, K-means plot, Collective outlier summary, collective outlier plot, global outlier summary, gobal outlier plot, help, and about.

5. Conclusion

Web-based applications for outlier detection of hotspots using k-means clustering algorithm and Shiny framework provides several major functions. First, summary and plot of the selected data. Second, clustering hotspots data with k-means algorithm. Third, visualization of the clustering results. Fourth, show the SSE of clustering result. Fifth, global and collective outliers detection. Last, visualization the outliers on the Riau Province map. All of main function that available on the application works well.

References

[1] Sizer N, et al. *Kebakaran hutan di Indonesia mencapai tingkat tertinggi sejak kondisi darurat kabut asap Juni*. 2014. Available: http://www.wri.org/blog/2014/03/kebakaran-hutan-di-indonesia-mencapai-tingkat-tertinggi -sejak-kondisi-darurat-kabut (URL).

[2] Adinugroho WC, Suryadiputra INN, Saharjo BH, Siboro L. *Panduan Pengendalian Kebakaran Hutan dan Lahan Gambut*. Bogor (ID): Wetlands International-Indonesia Programme and Wildlife Habibat Canada, 2005.

[3] Imas SS and Dhiya AMB. *Global and Collective Outliers Detection on Hotspot Data as Forest Fires Indicator in Riau Province, Indonesia*. Proceedings 2015 Second IEEE International Conference on Spatial Data Mining and Geographical Knowledge Services (ICSDM 2015), Fuzhou, China, Jul. 8-10, 2015, pp 66 – 70.

[4] Nida ZN. *Sequential Pattern Mining on Hotpsot data in Riau Province Using The PrefixSpanAlgorithm*. Proceedings of 2015 Third International Conference of Adaptive and Intelligent Agroindustry (ICAIA) 2015, pp 313-316.

[5] Tria A and Imas SS. *Sequential Patterns for Hotspots Occurences Based Weather Data Using Clospan Algorithm*. Proceedings of 2015 Third International Conference of Adaptive and Intelligent Agroindustry (ICAIA) 2015, pp 301-305.

[6] Imas SS, Razali Y, Norwati M, Ainuddin AN. *A Decision Tree Based on Spatial Relationship for Predicting Hotspots in Peatlands*. TELKOMINKA, pp 12(2): 511-518.

[7] Indry DN and Imas SS. *Classification Rules for Hotspots Occurences Using Spatial Entropy-based Decision Tree Algorithm*. The 1st International Symposium on LAPAN-IPB Sattelite for Food Security and Environmental Monitoring, Procedia Environmental Sciences, pp 24(2015): 120-126.

[8] Yaumil MK and Imas SS. *A Spatial Decision Tree Based on Topological Relationships for Classifying Hotspot Occurences in Bengkalis Riau Indonesia*. ICACSIS 2014, IEEE, pp 268-272.

[9] Imas SS. *Spatial Multidimensional Association Rules Mining in Forest Fire Data*. Journal of Data Analysis and Information Processing (JDAIP) 2013,1, pp 90-96.

[10] Imas SS, Razali Y, Norwati M, Ainuddin AN. *Predictive Models for Hotspots Occurence Using Decision Tree Algorithms and Logistic Regression*. Journal of Applied Science 2013, ISSN Asian Network for Scientific Information, pp 13(2): 251-261.

[11] Imas SS, Razali Y, Norwati M, Ainuddin AN. *Classification Model for Hotspot Occurences Using Spatial Decision Tree Algorithm*. Journal of Computer Science 2013, ISSN Science Purblicatons, pp 9(2): 244-251.

[12] Khairil A and Imas SS. *A Geographical Information System for Hotspot Occurences Classification in Riau Province Indonesia* . The 1st International Symposium on LAPAN-IPB Sattelite for Food Security and Environmental Monitoring, Procedia Environmental Sciences, pp 24(2015): 127-131.

[13] Beeley C. Web Application Development with R Using Shiny. Birmingham (UK): Packt, 2006.

[14] Han J, Kamber M, and Pei J. *Data Mining: Concepts and Techniques 3rd ed.* Massachusetts. US: Morgan Kaufman, 2012.

[15] Larose DT. Discovering Knowledge in Data: An Introduction to Data Mining. New Jersey (US): J Wiley, 2005.
[16] Zhang Y, Meratnia N, Havinga P. *Outlier detection techniques for wireless sensor networks*. IEEE Communications Surveys & Tutorials, Second Quarter 2010 pp 12(2): 1-12.

IOP Conf. Series: Earth and Environmental Science **31** (2016) 012004 doi:10.1088/1755-1315/31/1/012004

Cloud Computing Application for Hotspot Clustering Using Recursive Density Based Clustering (RDBC)

Aries Santoso*[1], Karlina Khiyarin Nisa[2]

[1, 2] Department of Computer Science, Bogor Agricultural University, INDONESIA.

E-mail: aries.santoso303@gmail.com, karlinanisa@gmail.com

Abstract. Indonesia has vast areas of tropical forest, but are often burned which causes extensive damage to property and human life. Monitoring hotspots can be one of the forest fire management. Each hotspot is recorded in dataset so that it can be processed and analyzed. This research aims to build a cloud computing application which visualizes hotspots clustering. This application uses the R programming language with Shiny web framework and implements Recursive Density Based Clustering (RDBC) algorithm. Clustering is done on hotspot dataset of the Kalimantan Island and South Sumatra Province to find the spread pattern of hotspots. The clustering results are evaluated using the Silhouette's Coefficient (SC) which yield best value 0.3220798 for Kalimantan dataset. Clustering pattern are displayed in the form of web pages so that it can be widely accessed and become the reference for fire occurrence prediction.

1. Introduction

Forest fires have become a major problem in Indonesia for many years. These are primarily triggered by the drought and irresponsible farmers who slash and burn trees. Forest fires are also caused by political economy of Indonesian forest mismanagement, where decades of reckless logging have left wide areas degraded and highly susceptible to fires[1]. The impact caused by forest fire is quite extensive and very complex, not only damages the environment, but also covers other areas such as economy, culture, even relations between states.

Monitoring hotspots through satellite remote sensing become a solution of forest fire disaster management. In this research, hotspot dataset used are collected using MODIS satellite which covers areas of Kalimantan island and South Sumatra province. Each hotspot contains spatial data (latitude, longitude), temporal data (time and date of monitoring data), and nonspatial data (air temperature, rainfall, etc.). Spreading hotspots are usually converge in a certain areas so that these data can be analyzed using clustering techniques.

Hotspot satellite data is a huge dataset. DBSCAN[2] is a density based algorithm which is robust enough to handle large spatial data. Previous research on web based hotspots clustering has been done using DBSCAN, which needs careful determination of parameter Eps and MinPts in order to yield good clustering results[3]. This research performs clustering on hotspot dataset using Recursive Density Based Clustering (RDBC)[4] that efficiently choose parameter for DBSCAN clustering.

In desktop based clustering application, access to such researchs is limited because it is only accessible on a local computing. The need for larger storage also become constraints when the system is growing. Therefore, the development of the system is done by making use of cloud computing based on R programming language with Shiny web framework. Shiny framework can create interactive web with R programming language so that it can be widely accessed through the internet.

2. Clustering Of Hotspots

2.1. Hotspot Data

This research uses hotspot data which taken by remote sensing satellite from Kalimantan island and South Sumatra Province in 2002-2003. This data was obtained from the Fire Information for Resource Management System (FIRMS)[5] are provided by the National Aeronautics and Space Administration (NASA). There are 4999 hotspots in Kalimantan island and 4821 hotspots in South Sumatera province. The initial dataset obtains 11 fields of data including spatial, non spatial, and temporal data. Then it is reduced into two fields of spatial data including latitude and longitude. Figure 1 and Figure 2 shows area in South Sumatra and Kalimantan covered in the dataset.

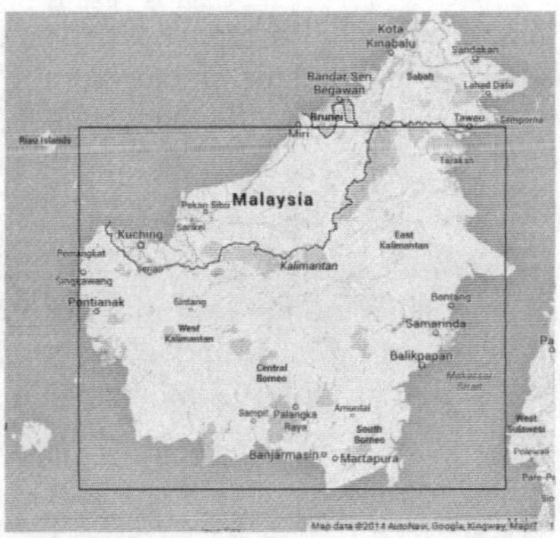

Figure 1. Dataset coverage area in Kalimantan island between 4.267 and -4.144 latitude, 118.763 and 108.913 longitude

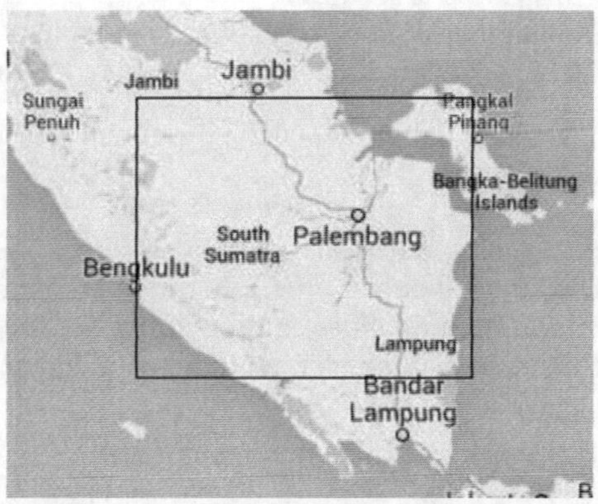

Figure 2. Dataset coverage area in South Sumatera province between -1.652 and -4.861 latitude, 106.053 and 102.259 longitude

2.2 RDBC Algorithm

RDBC is an extended form of DBSCAN, an algorithm to group neighboring objects of the database into clusters. Moreover, it does not require a predetermined cluster number to operate[6].

Input parameters for the RDBC algorithm is same as DBSCAN algorithm, those Eps and MinPts.

In RDBC algorithms, determining the distance threshold Eps and density threshold MinPts is performed recursively until the number of clusters is appropriate. MinPts is a minimum number of points that must be held to determine whether a point is classified into core point, a border point, or noise point. Meanwhile, Eps is the minimum distance between core point to another point around the core point. Point types in the density-based clustering can be seen in Figure 3, core points are those

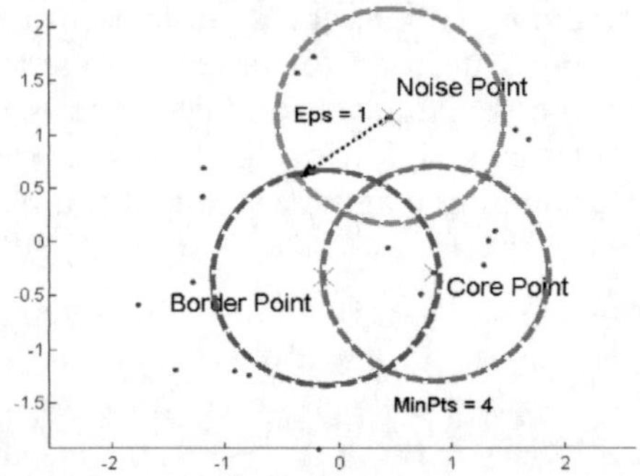

Figure 3. Types of point in density based clustering.

inside the circle, border points are points on the border of circle, while noise points are those outside the circle.

Different values of Eps and Mpts have been used in RDBC to identify core point set, CSet. Only after appropriate CSet is determined, the core points are clustered, and the remaining data points are then assigned to clusters according to their proximity to a particular cluster [6]. Figure 4 shows the pseudocode of RDBC algorithm.

If the number of core point set, Cset, is less than half the number of datasets, Eps and MinPts will be reduced, otherwise the next process is DBSCAN clustering using parameter values that have been reduced.

2.3 Cluster Validation

Validation is required for clustering results. This research uses Silhouette coefficient (SC) cluster validation. Silhouette coefficient combines the idea of cohesion and separation for each point. Cohesion measures how close relationships between objects in a cluster, while separation measures how much difference a cluster with another cluster. Here is the formula to calculate the value of Silhouette coefficient[7].

$$s(i) = \frac{b(i)-a(i)}{max\{a(i),b(i)\}} \tag{1}$$

```
Set initial values ε = ε1 and Mpts=Mpts1
Dataset = data_set
RDBC(ε, Mpts, Dataset)
//Use ε and Mpts to get the core points set Cset
   If size(Cset) > size(Dataset)/2
      DBSCAN(Dataset, ε, Mpts);
   Else
      ε =ε/2;
      Mpts=Mpts/4;
      RDBC(ε, Mpts, Dataset);//collect all points
around cluster
```

Figure 4. RDBC pseudocode

Based on formula(1), there is a point i in cluster A and s(i) is the silhouette of a point i, whereas a(i) represents the average rate of difference (dissimilarity) between the point i with another point and b(i) as average value of difference between the point i to point in the closest cluster. The final value of SC is obtained from the average of the overall silhouette on data set.

2.4 Cloud Computing Application using Shiny

Shiny is a web framework of RStudio that simplify the development of web applications directly from R programming language. Shiny is a perfect package for R, easy and simple to display the results of the analysis and graphs of R and users can interact via the web[8]. This package is classified as reactive programming or easily displaying the output corresponds to the input. The output value can be obtained without the need to re-load the page. Shiny composed of two components, the interface and the server side. The interface as a controller of the display of the application, while server side as the instructions for the computer to build the application.

3. Result and Discussion

In the application, RDBC algorithm is defined as a function with input parameters are datasets, Eps and MinPts. The differences between DBSCAN and RDBC is, that in RDBC, determination of core point is done separately from the clustering process. Figure 5 shows the implementation of the algorithm RDBC into R programming language.

```
rdbc <- function(data,eps,minpts){
     n <- nrow(data)
     cset <- cset(data,eps,minpts)
     if (cset > n/2)
       result <- dbscan(data,eps,minpts)
     else {
       eps <- eps/2
       minpts <- minpts/4
       result <- rdbc(data,eps,minpts)
     }
return(result);}
```

Figure 5. RDBC implementation using R

In RDBC algorithm, the number of core point is the main factor to produce minimal noise cluster. Therefore, the RDBC algorithm begins with the calculation of the number of core points. If the number of core points is less than half the number of points in the dataset, then the value of Eps and MinPts are reduced. The process is carried out recursively until the number of the core points qualified, i.e. more than half of the dataset. If the number of core points already qualifies or more than half of the dataset, then the value of Eps and MinPts is used to perform DBSCAN clustering on the dataset.

Table 1 is a comparison between clustering algorithms RDBC and DBSCAN using Kalimantan island hotspot dataset which consist of 4999 points, with initial parameters Eps = 0.009 and MinPts = 2. RDBC algorithm produces better clusters with higher value of SC because it choose optimal parameter values of Eps and Minpts based on the number of core points. RDBC algorithm considers the number of core points, when the number of cores point not yet reached half the number of datasets, the parameters Eps and MinPts will be reduced. The more core point are obtained then the less noises are generated. In DBSCAN algorithms, no constraint is applied in choosing the parameter, resulting in a cluster that is less good because the number of the core point may be fewer than the number of non-core point.

Table 1. Comparison of RDBC and DBSCAN

	RDBC	DBSCAN
MinPts	initial = 4, end = 1	4
Eps	initial = 0.009, end = 0.0045	0.009
Cluster	541	482
Noise	74	2887
Silhouette Coefficient	0.3220798	-0.2491384

Results of clustering process RDBC are then visualized in web application using Shiny web framework for R, as showed in Figure 5. In this application, the user can select datasets to be processed. Then, as early initialization, users can enter clustering parameter values Eps and Minpts. Using the initial value of Eps and Minpts, the algorithm will seek the best parameter values which generates core points more than half of the datasets.

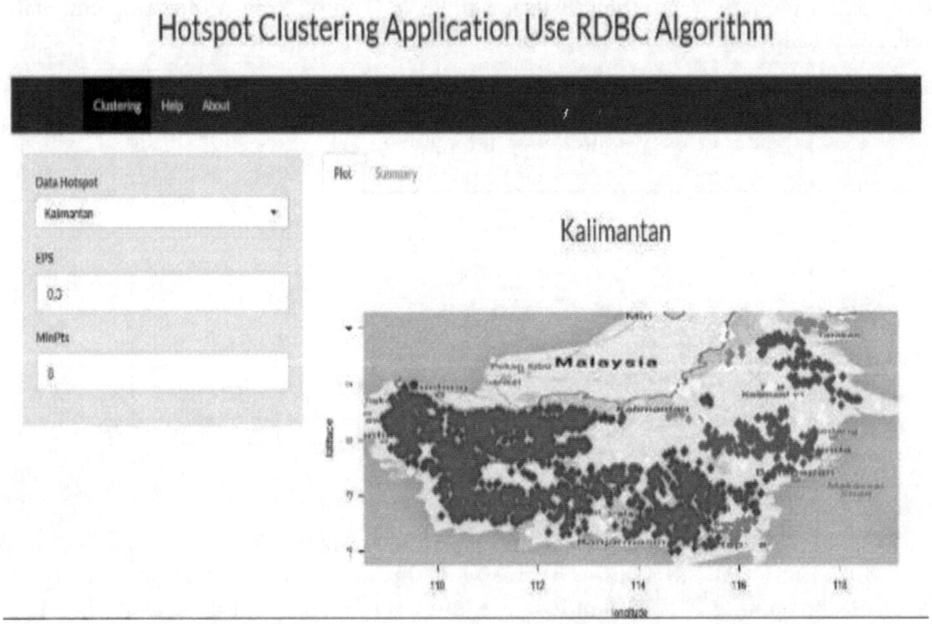

Figure 6. Web based application for clustering hotspot

IOP Conf. Series: Earth and Environmental Science **31** (2016) 012004 doi:10.1088/1755-1315/31/1/012004

4. Conclusion

In this study, RDBC algorithm is implemented in the language of R into a web application framework clustering with the help of Shiny. This algorithm identifies a core point separate from the clustering process to determine the optimal parameters of Eps and Minpts. Experiment result shows that RDBC yield better Silhouette coefficient compared with DBSCAN on hospot dataset. It is mainly because RDBC efficiently chooses oprimal parameter based on the number of core points obtained.

5. References

[1] P. Deauvergne, "The Political Economy of Indonesia's 1997 Forest Fires," Australian Journal of International Affairs, vol. 52, no. 1, 1998

[2] Ester, Martin, H.-P. K., "A Density-Based Algorithm for Discovering Clusters in Large Spatial Databases with Noise," Proceedings of 2nd International Conference on Knowledge Discovery and Data Mining (KDD-96), 1996, (pp. 226-231).

[3] K. Nisa, H. Adrianto, R. Mardhiyyah, "Hotspot Clustering Using DBSCAN Algorithm and Shiny Web Framework," Proceedings of 6th International Conference on Advanced Computer Science and Information System, 2014.

[4] M. Rehman, SA. Mehdi, "Comparison of density-based clustering algorithms".

[5] Fire Information For Resource Management System. [accessed Januari 2012]. Available at https://earthdata.nasa.gov/data/near-real-time-data/firms/

[6] Zhong S, Qiang Y, Hongjiang Z, Xiaowei X, Yuhen H, "Correlation-based document clustering using web logs department of computing science," Proceedings of the 34th Annual Hawaii International Conference on System Sciences, 2001; 2001 Jan 6; Maui, USA. Maui (US): IEEE. pp 5022.

[7] L. Kaufman, PJ. Rousseeuw, "Finding Groups in Data: An Introduction to Cluster Analysis," London (GB) : John Wiley and Sons.

[8] C. Beeley, "Web Application Development with R Using Shiny," Birmingham (GB) : Packt Publishing.

IOP Conf. Series: Earth and Environmental Science **31** (2016) 012005 doi:10.1088/1755-1315/31/1/012005

SOFish ver. 1.2 - A Decision Support System for Fishery Managers in Managing Complex Fish Stocks

A K Supriatna[1]*, A Sholahuddin[1], A P Ramadhan[1] and H Husniah[2]
[1]Department of Mathematics, *Padjadjaran University*, Sumedang 45363, INDONESIA
[2]Department of Industrial Engineering, Langlangbuana *University*, Bandung 40261, INDONESIA

E-mail: aksupriatna@gmail.com

Abstract. Sustainability is an important issue in a fishery industry. A manager of the fishery industry is responsible in deciding the best harvest that is able to sustain the industry while it should also guarantee the profitability of the industry. The most used concept in determining the best harvest in many fisheries industries is the Maximum Sustainable Yield (MSY). It represents the maximum amount of biomass that can be taken out from the fish population without harming the sustainability of the fishery. In other words, it is used to keep the population size stay over a threshold value of population level whenever harvesting activities is going on until indefinite time. In this paper we discuss a Decision Support System (DSS) for fishery managers in estimating the best harvest in a fishery industry. The best harvest is known as the Maximum Sustainable Yield (MSY) of the fishery. The DSS produces the MSY based on the discretization of some mathematical models of population growth, including the most popular models, such as Verhulst, Gompertz and Richards models. We also adding a biological complexity into the models, i.e. the presence of various degree of intra-specific competition of the population, which enhances the realism of the model and the DSS..

1. Introduction

Sustainability is an important issue in a fishery industry. A manager of the fishery industry is responsible in deciding the best harvest that is able to sustain the industry while it should also guarantee the profitability of the industry. The most used concept in determining the best harvest in many fisheries industries is the Maximum Sustainable Yield (MSY). It represents the maximum amount of biomass that can be taken out from the fish population without harming the sustainability of the fishery. In other words, it is used to keep the population size stay over a threshold value of population level whenever harvesting activities is going on until indefinite time.

The awareness among fisheries managers about the excessive of fish exploitation, its extinction issue, and fisheries population dynamic has created a demand for highly reliable software in the form of a decision support system (DSS). There are many examples of the use

* To whom any correspondence should be addressed.

of the DSS in fisheries industries, such as described in [1-5]. The software is usually directed for computing several complex mathematical processes that guides the user to take a proper decision that support population sustainability in managing the fisheries, e.g. by giving the right estimation of the MSY.

Knowing the amount of biomass of a fishery stock, at least its growth parameters, is regarded as an effective way in order to sustain the fish or aquatic populations in the fishery, since the MSY is a function of these growth parameters. In order to find the right value of the MSY, a DSS application needs a mathematical model used as the bases for the MSY calculation. In this paper we propose an integrated DSS using several different mathematical models as the bases for the MSY calculation. The models are continuous growth models, so to run the process in the computer, we discretize the models in terms of yield and effort variables to process the data input. There are two discrete forms for each models generated in the computation. These two discrete forms, in terms of yield and effort variables, are used as the bases for computing the intrinsic growth rate r and the carrying capacity K parameters. A Multiple linear regression with ordinary least square (OLS) method is needed to find these growth parameters, which are the main ingredient in obtaining the value of the MSY. The DSS is intended as the improvement of the previous version (SOFish ver. 1.0 [6] and SOFish ver. 1.1 [7]).

The rest of this paper consists of four sections and organized as follows. Section 2 introduces the mathematical models used in the development of the DSS. The main difference between the known mathematical models with the models in this paper is that here we consider a more complex biological interaction of the fish stock, where we assume that it has intra-specific competition between individuals in the fish stock. Section 3 discusses the design of the DSS followed by a case study using a known fishery data set in literature. Section 4 furnishes the concluding remarks of the research.

2. The Mathematical Models

It is known in literatures that to estimate the growth parameters in a continuous population model we can discretize the model in several different ways [1]. Here we consider a parameterization of four different well known models, the Verhulst growth model (1), the α-Verhulst growth model (2), the Richards growth model (3), the Gompertz model (4), the α-Gompertz growth model (5), and the generalized Gompertz growth model (6), as seen in the first two columns of Table 1. The appearence/disappearence of α in the different position reflects the different biological behaviour of the population, e.g.. the degree of intra-specific competition of the population. Here the growth parameters r and K , respectively, represent the intrinsic growth rate parameter and the carrying capacity parameter of the stock/population. Now, suppose there is a constant rate of harvesting C, then the dynamics of the exploited population for the Verhuslt growth model (1) is given by $dX/dt = rX(1-X/K)-C$ and for the Gompertz growth model (4) is given by $dX/dt = rX\ln(K/X)-C$. All other models follow the same pattern. In this new system, i.e. in the harvested system, sustainability can be achieved when the growth of the harvested population is non-negative. The easiest one is when the exploited population is at its equilibrium population size, i.e. when $dX/dt = 0$. This steady-state condition ensures the sustainability of the stock in the long term, and at the same time it gives the harvest as a function of stock abundance, i.e. $C = f(X,\alpha,r,K)$. The maximum value of C that is able to maintain the population at certain sustainable level is called the Maximum Sustainable Yield (MSY). Upon executing the same standard procedure, in the last two columns of Table 1, we obtain the MSY's for equations (1) to (6) given by expressions (7) to (12), respectively.

Table 1. Growth equations and their respective MSY's.

Growth equation	eqn. no.	MSY	eqn. no.
$dX/dt = rX\left(1-X/K\right)$	(1)	$\frac{rK}{4}$	(7)
$dX/dt = rX\left(1-X^{\alpha}/K\right)$	(2)	$r\left(\frac{K}{1+\alpha}\right)^{\frac{1}{\alpha}}\left(1-\frac{1}{1+\alpha}\right)$	(8)
$dX/dt = rX\left(1-X^{\alpha}/K^{\alpha}\right)$	(3)	$r\left(\frac{K^{\alpha}}{1+\alpha}\right)^{\frac{1}{\alpha}}\left(1-\frac{1}{1+\alpha}\right)$	(9)
$dX/dt = rX\ln\left(K/X\right)$	(4)	$\frac{rK}{e}$	(10)
$dX/dt = rX\ln\left(K/X^{\alpha}\right)$	(5)	$re^{\frac{1}{\alpha}\ln K-1}$	(11)
$dX/dt = rX\ln\left(K^{\alpha}/X^{\alpha}\right)$	(6)	$\alpha re^{\frac{1}{\alpha}\ln K-1}$	(12)

In practice, to obtain the value of the MSY directly from the continuous models is a difficult task. Because known fishery data is discrete and often measured in Catch per unit Effort (CPUE), and it is not measured in the population density X. Here we arrive at two complexities: the first one is dealing with transforming the continuous models into the discrete one, and the second is dealing with transforming the known data (Catch and Effort data) into the population density variable X. This task is challenging and involved some mathematical complexities and sophistication, which some managers try to avoid. We omit the detail derivation of the discretization, instead we propose a DSS that can be easily operated by fisheries managers to compute the MSY from the available catch-effort data they have at hand.

Some mathematical investigation have been undertaken resulted in fruitful theory of harvesting [8-25] and some still on-going pursuing some insight in many direction of research, e.g. by improving the realism of the model in terms of coupled sub-populations and in terms of intra-specific competition within sub-populations [26,27]. In this regards, our DSS is built in the endeavour to help the manager in deciding the best policy (i.e. determining the level of harvesting that can sustain the fishery in the long-run), without relying in a depth mathematical comprehension of the theory as the prerequisite. The DSS especially has a rich feature in terms of intra-specific competition of the underlying fishery populations, encompassing major well-known sigmoid shape population growth models.

3. The Decision Support System

The potential of a DSS as computerized tools in assisting decision-making process has been attracting many researchers for the last fifty years to improve decisions in many decision problems [28], especially for those complex problems. The issue has also been attracting many harvesters as the manager in fishery industries as explained in the introduction. In implementing the harvesting models in the DSS, here we develop discrete forms of the models shown in equations (1) to (6). The population density variable (*X*) in the model is transformed into the catch (*C*) and effort (*E*) variables via the relation *C=qEX*, where *q* denotes the catchability coefficient. The only data used as the input to run the DSS is the time series of *C* and *E*. This data is analysed using the ordinary least squares method for multiple linear regression to produce the MSY level for the respected model chosen in the DSS. The implementation of a prototype of the DSS was built in Visual C# for windows form application. In the previous study [1] we used several software that have a capability to perform the OLS in order to check the validity of some computational outputs. In order to make a good and fair comparison between the previous DSS prototype and the new one, we adopt the previous feature but make some improvement in several fields to the system, especially for the mathematical models or methods.

In the development of the DSS, the code used for the computational process of the OLS has been programmed manually. We do not use any library or any kind of plug-in that able to perform the multiple linear regression automatically. We choose this way in order to implement a step-by-step procedure of the mathematical process of the OLS, so that it will be easier to modify in the future when needed. We design the physical appearance of the DSS as simple as possible, where the main window is organized in single– compact sub-windows form with three major functional panels: Input data panel, regression panel and results panel as shown in Figure 1. The first step to do to use the DSS is to input the data needed in the MSY calculation into a number of rows. This can be done by pressing the "Insert Rows" button to reserve the number of rows depending the number of data we have (Figure 2). The automatic input button is available to use a pre defined data as an example in the MSY calculation with known MSY in the literature.

The next functional area is the input table. The availability of space depends on the number of rows that has been reserved in the previous step. In the table sheet, there are five columns with different headers that correspond to different categories: order or data id number, year, catch, effort, and Catch per unit Effort (CPUE). The DSS uses a Catch and Effort data in order to find the value of the MSY as a final consideration to the primary user. Catch and Effort data were required in order to continue the process of computation. The CPUE column would be filled automatically, right after one of mathematical model and method has been selected respectively in the option and the computation processes has been executed by the user. This characteristic makes the functionality of the table proceed as well as a spreadsheet. The user could execute them by clicking the MSY button and all required fields have completely filled before.

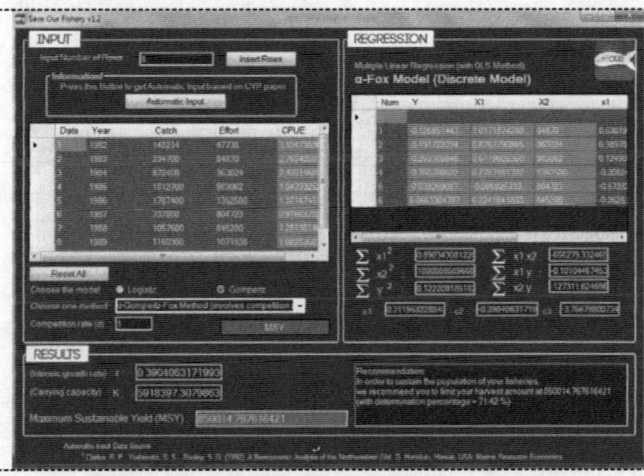

Figure 1. The DSS main window.

Figure 2. The user needs to enter the number of rows for the input table.

Figure 3. The MSY button.

The users also have to choose between two options of mathematical models, namely Verhulst and Gompertz family models. If the user considers to choose one of the Verhulst models in equations (1) to (3), there would be five options for the discrete approximation of Logistic model. There were Verhulst-Schaefer, α-Verhulst-Schaefer, Richards-Schaefer, α-Verhulst-Schnute and Richards-Schnute method on the selection box (Fig. 4.a). On the other hand, if the user considers to choose one of the Verhulst models in equations (4) to (6), then Gompertz-Fox, α-Gompertz-Fox, Modified- α-Gompertz-Fox, α-Gompertz-CYP and modified-α-Gompertz-CYP methods would come as the five options for the Gompertz model (Fig. 4.b).

The last four options for each model came as the improvement in this new version of the DSS, which is a distinct feature in the current version of the DSS and is not available in many known DSS model for fishery managers. In addition, almost all of the methods involve competition rate α to enhance the robustness of the model. The discrete approximation would be used for the computational processes of multiple linear regression with OLS in order to estimate the intrinsic growth rate *r* and carrying capacity *K*. The output of the OLS would be shown in the regression panel. In this panel, the user would find the regression coefficients that will be used to calculate the three growth parameters, r and K. The growth parameters would be used to find the MSY value. The value of each parameters and MSY would appear respectively in the results panel with the recommendation box that comes as suggestion for the user.

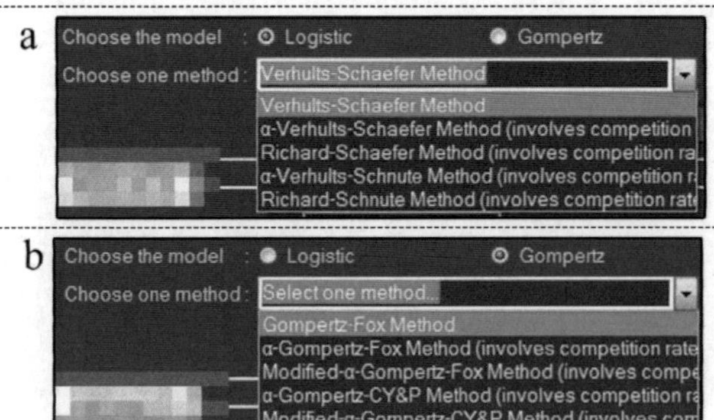

Figure 4. (a) Verhulst model with its discrete approximation; (b) Gompertz model with its discrete approximation.

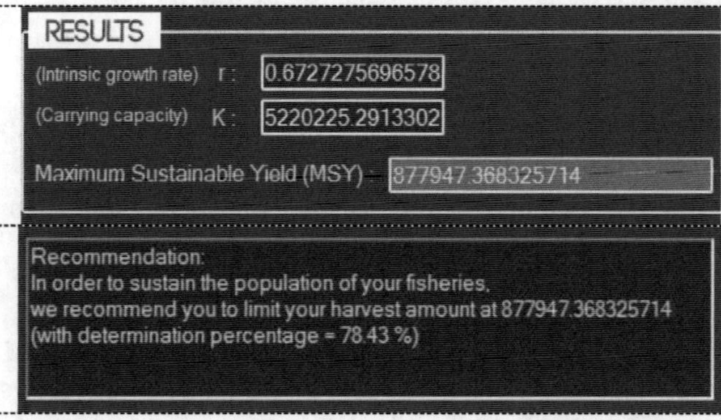

Figure 5. The results panel with the recommendation box.

4. Implementation and Discussion

In the DSS we use the discrete version of equation (1) derived by [14] (Verhulst-Schaefer method) and by [23] (Verhust-Schnute method). We follow the method in [14] and [23] to derive the discrete version of equation (2), to obtain the α-Verhulst-Schaefer and the α-Verhulst-Schnute methods. Analogously, we use the discrete version of equation (4) derived by [29] (Gompertz-Fox method) and by [24] (Gompertz-CYP method). We also follow the method in [29] and [24] to derive the discrete version of equation (5), to obtain the α-Gompertz-Fox and the α-Gompertz-CYP methods. The discrete equations can be seen in [7]. To test the software we use the data set in [24], because the MSY is already known in [24] for the reason of comparison. The results in Figure 5, uses the data set in [24], and the value of the resulting MSY in Figure 5 is exactly the same as in [24]. This indicates that the DSS is in agreement with known method, with the addition that in the DSS we can find the best α that fits the data, not necessarily 1 as in the known literature. This system need to improve in the future considering that there are some other biological complexities found in nature beside those treated in this paper, e.g. the metapopulation structure of the fish stocks. This is currently under investigation.

Acknowledgments

Authors wish to acknowledge financial support from the Indonesian Government through the scheme of Sinas Ristek 2015 and the travel grant from the ALG program.

References

[1] Alagappan M and Kumaran M 2013 Application of expert systems in fisheries sector – a review *Res. J. Anim. Vet. Fish. Sci.* **1**(8) 19-30
[2] Hughey K, Cullen R, Memon A, Kerr G and Wyatt N 2001 Developing a decision support system to manage fisheries externalities in New Zealand's exclusive economic zone *Microbehaviour and macroresults: Proc. Ten Bienn. Conf. Int. Institute Fish. Econ. and Trade*
[3] Lynch A J and Taylor W W 2013 Designing a decision support system for harvest management of Great Lakes Lake Whitefish in a changing climate *GLISA Project Reports* eds. Brown D, Bidwell D and Briley L (Available GLISA Center)
[4] Teniwut Y K and Marimin 2013 Decision support system for increasing sustainable productivity on fishery agroindustry supply chain *Int. Conf. Adv. Comp. Sci. Infor. Sys.* (ICACSIS)
[5] Truong T H, Rothschild B J and Azadivar F 2005 Decision support system for fisheries management *Proc. 2005 Winter Simul. Conf.* (IEEE Cat. No.05CH37732C), 1005

[6] Ramadhan A P, Setiawan K E, Amriyati P and Supriatna A K 2014 Software development in estimating parameters for Logistic and Gompertz population growth models arising from a fishery problem *E-Proc. 2nd Int. Conf. Artific. Intelli. Comp. Sci.* pp. 110-119 (E-ISBN: 978-967-11768-8-7)
[7] Supriatna A K, Ramadhan A P and Husniah H 2015 A decision support system for estimating growth parameters of commercial fish stock in fisheries industries *Procedia Comp. Sci. (elsevier)* **59** 331-339
[8] Cunningham S 1981 The evolution of the objectives of fisheries management during the 1970's *Ocean Manage.,* **6** 251-278
[9] Gordon H S 1954 The economic theory of a common-property resource: The fishery *J. Polit. Econ.* **62** 124-142
[10] Ricker W E 1954 Stock and recruitment *J. Fish. Res. Board of Canada* **11** 559-623
[11] Schaefer M B 1954 Some aspects of the dynamics of populations important to the management of the commercial marine fisheries *Bull. Inter-Amer. Trop. Tuna Comm.* **1** 27-56
[12] Clark C W 1976 *Mathematical Bioeconomics: The Optimal Management of Renewable Resources 1st Edn.* (New-York: John Wiley)
[13] Reed W J 1979 Optimal escapement levels in stochastic and deterministic harvesting models *J. Environ. Econ. Manage.* **6** 350-363
[14] Schaefer M B 1957 A study of the dynamic of fishery for yellowfin tuna in the Eastern Tropical Pacific Ocean *Bull. Inter-Amer. Trop. Tuna Comm.* **2** 247-285
[15] Conrad J M and Clark C W 1987 *Natural Resource Economics* (New-York: Cambridge University Press)
[16] Munro G R 1992 Mathematical bioeconomics and the evolution of modern fisheries economics *Bull. Math. Biol.* **54** 163-184
[17] Yodzis P 1994 Predator-prey theory and management of multispecies fisheries *Ecol. Appl.* **4** 51-58
[18] Botsford L W, Castilla J C and Peterson C H 1997 The management of fisheries and marine ecosystems *Science* **277** 509-515
[19] Mesterton-Gibbons M 1996 A technique for finding optimal two-species harvesting policies *Ecol. Model.* **92** 235-244
[20] Supriatna A K and Possingham H P 1998 Optimal harvesting for a predator-prey metapopulation *Bull. Math. Biol.* **60** 49-65
[21] Supriatna A K and Possingham H P 1999 Harvesting a two-patch predator-prey metapopulation *Nat. Resource Model.* **12** 481-498
[22] Tuck G N and Possingham H P 1994 Optimal harvesting strategies for a metapopulation *Bull. Math. Biol.* **56** 107-127
[23] Schnute, J 1977 Improved estimates from the Schaefer production model: theoretical consideration *J. Fish. Res. Board of Canada* **34** 583-603
[24] Clarke R P, Yoshimoto S S and Pooley S G 1992 A bioeconomic analysis of the Northwestern Hawaiian islands lobster fishery *Mar. Resource Econ.* **7** 115-140
[25] Verhulst P F 1838 Notice sur la loi que la population suit dans son accroissement *Corr. Mat. et Phys.* **10** 113–121
[26] Supriatna A K and Husniah H 2015 Sustainable harvesting strategy for natural resource having a coupled Gompertz production function *Interdisciplinary Behavior and Social Sciences* ed. Gaol L (Leiden: CRC Press Balkema) chapter 16 pp. 91-95
[27] Husniah H and Supriatna A K 2015 System dynamic approach in managing complex biological resources *ARPN J. Engineer. Appl. Sci.* **10** 1685-1690
[28] Mohemad R, Hamdan A R, Othman Z A and Noor N M 2010 Decision support systems (DSS) in construction tendering processes *Inter. J. Comp. Sci.* **7** 35-45
[29] Fox W 1970 Exponential surplus-yield model for optimizing exploited fish populations *Trans. Amer. Fish. Soc.* **99** 80-88

Bonus-Malus System with the Claim Frequency Distribution is Geometric and the Severity Distribution is Truncated Weibull

D N Santi[1], I G P Purnaba[2] and I W Mangku[3]
[1, 2, 3] Department of Mathematics, Bogor Agricultural University
Jalan Meranti Kampus IPB, Bogor, West Java, Indonesia

E-mail: [1]teresadiiaan5991@gmail.com, [2]purnaba@gmail.com
[3]wayan.mangku@gmail.com

Abstract. Bonus-Malus system is said to be optimal if it is financially balanced for insurance companies and fair for policyholders. Previous research about Bonus-Malus system concern with the determination of the risk premium which applied to all of the severity that guaranteed by the insurance company. In fact, not all of the severity that proposed by policyholder may be covered by insurance company. When the insurance company sets a maximum bound of the severity incurred, so it is necessary to modify the model of the severity distribution into the severity bound distribution. In this paper, optimal Bonus-Malus system is compound of claim frequency component has geometric distribution and severity component has truncated Weibull distribution is discussed. The number of claims considered to follow a Poisson distribution, and the expected number λ is exponentially distributed, so the number of claims has a geometric distribution. The severity with a given parameter θ is considered to have a truncated exponential distribution is modelled using the Levy distribution, so the severity have a truncated Weibull distribution.

1. Introduction

Bonus-Malus system is a system where in the next payment, policyholders who submitted one or more claim will be penalized by premium raise (malus), while policyholders who submitted no claim will be rewarded by premium reduction (bonus). Bonus-Malus system is said to be optimal if it is financially balanced for the insurance company (the total amount of bonus is equal to the total amount of malus) and fair to all policyholders, that is every policyholder pays premium proportionally with the risk [4].

In [1], it is discussed Bonus-Malus system based on the claim frequency having negative binomial distribution for modelling risk premium. While, [2] using claim frequency following Poisson inverse Gaussian distribution to minimize the risk of the insurer. Furthermore, Bonus-Malus system is not only based on the claim frequency but also based on the claim severity, were developed in [4] and [5]. This system uses a negative binomial distribution for claim frequency and Pareto distribution for claim severity. The other researchers, e.g. [3] used geometric distribution for claim frequency and Pareto distribution for claim severity on Bonus-Malus system. In the other hand, [7] is using negative binomial distribution in claim frequency and Weibull distribution for claim severity.

Previous research concern with the determination of the risk premium which applied to all of the severity that guaranteed by the insurance company. In fact, not all of the severity that proposed by policyholder may be covered by insurance company. When the insurance company sets a maximum bound of the severity incurred, so it is necessary to modify the model of the severity distribution into the severity bound distribution. Hence, in this paper, it is discussed Bonus-Malus system with the claim frequency distribution is geometric and the claim severity distribution is truncated Weibull, in this paper also compared the risk premium between the full severity and the severity with the maximum bound.

In Section 2, the process of deriving the risk premium for our models is discussed. The number of claims considered to follow a Poisson distribution, and the expected number λ is exponentially distributed, so the number of claims has a geometric distribution. The severity with a given parameter θ is considered to have a truncated exponential distribution is modelled using the Levy distribution, so the severity have a truncated Weibull distribution. The parameters are estimated using the quadratic error loss function [6]. The risk premium based on compound of geometric distribution and truncated Weibull distribution. In Section 3, as an application, the risk premium is calculated based on the claim frequency based on geometric distribution and the severity based on Weibull distribution and truncated Weibull distribution. In this section, we compared the risk premium between the full severity based on Weibull distribution and the severity with the maximum bound based on truncated Weibull. The conclusion of this paper is presented in Section 4.

2. Optimal Bonus-Malus System

In this section we discuss how to format the title, authors and affiliations. Please follow these instructions as carefully as possible so all articles within a conference have the same style to the title page. This paragraph follows a section title so it should not be indented.

The modelling of the claim frequency is the same as the one in [3]. The number of claims for given considered to follow a Poisson distribution, and the expected number of claims λ is exponentially distributed with parameter θ, so the number of claims has a geometric distribution with parameter $\theta(\theta + 1)^{-1}$. That is,

$$P(k)=\int_0^\infty P(k\mid\lambda)u(\lambda)d\lambda=\frac{\theta}{\theta+1}\left(\frac{1}{\theta+1}\right)^k,\quad k=0,1,2,\ldots \text{ and } 0<\frac{1}{\theta+1}<1. \tag{1}$$

Furthermore, by applying the Bayesian approach, the posterior distribution is given by,

$$u(\lambda\mid k_1,k_2,\ldots,k_t)=\frac{\lambda^K(\theta+t)^{K+1}e^{-\lambda(\theta+t)}}{\Gamma(K+1)},\lambda>0 \tag{2}$$

where $K=\sum_{i=1}^t k_i$ represents the total of claim frequency over t period with k_i present the number of claims in each period respectively. Using the quadratic loss function, the following choice for λ_{t+1}, the expected number of claims of a policyholder with a claim history $k_1, k_2, \ldots, k_t$, is :

$$\hat{\lambda}_{t+1}=\int_0^\infty \lambda u(\lambda\mid k_1,k_2,\ldots,k_t)d\lambda=\frac{K+1}{\theta+t}. \tag{3}$$

Equation (3) shows that, the risk premium payable at time $t+1$ depends on the claim history of the policyholder (K), time period (t) and the parameter of the exponential distribution (θ).

The modelling of the severity in [3] is using Weibull distribution. The expected number of severity for the next period is

$$\hat{\theta}_{t+1} = \frac{B_{K-\frac{3}{2}}\left(c\sqrt{M}\right)}{B_{K-\frac{1}{2}}\left(c\sqrt{M}\right)}\left(\frac{2\sqrt{M}}{c}\right), \tag{4}$$

where M is the total severity, K is the total of claim, c is the parameter of Weibull distribution.

So that the risk premium based on compound refer to (3) and (4) is given by,

$$\text{Premium}_{t+1} = \frac{K+1}{\theta+t}\left(\frac{B_{K-\frac{3}{2}}\left(c\sqrt{M}\right)}{B_{K-\frac{1}{2}}\left(c\sqrt{M}\right)}\left(\frac{2\sqrt{M}}{c}\right)\right). \tag{5}$$

When the insurance company sets a maximum bound of the severity incurred, so it is necessary to modify the model of the severity distribution into the severity bound distribution. Now, suppose that the amount of claim x is distributed according to the truncated exponential distribution with a given parameter θ. The cumulative distribution function is given by

$$F\left(X=x\,|\,\theta\right)=\begin{cases}1-e^{-\theta x}, 0<x<u\\ 1 \quad , x\geq u\end{cases}. \tag{6}$$

The parameter θ is referred as stable Levy distribution. Then the probability density function can be described as follows:

$$\pi\left(\Theta=\theta\right)=\frac{c}{2\sqrt{\pi\theta^3}}e^{-\frac{c^2}{4\theta}},\ \theta>0. \tag{7}$$

So the unconditional cumulative distribution function for $0<x<u$ is,

$$F(x)=1-\frac{2}{\sqrt{\pi}}\int_0^\infty e^{-\left(\delta^2+\frac{a}{\delta^2}\right)}d\delta,\ \ \frac{c^2x}{4}=a\geq 0. \tag{8}$$

The next task is solving refer to (8). First assume that,

$$\begin{aligned} I &= \int_0^\infty e^{-\left(\delta-\frac{\sqrt{a}}{\delta}\right)^2} d\left(\delta-\frac{\sqrt{a}}{\delta}\right) \\ &= \int_0^\infty e^{-\left(\delta-\frac{\sqrt{a}}{\delta}\right)^2} d\delta + \int_0^\infty e^{-\left(\delta-\frac{\sqrt{a}}{\delta}\right)^2}\left(\frac{\sqrt{a}}{\delta}\right)d\delta. \end{aligned} \tag{9}$$

Using substitution, let $g=\delta-\frac{\sqrt{a}}{\delta}\in\left(-\infty,\infty\right)$ and $\frac{dg}{d\delta}=1+\frac{\sqrt{a}}{\delta^2}>0$, then g is a monotonically increasing function from -∞ to ∞. So that refer to (9) becomes

$$I=\int_{-\infty}^{\infty} e^{-g^2}\left(1+\frac{\sqrt{a}}{\delta^2}\right)\frac{1}{\left(1+\frac{\sqrt{a}}{\delta^2}\right)}dg=\sqrt{\pi} \tag{10}$$

The form obtained from solving refer to (9) is needed to have $\int_0^\infty e^{-\left(\delta-\frac{\sqrt{a}}{\delta}\right)^2}d\delta$.

Let $\varepsilon=-\frac{\sqrt{a}}{\delta^2}$ and $d\varepsilon=-\frac{\sqrt{a}}{\delta^2}d\delta$. Since δ = 0 implies ε = ∞ and δ = ∞ implies ε = 0, refer to (9) can be written as

$$I=\int_0^\infty e^{-\left(\delta-\frac{\sqrt{a}}{\delta}\right)^2}d\delta+\int_0^\infty e^{-\left(\delta-\frac{\sqrt{a}}{\delta}\right)^2}\left(\frac{\sqrt{a}}{\delta}\right)d\delta$$

$$=\int_0^\infty e^{-\left(\delta-\frac{\sqrt{a}}{\delta}\right)^2}d\delta+\int_0^\infty e^{-\left(\frac{\sqrt{a}}{\varepsilon}-\varepsilon\right)^2}\left(\frac{\sqrt{a}}{\delta^2}\right)\left(-\frac{\delta^2}{\sqrt{a}}\right)d\varepsilon$$

$$=\int_0^\infty e^{-\left(\delta-\frac{\sqrt{a}}{\delta}\right)^2}d\delta+\int_0^\infty e^{-\left(\varepsilon-\frac{\sqrt{a}}{\varepsilon}\right)^2}d\varepsilon$$

$$=2\int_0^\infty e^{-\left(\delta-\frac{\sqrt{a}}{\delta}\right)^2}d\delta \tag{11}$$

Because $I=2\int_0^\infty e^{-\left(\delta-\frac{\sqrt{a}}{\delta}\right)^2}d\delta$ and $I=\sqrt{\pi}$ then we have $\int_0^\infty e^{-\left(\delta-\frac{\sqrt{a}}{\delta}\right)^2}d\delta=\frac{\sqrt{\pi}}{2}$. Equation (6) can be described as follows

$$F(x)=1-\frac{2}{\sqrt{\pi}}\int_0^\infty e^{-\left(\delta^2+\frac{a}{\delta^2}\right)}d\delta$$

$$=1-\frac{2}{\sqrt{\pi}}\int_0^\infty e^{-\left(\delta-\frac{\sqrt{a}}{\delta}\right)^2-2\sqrt{a}}d\delta$$

$$=1-e^{-\left(\frac{x}{c^{-2}}\right)^{\frac{1}{2}}} \tag{12}$$

So we have,

$$F(x)=\begin{cases}1-e^{-\left(\frac{x}{c^{-2}}\right)^{\frac{1}{2}}} & ,0<x<u\\ 1 & ,x\geq u\end{cases} \tag{13}$$

Thus, the cumulative distribution function above is the truncated Weibull distribution with parameter (c^{-2}, 0.5).

If x_i, $i = 1,2,\ldots,K$ denotes the amount of *i-th* claim, then the total amount of claimed for a policyholder over t period will be equal to $\sum_{i=1}^{K}x_i$.. We assumed that there is n claim with $0<x_i<u$, and the amount of claim for $x_i\geq u$ is $K-n$ claim. Suppose N is the total of amount claimed for $0<x_i<u$ that is $N=\sum_{i=1}^{n}x_i$. The total amount of K claim becomes $\sum_{i=1}^{K}x_i=\sum_{i=1}^{n}x_i+u(K-n)=N+u(K-n)$. Hence, the posterior distribution for θ can be obtained by using Bayesian approach as follows

$$\pi(\theta\mid x_1,x_2,\ldots,x_K)=\frac{\theta^{K-\frac{3}{2}}e^{-\left(\frac{c^2}{4\theta}+\theta(N+u(K-n))\right)}}{\int_0^\infty\theta^{K-\frac{3}{2}}e^{-\left(\frac{c^2}{4\theta}+\theta(N+u(K-n))\right)}d\theta}. \tag{14}$$

By slightly modifying the variables refer to (14), it can be rewritten as,

$$\pi\left(\theta \mid x_1, x_2, \ldots, x_K\right)=\frac{\left(\frac{c}{2\sqrt{N+u(K-n)}}\right)^{-\left(K-\frac{1}{2}\right)} \theta^{K-\frac{3}{2}} e^{-\left(\frac{c^2}{4\theta}+\theta(N+u(K-n))\right)}}{\int_0^{\infty}\left(\frac{2\theta\sqrt{N+u(K-n)}}{c}\right)^{K-\frac{3}{2}} e^{-\left(\frac{c\sqrt{N+u(K-n)}}{2}\left(\frac{2\theta\sqrt{N+u(K-n)}}{c}+\frac{c}{2\theta\sqrt{N+u(K-n)}}\right)\right)} d\left(\frac{2\sqrt{N+u(K-n)}}{c}\theta\right)}. \tag{15}$$

The integral on the denominator can be transformed to a modified Bessel function, whose integral representation is given as follow

$$B_v(x)=\frac{1}{2}\int_0^{\infty} y^{v-1} e^{-\frac{x}{2}\left(y+\frac{1}{y}\right)} dy,\ B_v(x)>0, \forall x>0, v\in R. \tag{16}$$

Then the posterior distribution refer to (15) can be written below,

$$\pi\left(\theta \mid x_1, x_2, \ldots, x_K\right)=\frac{\left(\frac{c}{2\sqrt{N+u(K-n)}}\right)^{-\left(K-\frac{1}{2}\right)} \theta^{K-\frac{3}{2}} e^{-\left(\frac{c^2}{4\theta}+\theta(N+u(K-n))\right)}}{2B_{K-\frac{1}{2}}\left(c\sqrt{N+u(K-n)}\right)}. \tag{17}$$

Using the quadratic loss function, the optimal choice for θ_{t+1} for a policyholder having amount of claim reports x_i, $i = 1,2,\ldots, K$ over t period is estimated as

$$\hat{\theta}_{t+1}=\frac{B_{K-\frac{3}{2}}\left(c\sqrt{N+u(K-n)}\right)}{B_{K-\frac{1}{2}}\left(c\sqrt{N+u(K-n)}\right)}\left(\frac{2\sqrt{N+u(K-n)}}{c}\right). \tag{18}$$

The risk premium to be paid at time $t+1$ for a policyholder whose number of claims history is k_1, $k_2, \ldots, k_t$ and whose amount claim history is $x_1, x_2, \ldots, x_k$ can be calculated according to the net premium as

$$\text{Premium}_{t+1}=\frac{K+1}{\theta+t}\left(\frac{B_{K-\frac{3}{2}}\left(c\sqrt{N+u(K-n)}\right)}{B_{K-\frac{1}{2}}\left(c\sqrt{N+u(K-n)}\right)}\left(\frac{2\sqrt{N+u(K-n)}}{c}\right)\right). \tag{19}$$

The risk premium in (19) is compound of refer to (3) and (18). It is shows that the risk premium depends on the claim frequency, time period, and total severity.

The calculation of risk premium with Weibull distribution is refer to (5) and the risk premium based on truncated Weibull distribution is show in (19). It is give varying results according to the proposed huge losses. At the beginning and when the policyholders have no claim, the risk premium paid are equal. In other words, when the total severity is $M \leq N + u\,(K - n)$, the risk premium based on Weibull distribution will be equal to the one based on truncated Weibull distribution. Whereas, when the total severity is $M > N + u\,(K - n)$, the risk premium based on the Weibull distribution is more expensive than the one based on the truncated Weibull distribution.

Let suppose the ratio of two Bessel functions is

$$Q_K\left(c\sqrt{N+u(K-n)}\right)=\frac{B_{v-1}\left(c\sqrt{N+u(K-n)}\right)}{B_v\left(c\sqrt{N+u(K-n)}\right)}$$

$$=\frac{B_{K-\frac{3}{2}}\left(c\sqrt{N+u(K-n)}\right)}{B_{K-\frac{1}{2}}\left(c\sqrt{N+u(K-n)}\right)}. \tag{20}$$

When $K = 1$ and $B_v(x)=B_{-v}(x)$, then we have

$$Q_1\left(c\sqrt{N+u(K-n)}\right)=\frac{B_{-\frac{1}{2}}\left(c\sqrt{N+u(K-n)}\right)}{B_{\frac{1}{2}}\left(c\sqrt{N+u(K-n)}\right)}=1. \tag{21}$$

Because $B_{v+1}(x)=\frac{2v}{x}B_v(x)+B_{v-1}(x)$, so the recursive function Q_k is

$$\frac{1}{Q_{K+1}\left(c\sqrt{N+u(K-n)}\right)}=\frac{2\left(K-\frac{1}{2}\right)}{c\sqrt{N+u(K-n)}}+Q_K\left(c\sqrt{N+u(K-n)}\right). \tag{22}$$

In (19), when the total severity is $N + u(K - n) = 0$, it is means that no claim filed ($K = 0$), it shows that the risk premium is undefined. Hence, we redefine the risk premium when $K = 0$. Then the risk premium for the following period if no claim filed will be given by,

$$\text{Premium}_{t+1\mid N+u(K-n)=0}=\frac{1}{t+\theta}\left(\frac{2}{c^2}\right). \tag{23}$$

This means that when there are no claim reported, the greater of the period time will cause the smaller of the risk premium must be paid.

3. Numerical Illustration

3.1. Bonus-Malus System without Truncated Distribution

In this set up the number of claims is assumed to follow a geometric distribution with parameter θ $(\theta + 1)^{-1}$, $\theta = 1.5$. Total amount of claim is assumed to follow a truncated Weibull distribution with parameter c^{-2}, $c = 0.052$. The calculation of the risk premium to be paid by each policyholder if no claim filed refer to (23), whereas if there is a claim filed refer to (5). Total of claim severity used are 8000 and 10000. The risk premium are calculated for *t-th* period, $t = 0,1,2,\ldots,7$ and for the number of claims $K = 0,1,2,\ldots,5$.

When there is no change in the number of claims filed, the risk premium gets cheaper along with increasing time. For example, if a policyholder submits one claim with total loss is 8000 in the first period of observation, the amount of risk premiums have to be paid is 2752, it can be seen in Table 1. Furthermore, in the second period, the policyholder submits another claim with loss is 2000. It makes the number of claims during the two periods are 2 claims and the total loss is 10000, so the policyholder have to pay for 2765 that can be seen in Table 2.

Table 1. The result of risk premium with total severity is 8000

Period	Number of claim (K)					
(t)	0	1	2	3	4	5
0	493					
1	296	2752	3398	3749	3918	3980
2	211	1966	2427	2678	2798	2843
3	164	1529	1888	2083	2176	2211
4	134	1251	1544	1704	1781	1809
5	114	1058	1307	1442	1507	1531
6	99	917	1133	1250	1306	1327
7	87	809	999	1103	1152	1171

Table 2. The result of risk premium with total severity is 10000

Period	Number of claim (K)					
(t)	0	1	2	3	4	5
0	493					
1	296	3077	3871	4347	4612	4744
2	211	2198	2765	3105	3294	3389
3	164	1709	2151	2415	2562	2636
4	134	1399	1760	1976	2096	2156
5	114	1183	1489	1672	1774	1825
6	99	1026	1290	1449	1537	1581
7	87	905	1139	1279	1356	1395

3.2. Bonus-Malus System with Truncated Distribution

In order to see the effect of the truncated distribution on claim severity, the calculation of the risk premium refer to (23) if no claim filed, whereas if there is a claim filed refer to (19). The maximum bound of premiums that can be covered by insurance companies is u = 2300. Claim severity x_i is assumed to be equal to the average of the claim size, that is 1000.

Table 3 shows that the risk premium prices depend on the claim frequency variation. For example, in the beginning the policyholder pays the premium for 493. If in the first period there is no accident, there will be a premium reduction to the level 296. But, if a policyholder have one accident, the policyholder have to pay 1476 if the amount of claim is more than 2300, otherwise have to pay 973.

Table 3. The result of risk premium with $u = 2300$

Period (t)	Number of claim (K)									
	0	1		2			3			
		Number of claim with the severity less than u (n)								
		0	1	0	1	2	0	1	2	3
0	493									
1	296	1476	973	2439	1986	1443	3393	2940	2441	1879
2	211	1054	695	1742	1419	1031	2424	2100	1744	1342
3	164	820	541	1355	1104	802	1885	1633	1356	1044
4	134	671	442	1109	903	656	1542	1336	1110	854
5	114	568	374	938	764	555	1305	1131	939	723
6	99	492	324	813	662	481	1131	980	814	626
7	87	434	286	717	584	425	998	865	718	553

Table 3. The result of risk premium with $u = 2300$ (continue)

Period (t)	Number of claim (K)										
	4					5					
	Number of claim with the severity less than u (n)										
	0	1	2	3	4	0	1	2	3	4	5
0											
1	4341	3881	3393	2868	2299	5284	4818	4331	3820	3282	2711
2	3101	2772	2423	2049	1642	3774	3441	3093	2729	2344	1936
3	2412	2156	1885	1594	1277	2936	2677	2406	2122	1823	1506
4	1973	1764	1542	1304	1045	2402	2190	1969	1736	1492	1232
5	1670	1493	1305	1103	884	2032	1853	1666	1469	1262	1043
6	1447	1294	1131	956	766	1761	1606	1444	1273	1094	904
7	1277	1142	998	844	676	1554	1417	1274	1124	965	797

4. CONCLUSION

This paper consider Bonus-Malus system with the claim frequency distribution is geometric that compound distribution of Poisson distribution and exponentially distribution and the claim severity distribution is truncated Weibull which compound distribution of truncated exponentially distribution and Levy distribution. Based on the results, it can be shown that the risk premium depend on the level of risk. When the total severity are less than the maximum bound of severity or $M \leq N + u\,(K - n)$, the risk premium based on Weibull distribution will be equal to the one based on truncated Weibull distribution. Where M is the total severity, u is the maximum bound that covered by insurance company, N is the total severity which the severity are lower than bound u, K is the total claim frequency, and n is the claim frequency which the frequency are lower than bound u. In other hand, when the total of severity are greater than the maximum bound of the severity or $M > N + u\,(K - n)$, the risk premium based on the Weibull distribution is more expensive than the one based on truncated Weibull distribution.

IOP Conf. Series: Earth and Environmental Science **31** (2016) 012006 doi:10.1088/1755-1315/31/1/012006

5. References

[1] Georges D and Charles V 1989 A generalization of automobile insurance rating models: the negative binomial distribution with a regression component *J. ASTIN Bulletin* **19** 199-212.

[2] Luc T 1992 Using the poisson inverse gaussian in bonus-malus systems *J. ASTIN Bulletin* **22** 97-106.

[3] Mehmet M and Yasemin S 2005 On a bonus malus system where the claim frequency distribution is geometric and the claim severity distribution is pareto *J. Hacettepe Journal of Mathematics and Statistics* **34** 75-81.

[4] Nicholas E F and Spyridon D V 2011 Design of optimal bonus-malus systems with a frequency and a severity component on an individual basis in automobile insurance *J. ASTIN Bulletin* **31** 1-22.

[5] Rahim M and Hossein H 2009 Generalized bonus-malus systems with a frequency and a severity component on an individual basis in automobile insurance *J. ASTIN Bulletin* **39** 307-315.

[6] Stuart A K, Harry H P and Gordon E W 2012 *Loss Models* (United States of America: John Wiley & Sons).

[7] Weihong N, Corina C and Athanasios A P 2014 Bonus-malus systems with weibull distributed claim severities *J. Annals of Actuarial Science* **8** 217-233.

IOP Conf. Series: Earth and Environmental Science **31** (2016) 012007 doi:10.1088/1755-1315/31/1/012007

Constant optimization of oral drug absorption kinetics in the compartment absorption and transit models using particle swarm optimization algorithm

K Prabowo[1], **T Sumaryada**[2,3], **A Kartono**[3]
[1] Department of Physics, Bogor Agricultural University, Jalan Meranti Kampus Dramaga IPB, 16880, INDONESIA
[2]Biopharmaca Research Center, Bogor Agricultural university, Jalan Taman Kencana No.3, Bogor 16128, INDONESIA
[3]Computational Biophysics and Molecular Research Group (CBMoRG), , Department of Physics, Bogor Agricultural University, Jalan Meranti Kampus Dramaga IPB, 16880, INDONESIA

E-mail: kemal.prabowo@gmail.com

Abstract. Simulation of predictive modeling oral drug namely Compartment Absorption and Transit (CAT) using Particle Swarm Optimization (PSO) algorithm has been performed. This research will be carried out optimization of kinetic constant value oral drug use PSO algorithm to obtain the best global transport constant values for CAT equation that can predict drug concentration in plasma. The value of drug absorption rate constant for drug atenolol 25 mg is k_{10}, k_{12}, k_{21}, k_{13} and k_{31} with each value is 0.8562 , 0.3736 , 0.2191 , 0.4334 and 1.000 have been obtained thus raising the value of the coefficient of determination of a model CAT. From the experimental data plasma drug concentrations used are Atenolol, the coefficient of determination (R^2) obtained from simulations atenolol 25 mg (PSO) was 81.72% and 99.46%. Better correlation between the dependent variable as the drug concentration and explanatory variables such as mass medication, plasma volume, and rate of absorption of the drug has increased in CAT models using PSO algorithm. Based on the results of CAT models fit charts can predict drug concentration in plasma.

1. Introduction

Modern drug design is not only focused on the pharmacological activity but also reach the area where the drug is working and the ability of the drug to be absorbed by the body. A few of mathematical models that can predict the oral bioavailability of a drug, where bioavabilitas is the extent to which the drug is absorbed in the body. Predictive modeling oral drug that made Yu et al namely Compartment Absorption and Transit (CAT) has provided an understanding of oral drug absorption since the 1995's until the present[1-2]. The mathematical model is an interesting tool that provides an understanding of oral drug absorption since the 1650s until now.

Particle Swarm Optimization (PSO) is a method of optimization solutions that is adapted from the behavior of animals such as birds flock movements are then each object animals consider to be a particle. A particle in D-dimensional space has a position that is encoded as a vector coordinates. This position vector is considered as a condition of the particle in the search space. Each position in the search space is an alternative solution that can be evaluated using the objective function. Each particle moves with a certain speed. Particle Swarm Optimization (PSO) to apply the nature of each individual in a large group. Then combine these properties to solve problems. Particle Swarm Optimization was introduced in 1995, since then that the researchers who developed the method of PSO. The unique thing of the PSO algorithm is heuristic particle speed settings and probabilistic. If a particle has a constant speed so if the position of a particle visualized trail will form a straight line. With the external factors that alter the line will then move the particle in the search space it is expected that the particles can lead, approached, and ultimately achieve an optimal point. External factors in question include the best position ever skipped a particle and the best position of the entire particles (each particle is assumed to know the best position of any other particles).

In this simulation, to optimize the absorption of orally administered drugs in the small intestine are used PSO algorithm. Given the effectiveness of the concept and simple, PSO optimizer has become popular and has been widely applied in practical problem solving. Thus, the theoretical study and improvement of the performance of the algorithm has become an important and attractive[3-18]. This research will be carried out optimization of kinetic constant value oral drug use PSO algorithm to obtain the best global transport constant values for CAT equation that can predict drug concentration in plasma.

2. Theory

2.1. Particle Swarm Optimization

There are n particles in a space dimension D, the position of the i-th particle in a room is $r_i = (r_{i1}, r_{i2}, ..., r_{iD})$ $, i = 1,2,3,...,n$, the particle velocity is $\dot{r}_i = (\dot{r}_{i1}, \dot{r}_{i2}, ..., \dot{r}_{iD})$ $, i = 1,2,3,...,n$ and the best position of a particle is $P_i = (P_{i1}, P_{i2}, ..., P_{iD})$ $, i = 1,2,3,...,n$. The position of the particle swarms is $P_g = (P_{g1}, P_{g2}, ..., P_{gD})$ $, i = 1,2,3,...,n$, where the PSO algorithm is as follows:

$$r_{id} = r_{id} + \dot{r}_{id} \tag{1}$$

$$\dot{r}_{id} = \dot{r}_{id} + c_1\beta_1(p_{id} - r_{id}) + c_2\beta_2(p_{gd} - r_{id}) \tag{2}$$

Where, $i=1,2,...,n$, $d = 1,2,...,D$, c_1 and c_2 is the coefficient identifier on each swarm and are positive. The relative coefficient value expressing the degree of the relative importance of Pi and Pg with developments. β1 and β2 is a random value between 0-1, is determined by the user, equation (1) and (2) can be modified into

$$r_{id} = \gamma\, r_{id} + \dot{r}_{id} \tag{3}$$

$$\dot{r}_{id} = \omega\, \dot{r}_{id} + c_1\beta_1(p_{id} - r_{id}) + c_2\beta_2(p_{gd} - r_{id}) \tag{4}$$

Where γ and ω each is a constant factor and the factor of inertia. This model will be simulated in space with specific dimensions with a number of iterations so that in each iteration, the particle's position will increasingly lead to the intended target (minimizing or maximizing functions). This is done until the maximum iteration can also be used to achieve or other termination criteria.

2.2. Compartement Absorption and Transit (CAT) models

The distribution of the different compartments have been used to simulate and explain the oral drug absorption. therefore it is necessary defining the number of compartments closest to the small intestine absorption characteristics in humans. CAT models created to describe the transit flow of drugs in the human intestine. This model can explain the problem and estimate the plasma concentrations of oral absorption of oral medications. This model defines that the drug crosses the small intestine passes through a series of compartments. Each compartment has a linear transfer kinetics. Each compartment is smoothly flow rate and volume are different but have Kt (constant transit) are the same.. In this model the digestive system is divided into three segments, namely the stomach, small intestine, and large intestine. Human small intestine is divided into seven compartments with drug transfer from one compartment to the other compartment occurs in the first order.

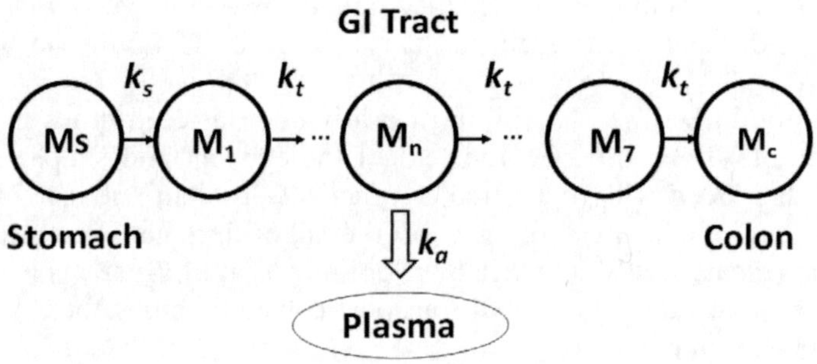

Figure 1 CAT diagram with linear transit and absorption kinetics[1]

Figure 1 illustrates the CAT models to calculate the flow of transit in the stomach, duodenum, jejenum, ileum and colon. Drugs should not be described soon after entry into the human body where M_s, M_c and M_a are the mass of the drug in the stomach, small intestine and colon, K_s, K_a and K_t are rate constant of gastric emptying, intrinsic absorption and small intestine transit. So that modeling transit and gastrointestinal absorption can be formulated as follows.

Stomach:

$$\frac{dM_s}{dt} = -K_s M_s, \qquad \text{where, } K_s = \frac{1}{T_{ge}} \tag{5}$$

Small Intestine:

$$\begin{aligned}
\frac{dM_1}{dt} &= K_s M_s - K_t M_1 - K_a M_1, \\
\frac{dM_2}{dt} &= K_s M_1 - K_t M_2 - K_a M_2, \\
\frac{dM_3}{dt} &= K_s M_2 - K_t M_3 - K_a M_3, \\
\frac{dM_4}{dt} &= K_s M_3 - K_t M_4 - K_a M_4, \\
\frac{dM_5}{dt} &= K_s M_4 - K_t M_5 - K_a M_5,
\end{aligned} \tag{6}$$

$$\frac{dM_6}{dt} = K_s M_5 - K_t M_6 - K_a M_6,$$

$$\frac{dM_7}{dt} = K_s M_6 - K_t M_7 - K_a M_7,$$

Where, $K_a = 2/(P_{eff} * R)$, $K_t = 7/T_{si} = 7/3.32\, jam$ and $K_t M_0 = K_s M_s$

Colon:

$$\frac{dM_c}{dt} = -K_t M_a, \tag{7}$$

The amount of drug absorbed by the small intestine:

$$\frac{dM_a}{dt} = K_a \sum_{a=1}^{7} M_n, \quad n = 1,\dots,7 \tag{8}$$

Yu et al provide modeling compartment as follows[1]:

$$\frac{dC_1}{dt} = \frac{1}{V_1}\frac{dM_a}{dt} - (k_{12} + k_{13} + k_{10})C_1 + k_{21}C_2 + k_{31}C_3$$

(9)

$$\frac{dC_2}{dt} = k_{12}C_1 + k_{21}C_2 \tag{10}$$

$$\frac{dC_3}{dt} = k_{13}C_1 + k_{31}C_3 \tag{11}$$

where V_1 is the volume of the central compartment, k_{10}, k_{12}, k_{21}, k_{13} and k_{31} are the value of drug absorption rate constant.

3. Result and Discussion

Parameter values such as k_{10}, k_{12}, k_{21}, k_{13} and k_{31} is used as a constant transport of (Compartment Absorption and Transit) CAT models can be optimized using (Particle Swarm Optimization) PSO algorithm. The value of drug absorption rate constant for drug atenolol 25 mg is k_{10}, k_{12}, k_{21}, k_{13} and k_{31} with each value is 0.8562 , 0.3736 , 0.2191 , 0.4334 and 1.000 have been obtained thus raising the value of the coefficient of determination of a model CAT. From the experimental data plasma drug concentrations used are Atenolol, the coefficient of determination (R^2) obtained from simulations atenolol 25 mg and 25 mg (PSO) was 81.72% and 99.46%. Based on the picture, the drug experienced three phase process in which plasma absorption, post absorption phase and elimination phase. Undergo drug absorption phase indicated by the increase in concentrations of the drug, and the drug experienced post absorption phase is indicated by a decrease in drug concentration. The last stage is the elimination phase which is characterized by a decrease in medication to near zero. Based on the results of CAT models fit charts can predict drug concentration in plasma.

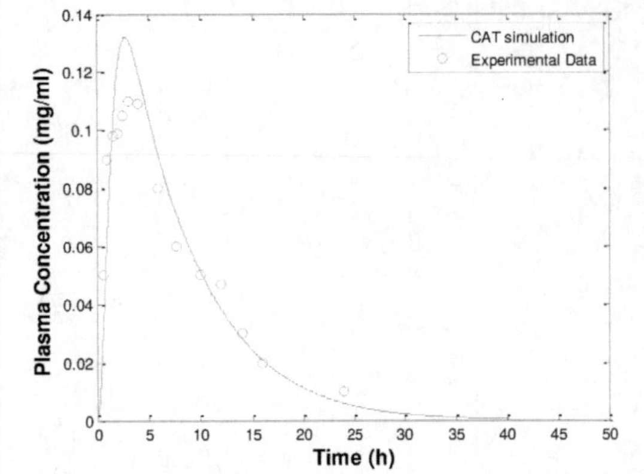

Figure 2. CAT models without PSO (Atenolol 25mg), R^2=81,72%

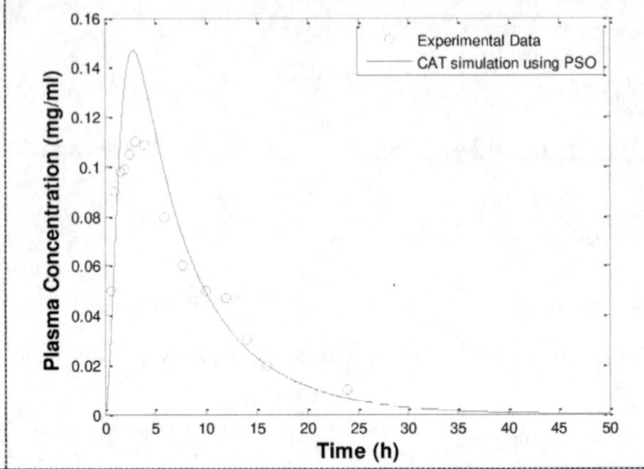

Figure 3. CAT models with PSO (Atenolol 25mg). R^2=99,46%

Table 1. CAT model parameters for drug atenolol 25 mg and 25 mg (PSO)

Parameters	Doses (mg)	
	25 (PSO)	25
F	0.56 [a]	0.56 [a]
P_{eff}(cm/hr)	0.19 [a]	0.19 [a]
T_{ge}(hr)	2 [a]	2 [a]
K_a(hr^{-1})	0.2171 [a]	0.2171 [a]
K_s(hr^{-1})	0.5 [a]	0.5 [a]
K_{12}(hr^{-1})	0.3736 [c]	0.05 [b]
K_{13}(hr^{-1})	0.4334 [c]	0.9 [b]
K_{21}(hr^{-1})	0.2191 [c]	11.32 [b]
K_{31}(hr^{-1})	1.0 [c]	0.39 [b]
K_{10}(hr^{-1})	0.8562 [c]	0.9 [b]

[a] Yu et al (1999).
[b] estimates.
[c] Using PSO with upper and lower bound that has been determined from 0 to 1.

4. Conclusion

This study reported that the PSO algorithm can determine the rate constant of drug absorption and also increases the coefficient of determination of CAT models in order to estimate the oral drug absorption in plasma

5. References

[1] Yu LX, Amidon GL. A compartmental absorption and transit model for estimating oral drug absorption. Int J Pharm. 1999;186:119–25.

[2] Yu LX, Lipka E, Crison JR, Amidon GL. Transport approaches to the biopharmaceutical design of oral drug delivery systems: prediction of intestinal absorption. Adv Drug Deliv Rev. 1996;19:359–76.

[3] R. C. Eberhart and Y. Shi, "Guest editorial," *IEEE Trans. Evol. Comput.—Special Issue Particle Swarm Optimization*, vol. 8, no. 3, pp. 201–203, Jun. 2004

[4] J. Kennedy and R. C. Eberhart, "Particle swarm optimization," in

Proc. IEEE Int. Conf. Neural Netw., Perth, Australia, 1995, vol. 4, pp. 1942–1948.
[5] R. C. Eberhart and J. Kennedy, "A new optimizer using particle swarm theory," in *Proc. 6th Int. Symp. Micromachine Human Sci.*, Nagoya, Japan, 1995, pp. 39–43
[6] J. Kennedy, R. C. Eberhart, and Y. H. Shi, *Swarm Intelligence*. San Mateo, CA: Morgan Kaufmann, 2001.
[7] R. C. Eberhart and Y. H. Shi, "Particle swarm optimization: Developments, applications and resources," in *Proc. IEEE Congr. Evol. Comput.*, Seoul, Korea, 2001, pp. 81–86.
[8] X. D. Li and A. P. Engelbrecht, "Particle swarm optimization: An introduction and its recent developments," in *Proc. Genetic Evol. Comput. Conf.*, 2007, pp. 3391–3414.
[9] R. A. Krohling and L. dos Santos Coelho, "Coevolutionary particle swarm optimization using Gaussian distribution for solving constrained optimization problems," *IEEE Trans. Syst., Man, Cybern. B, Cybern.*, vol. 36, no. 6, pp. 1407–1416, Dec. 2006.
[10] N. Franken and A. P. Engelbrecht, "Particle swarm optimization approaches to coevolve strategies for the iterated prisoner's dilemma," *IEEE Trans. Evol. Comput.*, vol. 9, no. 6, pp. 562–579, Dec. 2005.
[11] S.-Y. Ho, H.-S. Lin, W.-H. Liauh, and S.-J. Ho, "OPSO: Orthogonal particle swarm optimization and its application to task assignment problems," *IEEE Trans. Syst., Man, Cybern. A, Syst., Humans*, vol. 38, no. 2, pp. 288–298, Mar. 2008.
[12] B. Liu, L. Wang, and Y. H. Jin, "An effective PSO-based memetic algorithm for flow shop scheduling," *IEEE Trans. Syst., Man, Cybern. B, Cybern.*, vol. 37, no. 1, pp. 18–27, Feb. 2007.
[13] R. C. Eberhart and Y. Shi, "Guest editorial," *IEEE Trans. Evol. Comput.—Special Issue Particle Swarm Optimization*, vol. 8, no. 3, pp. 201–203, Jun. 2004.
[14] I. C. Trelea, "The particle swarm optimization algorithm: Convergence analysis and parameter selection," *Inf. Process. Lett.*, vol. 85, no. 6, pp. 317–325, Mar. 2003.
[15] K. Yasuda, A. Ide, and N. Iwasaki, "Stability analysis of particle swarm optimization," in *Proc. 5th Metaheuristics Int. Conf.*, 2003, pp. 341–346.
[16] V. Kadirkamanathan, K. Selvarajah, and P. J. Fleming, "Stability analysis of the particle dynamics in particle swarm optimizer," *IEEE Trans. Evol. Comput.*, vol. 10, no. 3, pp. 245–255, Jun. 2006.
[17] F. van den Bergh and A. P. Engelbrecht, "A study of particle optimization particle trajectories," *Inf. Sci.*, vol. 176, no. 8, pp. 937–971, Apr. 2006.
[18] M. Clerc and J. Kennedy, "The particle swarm-explosion, stability and convergence in a multidimensional complex space," *IEEE Trans. Evol. Comput.*, vol. 6, no. 1, pp. 58–73,

Elucidation of GB1 Protein Unfolding Mechanism via a Long-timescale Molecular Dynamics Simulation

T Sumaryada[1, 2,*], J Hati[1], S T Wahyudi[1], N D Malau[1] and K N Sawitri[1]
[1]Computational Biophysics and Molecular Modelling Research Group (CBMoRG), Department of Physics, Bogor Agricultural University, Jalan Meranti Kampus IPB Dramaga Bogor 16680, Indonesia
[2]Biopharmaca Research Center, Bogor Agricultural University, Jalan Taman Kencana No. 3, Bogor 16128, Indonesia

E-mail: tsumaryada@ipb.ac.id

Abstract. This study investigates the unfolding mechanism of 1GB1 protein at various simulation temperatures using a long-timescale molecular dynamics simulation. Analysis of structural parameters of molecular dynamics simulation have indicated that the unfolding process of GB1 protein has started at 95 ns for 475 K simulation, and at 745 ps for 500 K simulation. The unfolding process in this simulation exhibit the feature of hydrophobic core collapse model, in which the beta-hairpin destruction precedes the α-helix to coil transition. The unfolding was started with the increasing flexibility of the beta-sheets and hydrophobic core region, continued with beta-hairpins destruction, and ended with α-helix to coil and turn transition. The final structures of GB1 protein after unfolding, suggest an unfinished denaturation of protein as seen from the small remains of α-helix structure.

1. Introduction

The folding/unfolding mechanism of protein is still one of the hot topic in life sciences. The ability to predict the pathway of folding/unfolding event and mechanism of protein could lead to the breakthrough and revolution in biotechnology, especially in biomedical sciences and drug design engineering. Molecular dynamics (MD) simulation that basically rely on the first principal of physics laws can be used to simulate the folding/unfolding process of proteins and and elucidate its mechanism. The typical time scale of folding and unfolding event in nature usually in microseconds or milliseconds time frame. To realistically mimic that event through a computer simulation, a computationally expensive process and time consuming effort must be done. This burden sometimes hinder scientists from using a full microseconds or milliseconds time frame simulation. However, the advance of computational algorithm, computer software and hardware nowadays has make long-timescale MD possible [1].

B1 domain of protein G (pdb code 1GB1) [2] is one of the small proteins which extensivelly studied to unveil the folding/unfolding mechanism of protein. The small number of residues (56 residues) and a highly symmetric structure has made 1GB1 protein as a as one of the favorite choices for scientists to study the folding/unfolding process and to unveil the most responsible interactions in maintaining protein in its native state. The high thermostability of beta hairpin fragment of GB1 protein have been well studied using MD simulation [3, 4-7]. The simulation of a full 56-residues of GB1 protein has also been done by some researchers [8-12]. Unlike beta-hairpin, simulation of a full

GB1 protein is more difficult and complex due to more secondary structures and molecular interactions involved.

Two models that commonly used in explaining the folding/unfolding mechanism of protein, the zipper model, and the hydrophobic core collapse model. In the zipper model [13] the folding process is dominated by hydrogen bond force [7, 14, 15], while in the hydrophobic collapse model the folding process is dominated by the hydrophobic force from the hydrophobic residues buried in the core of protein [5,17].

In this research, long-timescale (sub-microseconds) molecular dynamics simulations of GB1 protein have been conducted at three different temperatures, 450 K, 475 K, and 500 K for 0.10 μs to determine the pathway process of unfolding event and to reveal the unfolding mechanism of a GB1 protein. Some parameters that were analyzed including Root Mean Squared Deviation (RMSD), Root Means Squared Fluctuations (RMSF), secondary structure, and cartoon representation (snapshot) of the protein.

2. Materials and Methods

There are three important parts in this research. The first part is protein data preparation, the second is running the molecular dynamics simulation, and the last part is data analysis.

2.1. Preparation of protein file

The structure of B1 domain of protein G was taken from the Protein Data Bank with pdb code name 1GB1. This original file still contains of predicted coordinates of hydrogen atoms which have to be removed. Protein must be prepared in a simulation box filled with the solvent atoms, TIP3P water box [18] with the size of 80x80x80Å^3. The sodium and chloride counterions were used to neutralize the protein during simulation. All those preparations were done by using Virtual Molecular Dynamics program (VMD) [19].

2.2. Molecular dynamics simulations

The MD simulations were performed using NAMD v.2.9 simulator [20] with CHARMM++ used as the force field. The Particle Mesh Ewald (PME) and Periodic Boundary conditions (PBC) methods were used to calculate the interactions between atoms. Simulation can be divided into two parts. The first part is minimizations, heating and equilibration. Minimizations were run in four steps and are aimed to minimize the energy of the protein. After minimizations, protein was heated from 0 K to the final temperatures (three different simulations with three different final temperatures i.e 450 K, 475 K and 500 K) then equilibrated at constant temperature using Langevin protocol. The second part is production run, where protein held at constant temperature for 100 ns. All simulations was performed in single CPU powered by 3.4 GHz Intel® core i7 processor with 12 GB of RAM and using Ubuntu 12.04 Linux platform..

2.3. Data analysis

For the data analysis, the MD output (dcd type of files) were analyzed using VMD to produce the output such as RMSD, RMSF, secondary structures, and cartoon representation of protein structure. Some of the MD data need to be smoothed first using moving average method to be properly displayed in this paper. The data obtained after the unfolding event were cut and not included in the analysis.

3. Results and Discussions

The dynamics of protein structure and conformation can be observed through the RMSD plot. The RMSD change above 2.50 Å is commonly used as an indicator for significant conformational changes in protein structure, and RMSD above 8.0 Å indicates an unfolding event. For 450 K simulation, in general there was no significant change in RMSD occured, with the average RMSD lied within 1.0 Å to 2.0 Å range, except for some fluctuations at 33 ns and 65 ns (see figure 1(a)). For 475 K simulation, a significant structural change occured within 80.0 to 90.0 ns simulation time, and followed by a drastic change of RMSD at 95 ns (see figure 1(a)). This drastic change of RMSD up to 12 Å indicates a possible unfolding event happened at 475 K. For 500 K simulation the RMSD value jumps drastically from 1.0 Å to 9.5 Å at 745 ps which indicates the transition from a folded to an unfolded structure as seen in figure 1(b). This sharp transition is commonly expected from a typical two-state folding kinetics [13]. Notice that the result of RMSD at T=500 K is shown in the separate plot due to different time scaling.

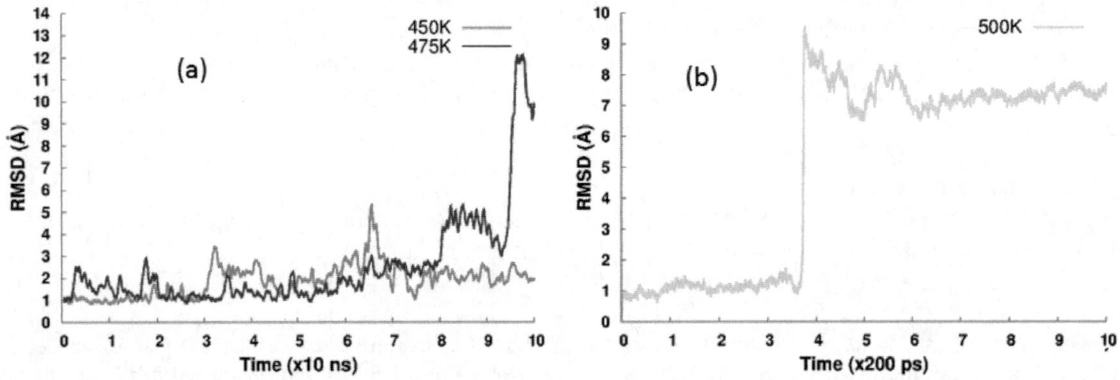

Figure 1. RMSD as a function of simulation time for (a) simulation temperatur at 450 K and 475 K and (b) at 500 K.

The RMSF value shows the flexibility of each residue during the simulation. The RMSF of each residue is consistenly increased as the temperature is raised. At 450 K the flexible residues were mostly found in coil (M1.E56) and turn (T11, A48) regions. At 475 K more flexibility were found in coil (M1, D22, G38, G41, E56) and in turn (T11, A48, K50) as seen from figure 2 (a). In contrast, when unfolding occurred at 500 K the flexibility were shifted to secondary structures, α-helix (E27, V29, Y33, D36), and beta-sheet (L5, I6, L7, G14, T16, E42) regions (figure. 2(b)).

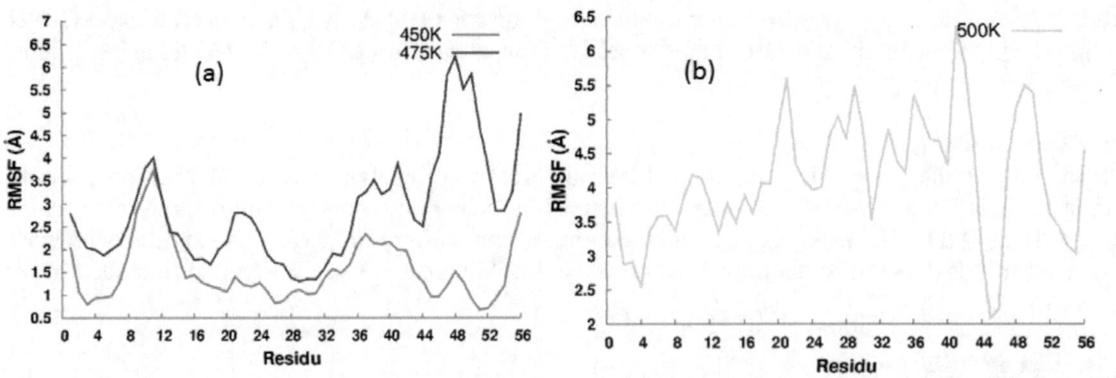

Figure 2. RMSF as a function of simulation time for (a) simulation temperatur at 450 K and 475 K and (b) at 500 K

The increasing flexibility, or a rigidity weakening, of secondary structures indicates the beginning of unfolding process. We also found that, of five hydrophobic core residues in the beta-

IOP Conf. Series: Earth and Environmental Science **31** (2016) 012008 doi:10.1088/1755-1315/31/1/012008

hairpins region (L5, L7, G9, F52, V54), two of them (L5, L7) have shown a significant increasing of flexibility at transition point. This found emphasizes the domination of hydrophobic core collapse process and beta-hairpins destruction upon helix to coil and turn transformation. The above results are in agreement with findings from the other groups [5,10-12,17].

The secondary structures of protein at various temperatures were shown in figure 3. There is no significant change observed at temperature of 450 K and 475 K. At the end of 475 K simulation, about one third of α-helix transform into a turn structure, while the beta-sheets remain intact. The unfolding signature of GB1 protein occurred at 500 K as indicated by the disappearance of beta-sheets and some of α-helix structure which transformed into 3_{10}-helix, coil and turn structures. The cartoon representation of protein secondary structures (figure 4) suggest a complete destruction (collapse) of the hydrophobic core in the beta-sheet region at T=500 K.

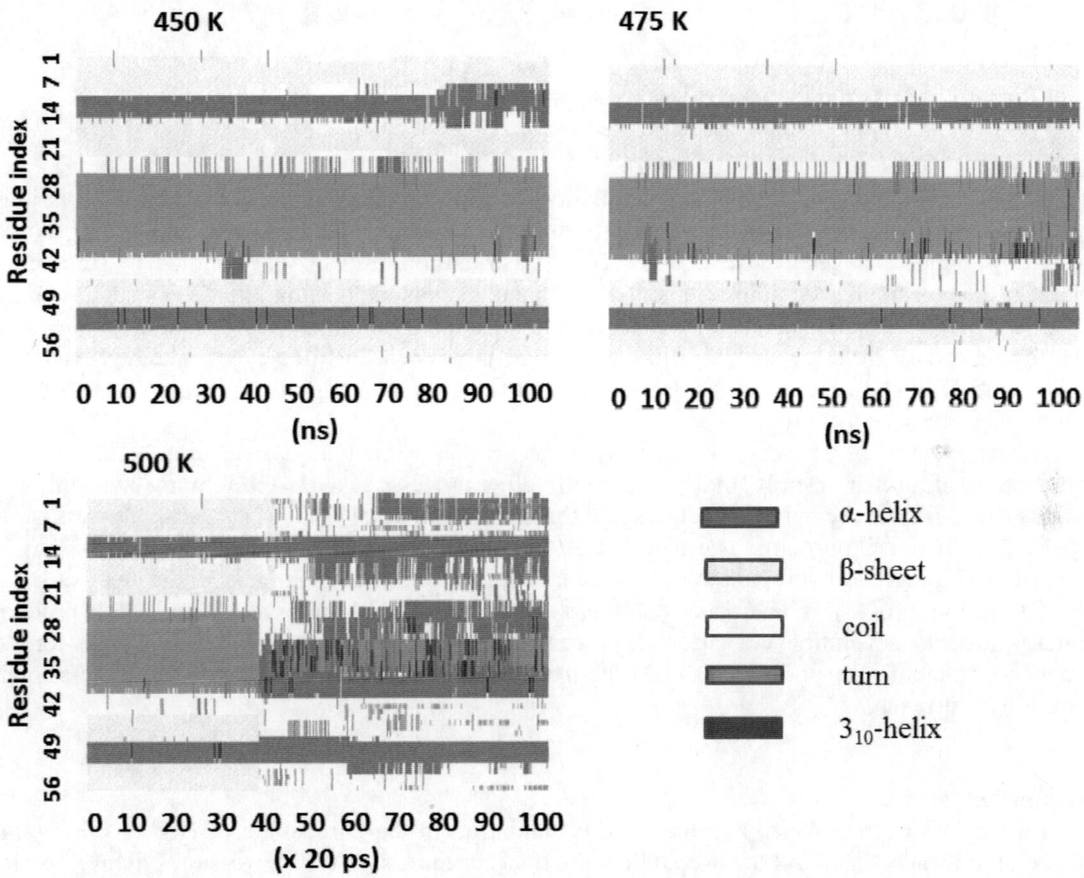

Figure 3. Evolution of secondary structures at various simulation temperatures.

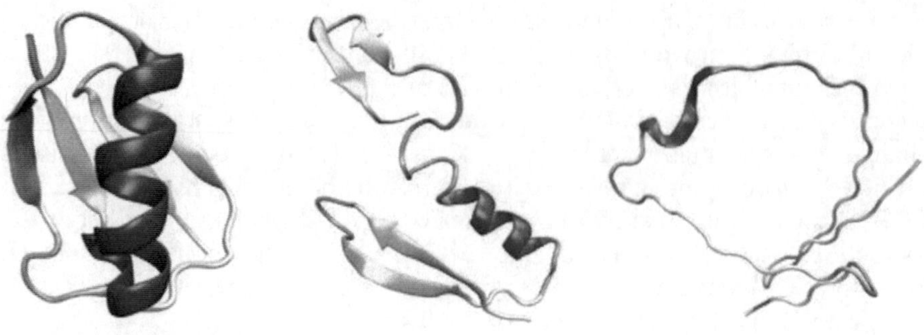

Figure 4. Cartoon representation of protein final structures at various temperatures

4. Conclusions

The long-time scale molecular dynamics simulation of GB1 protein has been performed at various temperatures. The unfolding signatures were found at transition point (T=500K and t=745 ps) as indicated by sudden changes in RMSD and secondary structure profiles. Analysis of MD simulation results has shown that the unfolding transition exhibited a two-state folding kinetics as suggested by other groups [13, 21-23]. Based on RMSD value, the indication of unfolding also found at 95 ns at 475 K, but the secondary structure analysis did not support the existence of this event. The high value of RMSD at 475 K likely correspond to a transformation from a compact globular into an open tertiary structure as can be seen from figure 4.

The structural parameters changes such as RMSF, secondary structures, and cartoon representation of protein clearly indicate an unfolding process of GB1 protein by showing the increasing flexibility of secondary structures and the disappearance of beta-sheet and some parts of α-helix structure at transition point. The fact that at 500 K there are six the most flexible residues in beta-sheet (including two hydrophobic core residues) compare to four in α-helix, indicates that beta-sheet destruction process initiate and dominate the overall unfolding process of GB1 protein. Secondary structures evolution and cartoon representation of protein emphasize a clear support for the hydrophobic core collapse model, as can be seen from the disappearance of all beta-sheets and a small part of α-helix structure.

5. Acknowledgment

The authors express their gratitude for the funding from the Directorate of Higher Education, Ministry of Education and Culture Republic of Indonesia through BOPTN research grant contract number 237/IT3.41.2/L2/SPK/2013 code 2013.089.521219.

6. References

[1] Klepeis J L, Lindorff-Larsen K, Dror R O and Shaw D E 2009 *Curr. Op. Struct. Bio.* **19** (2) 120
[2] Gronenborn A M, Fillipula D R, Essig N Z, Achari A, Whitlow M, Wingfield P T, Clore G M 1991 *Science*. **253** 657
[3] Pande V S and Rokhsar D S 1999 *Proc. Natl. Acad. Sci.USA*. **96** 9062
[4] Roccatano D, Amadei A, Di Nola A, Berendsen H J, 1999 *Protein Sci.* **8** (10) 2130
[5]. Lee J and Shin S 2001 *Biophys. J.* **5** 2507

IOP Conf. Series: Earth and Environmental Science **31** (2016) 012008 doi:10.1088/1755-1315/31/1/012008

[6] Klimov D K and Thirumalai D, 2000 *Proc.Natl. Acad. Sci.USA*. **97** (6) 2544
[7] Ma B and Nussinov R 2000 *J. Mol. Biol*. **296** (4) 1091
[8] Bui J M, Gsponer J, Vendruscolo M, Dobson C M 2009 *Biophyics Journal*. **97** 2513
[9] Kouza M, Hansmann U H E 2012 *J. Phys. Chem. B* **116** 6645
[10] Ceruso M A, Amadei A and Di Nola A 1991 *Protein Science* **8** 147
[11] Sheinerman F B and Brooks C L 1998 *J. Mol. Biol*. **278** (2) 439
[12] Sheinerman F B and Brooks C L 1997 *Proteins Struct. Funct. Gen*. **29** (2) 193
[13] Muñoz V, Thompson P A, Hofrichter J and Eaton W A 1997 *Nature* **390** 196
[14] Zhou R, Berne B J and Germain R 2001 *Proc. Natl. Acad. Sci*. **98** (26) 14931
[15] Zagrovic B, Sorin E J and Pande V 2001 *J. Mol. Biol*. **313** (1) 151
[16] Tanford C 1980 *The hydrophobic Effect* 2nd ed (New York :Wiley)
[17] Garcia A E, Sanbonmatsu K Y 2001 *Proteins* **42** (3) 345
[18] Jorgensen W L, Chandrasekhar J, Madura J D, Impey R W and Klein M L 1983 *J. Chem.Phys*.**79** 926
[19] Humprey W, Dalke A and Schulten K 1996 *J.Molec.Graphics* **14** 33
[20] Phillips J C, Braun R, Wang W, Gumbart J, Tajkhorshid E, Villa E, Chipot C, SkeelR D, Kalé L and Schulten K 2005 *Journal of Computational Chemistry*, **26** (16) 1781
[21] Blanco F J, Rivas G and Serano L 1994 *Nat.Struct.Biol*.**1** 584
[22] Alexander P, Orban J and Bryan P *Biochemistry* **31**(32) 7243
[23] Chung H S, Louis J M and Eaton W A 2010 *Biophys. J*. **98** 696

Molecular Docking Studies of Catechin and Its Derivatives as Anti-bacterial Inhibitor for Glucosamine-6-Phosphate Synthase

H Fikrika[1], L Ambarsari[1], T Sumaryada[2,3]*

[1]Department of Biochemistry, Bogor Agricultural University, Jalan Agatis Kampus Dramaga IPB, 16880, Indonesia

[2]Computational Biophysics and Molecular Modeling Research Group (CBMoRG), Department of Physics, Bogor Agricultural University, Jalan Meranti Kampus IPB Dramaga Bogor 16680, Indonesia

[3]Biopharmaca Research Center, Bogor Agricultural University, Jalan Taman Kencana No. 3, Bogor 16128, Indonesia

E-mail: tsumaryada@ipb.ac.id

Abstract. Molecular docking simulation of catechin and its derivatives on Glucosamine-6-Phosphate Synthase (GlmS) has been performed in this research. GlmS inhibition by a particular ligand will suppress the production of bacterial cell wall and significantly reduce the population of invading bacteria. In this study, catechin derivatives i.e epicatechin, galloatechin and epigalloatechin were found to have stronger binding affinities as compared to natural ligand of GlmS, Fructose-6-Phosphate (F6P). Those three ligands were docked on the same pocket in GlmS target as F6P, with 70% binding sites similarity. Based on the docking results, gallocatechin turns out to be the most potent ligand for anti-bacterial agent with ΔG= -8.00 kcal/mol. The docking between GlmS and catechin derivatives are characterized by a constant present of a strong hydrogen bond between functional group O3 and Ser-349. This hydrogen bond most likely plays a significant role in the docking mechanism and binding modes selection. The surprising result is catechin itself exhibited a quite strong binding with GlmS (ΔG= -7.80 kcal.mol), but docked on a completely different pocket compared to other ligands. This results suggest that catechin might still have a curing effect but with a completely different pathway and mechanism as compared to its derivatives.

1. Introduction

Tea is the most popular beverage in the world. One kind of tea is green tea, which has a high level of flavonoid compounds due to the absence of the oxidation process [1]. Those compounds, catechin and its derivatives (epicatechin, gallocatechin, and epigallocatechin) [2], have been known for a long time to have anti-oxidant, anti-diabetic, anti-bacterial, anti-inflamation, anti-virus, and anti-cancer potential [3].

The Glucosamine-6-phosphate synthase (GlmS) enzyme plays an important role in the synthesis of Bacterial cell wall. GlmS help the enzymatic reaction between fructose-6-phosphate (F6P) and L-glutamine to produce glucosamine-6-phosphate which lead to the formation of UDP-N acetylglucosamine which plays an important role in building peptidoglycan of bacterial cell wall. Inhibiting GlmS with a particular ligand will suppress the production of glucosamine-6-phosphate and kill the bacteria [4]. Inhibition of GlmS for anti-microbial purpose have been conducted by

Balachandran *et al* [5]. In that paper N-benzyl-2,2,2 trifluoroacetamide was used as a ligand, and had been proven to be a potential anti-microbial drugs.

The potential of flavonoid compounds of green tea have been extensively studied by [6,7,8,9]. In this paper, the potential of catechin and its derivatives for anti-bacterial purposes were explored through a molecular docking method. Docking simulations were performed between catechin (and its derivatives) and GlmS enzyme as the target. Fructose-6-Phosphate (F6P), one of the natural inhibitors of GlmS, was used as the standard ligand. The docking of F6P to GlmS were important to the production of glucosamine-6-phosphate, which is crucial to the formation of bacterial cell wall. By comparing the inhibition profiles of catechin and its derivatives with F6P, we can define their potential as anti-bacterial agents.

2. Materials and Methods

There are three important parts in this research. The first part is preparation of target protein and ligands, the second is running the molecular docking simulation, and the last part is data analysis.

2.1. Receptor Preparation

Three dimensional structure of GlmS (pdb code 2VF5) was downloaded from www.rcsb.org (see figure 1) [10]. Receptor data were downloaded and opened using Discovery Studio Visualizer to remove the natural ligand that still attached to it. Before the docking process, the non polar hydrogen atoms were added, followed by Gasteiger charger calculation using Autodock tools (ADT) 1.5.6. [11]. The protein file then saved in pdbqt format and ready to be used for docking.

2.2. Ligand Preparation

There are five ligands prepared for docking simulation, one natural inhibitors as a control (or standard) and four catechin and its derivatives (as test ligands). Three dimensional structure of ligands were downloaded from PubChem (see figure 1), opened and saved in pdb format using Marvin View 6.0 program from ChemAxon (http://www.chemaxon.com). Polar hydrogen atoms were added and Gasteiger charges were assigned using Discovery Studio 3.5 Client. All ligands has to be saved in pdbqt format.

2.3. Docking Methods

AutoDock vina [12] program was used to dock ligand with GlmS. Targeted docking method was used in this research with the coordinate of origin was set at x=30.054, y=22.75 and z= 4.171. This set of numbers was based on the location of F6P binding with Glms from protein data bank file. The box size was set at x=40, y=40 and z=40. Docking simulation were run with the number of modes set to 20 to get more accurate results.

2.4. Docking Analysis

Analysis of binding affinity and the binding sites were performed in this research. LigPlot 1.4.5 [13] was used to produce two dimensional docking representations of all ligands. There are two types of interaction analyzed, hydrogen bond and hydrophobic interaction. Interaction occured between amino acid residue on the target side and functional group on the ligand side. Binding site similarity analysis between test ligands and standard (natural) ligand were performed in this research to show the potential of catechin and its derivatives as anti-bacterial agent.

3. Results and Discussions

The docking simulation results are shown in Table.1. F6P as standard ligand binds with GlmS with binding affinity of -6.00 kcal/mol (see figure.2). The binding sites of F6P on GlmS target includes five residues involved in the hydrogen bond and six residues involved in hydrophobic interactions. The strongest binding occurred at Ser-303 in which two hydrogen bonds were formed with functional group O3 (bond length of 2.89 Å) and O1(bond length of 3.24 Å).

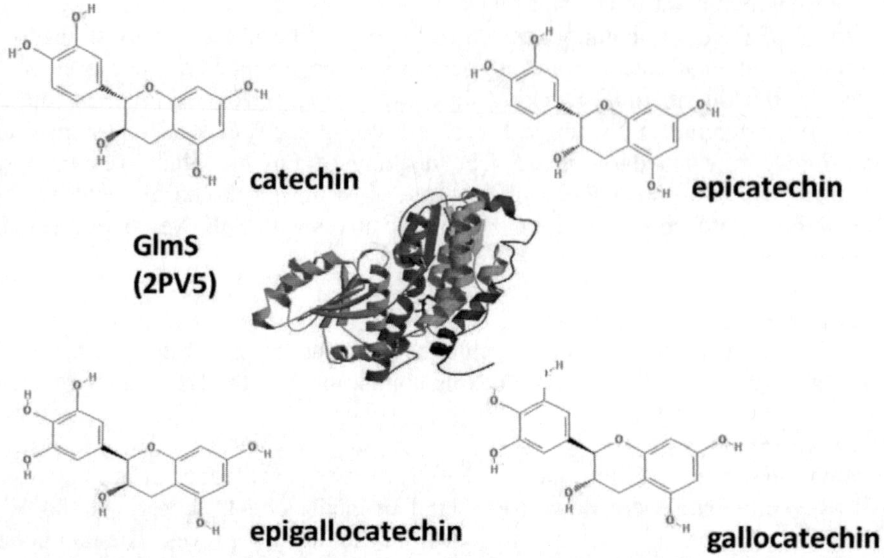

Figure1. Structure of target protein GlmS (pdb code 2PV5) and test ligands. Protein structure downloaded from www.rcsb.org and ligand's structures were downloaded from http://pubchem.ncbi.nlm.nih.gov

Table.1 shows that catechin were docked at completely different sites as compared to other ligands, as can be seen from zero binding site similarity. The binding affinity of catechin on GlmS is found to be -7.80 kcal/mol. This strong binding is mostly due to strong hydrogen bonds between Val-567 with functional group O5 (bond length of 2.72 Å) and O6 (bond length of 2.75 Å) as seen in figure 2. There are four residues involved in hydrogen bond and nine in hydrophobic interactions.

Epicatechin docked on the same region as F6P. Functional group O3 made two hydrogen bonds with Ser-348 and Ser-349. Ten hydrophobic interactions surround epicatechin inside the binding pocket (figure 3). The binding site similarity of epicatechin to F6P is 70.0% (7 out of 10 residues). The binding site similarity of all ligands as compared to F6P is also shown by highlighted residues in Table.1. The binding affinity of epicatechin is -7.50 kal/mol, with the strongest interaction found in hydrogen bond between O3 and Gln-348. Two residues,Thr-352 and Val-399, that previously involved in hydrogen bond with F6P, now changed into hydrophobic interactions with epicatechin.

Gallocatechin has the strongest binding affinity as compared to other ligands with $\Delta G= -8.00$ kcal/mol. This strong binding is due to the existence of four hydrogen bonds, with three of them have bond length < 3.00 Å. There are eight GlmS residues involved in hydrophobic interactions, as seen in figure 3. Two residues, Thr-352 and Glu-488, that previously involved in hydrogen bond with F6P now switched into hydrophobic interactions with gallocatechin. The binding site similarity of gallocatechin as compared to F6P is 70% (7 similar residues out of 10 GlmS residues docked by F6P).

Epigallocatechin has the binding affinity of -7.90 kcal/mol with similar type of hydrogen bond as epicatechin has (Gln-348 and Ser-349 hydrogen bond with functional group O3). One more hydrogen bond is also formed between Gln-348 with functional group O2. Ten hydrophobic interactions with identical sites as compared to epicatechin surround epigallocatechin (figure 3). The binding site similarity of epigallocatechin as compared to F6P is also 70%. Gln-348 which previously involved in hydrophobic interaction with F6P now changed into hydrogen bond with epigallocatechin.

Ser-349 is found to be a continuously present hydrogen bond in F6P, epicatechin, gallocatechin and epigallocatechin. In catechin derivatives, all hydrogen bonding with Ser-349 is occurred in functional group O3. Interaction between functional group O3 in catechin derivatives and Ser-349 residue in GlmS seems important and play a crucial role in determining the potential of catechin derivatives as anti-bacterial agent. The existence of Glu-488 binding site in the docking of F6P and catechin derivatives is in agreement with the results from Balachandran *et.al* [5] that used N-benzyl-2,2,2 trifluoroacetamide ligand for anti-microbial purpose. This finding emphasizes the results of our docking simulation is on the right track in searching for anti-bacterial and anti-microbial agents.

Table 1. Docking simulation results

Test ligand	Hydrogen bond length (Å)	Residue involved in H-bonding	Functional group of ligand	Residue involved in Hydrophobic interaction	Number of binding site similarity to standard ligand	Binding Affinity ΔG (kcal/mol)
Fructose-6-Phosphate (standard ligand)	2.89	Ser-303	O3	Thr-302, Ser-347, Gln-348, Ala-400, Ser-401, Ala-602	All	-6.00
	3.24	Ser-303	O1			
	3.22	Ser-349	O2			
	3.18	Thr-352	O9			
	3.02	Val-399	O8			
	3.09	Glu-488	O8			
Catechin	3.16	Arg-472	O3	Ala-520, Pro-521, Leu-525,Asp-548, His-566, Glu-569, Ala-572, Pro-573, Tyr-576	0	-7.80
	2.93	Asp-474	O3			
	3.23	Asn-522	O4			
	2.72	Val-567	O5			
	2.75	Val-567	O6			
	3.25	Val-567	O6			
Epicatechin	2.93	Gln-348	O3	Cys-300, Gly-301, Ser-347, Thr-352, Val-399, Leu-484, Glu-488, Leu-601, Ala-602, Lys-603	7	-7.50
	3.09	Ser-349	O3			
Gallocatechin	2.80	Thr-302	O5	Cys-300, Gly-301, Ser-347, Thr-352, Glu-488, Leu-601, Lys-603, Ser-604	7	-8.00
	2.93	Gln-348	O3			
	2.92	Ser-349	O3			
	3.11	Ala-602	O7			
Epigallocatechin	3.15	Gln-348	O2	Cys-300, Gly-301, Ser-347, Thr-352, Val-399, Glu-488, Leu-484,Leu-601, Ala-602, Lys-603	7	-7.90
	3.00	Gln-348	O3			
	3.08	Ser-349	O3			

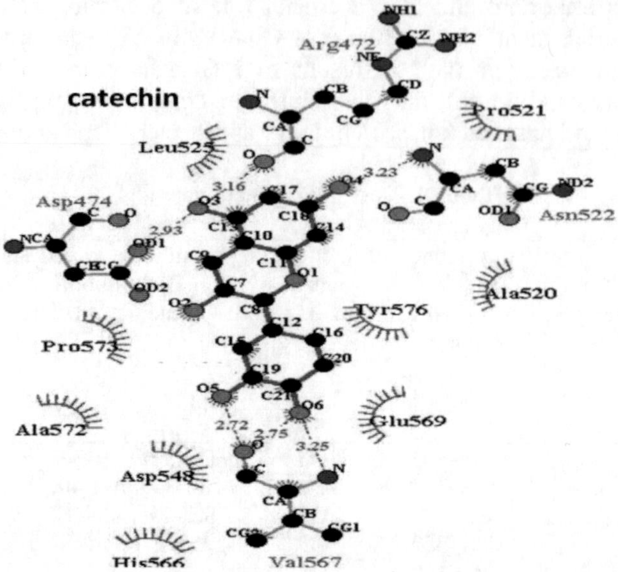

Figure 2. Docking representation of catechin on GlmS target.

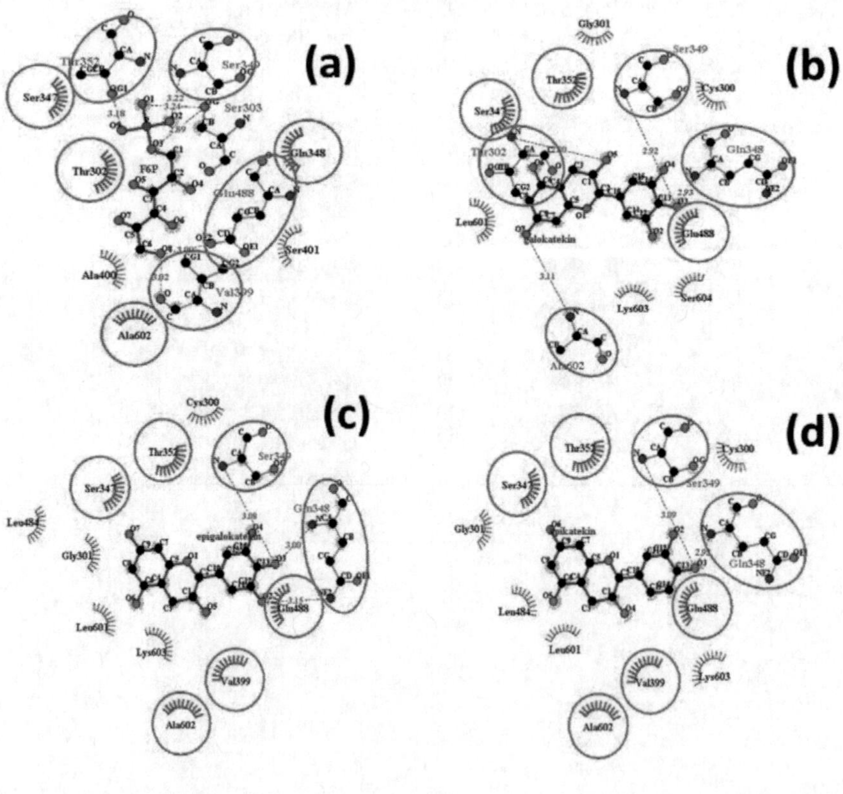

Figure 3. Comparison of docking (binding) sites of (a) Fructose-6-Phosphate, (b) gallocatechin, (c) epigallocatechin and (d) epicatechin on GlmS target. Note that all ligands docked on almost identical region (pocket).

4. Conclusions

Molecular docking simulation of catechin and its derivatives on GlmS target has been performed. All catechin derivative ligands exhibited a strong binding with GlmS with gallocatechin shows the best potential to be developed into an anti-bacterial drug. Except for catechin, all ligands in principal docked on the same pocket (binding sites) which indicates the same biochemistry pathway as compared to F6P, the natural ligand of GlmS. Catechin, on the other hand has also shown a strong binding to GlmS, but docked on a completely different pocket as compared to F6P. This result arise a new question whether catechin still have the remedy effect, but following a different pathway and mechanism as compared to F6P. Further study is clearly needed here.

5. Acknowledgment

The authors express their gratitude for the funding from the Directorate of Higher Education, Ministry of Education and Culture Republic of Indonesia through BOPTN research grant contract number 237/IT3.41.2/L2/SPK/2013 code 2013.089.521219.

6. References

[1] Banerjee S and Chatterjee J. 2014 *J.Food Sci.Technol* **487**-3
[2] Gramza A, Korczak J and Amarowicz R. 2005 *A Review. Pol. J. Food Nutr. Sci.* **14**(3) 219-235
[3] Dubreuill J D 2013 *Toxins.* **5** 2009-2041
[4] H Chmara, E Borowski 1986 *Acta. Microbiol. Pol* **35** 15-27
[5] Balachandran C et al 2015 Appl.Nanosci **5** 207-216
[6] Kumar D et al 2015 *J. Korean Soc Appl.Biol.Chem* ***58***(4) 581-585
[7] Reygaert W C 2014 *Frontiers in Microbiology* **5** (434) 1-8
[8] Taylor P W, Hamilton-Miller J M T, Stapleton P D. 2005 *Food.Sci.Technol.Bull* **2** 71-81
[9] Lee L S, Kim S H, Kim Y B and Kim Y C 2014 *Molecules.***19** 9173-9186
[10] Mouilleron S, Badet D M A, Golinell P B 2014 *J Mol Biol.* **337**:1174
[11] Morris G M, Huey R, Lindstrom W, Sanner M F, Belew R K, Goodsell D S and Olson A J 2009 *J.* Computational Chemistry **16** 85-91
[12] Trott O, Olson A J, 2010 Journal of Computational Chemistry **31** 455-461
[13] Laskowski R A, Swindells M B 2011 *J Chem.Inf.Model.* **51**(10) 2778-2786

High Temperature Unfolding and Low Temperature Refolding Pathway of Chymotrypsin Inhibitor 2 Using Molecular Dynamics Simulation

N D Malau[1] and T Sumaryada[1,2]
[1]Computational Biophysics and Molecular Modeling Research Group (CBMoRG), Department of Physics, Bogor Agricultural University, Jalan Meranti Kampus IPB Dramaga Bogor 16680, Indonesia
[2]Biopharmaca Research Center, Bogor Agricultural University, Jalan Taman Kencana No. 3, Bogor 16128, Indonesia

E-mail : tsumaryada@ipb.ac.id

Abstract. The mechanism that explains the unfolding/refolding process of the protein is still a major problem that has not been fully understood. In this paper we present our study on the unfolding and refolding pathway of Chymotrypsin Inhibitor 2 (CI2) protein through a molecular dynamics simulation technique. The high temperature unfolding simulation were performed at 500 K for 35 ns. While the low temperature refolding simulation performed at 200 K for 35 ns. The unfolding and refolding pathway of protein were analysed by looking at the dynamics of root mean squared deviation (RMSD) and secondary structure profiles. The signatures of unfolding were observed from significant increase of RMSD within the time span of 10 ns to 35 ns. For the refolding process, the initial structure was prepared from the structure of unfolding protein at t=15 ns and T=500 K. Analysis have shown that some of the secondary structures of CI2 protein that have been damaged at high temperature can be refolded back to its initial structure at low temperature simulation. Our results suggest that most of α-helix structure of CI2 protein can be refolded back to its initial state, while only half beta-sheet structure can be reformed.

1. Introduction

Unfolding and refolding of protein are rapid event which involve complex natural processes. The simplest unfolding pathway is two-states model, in which the denatured and native states are separated distinctly at transition point.

Chymotrypsin inhibitor 2 (CI2) was among the first protein that demonstrate the two-state folding mechanism, and it has since been the focus of a various number of experimental and theoretical studies [1,2,3,4,5]. CI2 protein consist of 64-residue with one α-helix and a three-stranded β-sheets. The main hydrophobic core of this protein is formed by the packing of the α-helix against the β-sheet [2]. The unfolding of CI2 was first characterized by simulation at high temperature of 498 K [3]. The use of artificially high temperatures simulation causes the unfolding process occurres more quickly without changing the unfolding pathway too much [4] and significantly reduce the usage of computational resources. Similar to unfolding, the protein refolding is also a complex process which bring the unfolded protein back to its native state. In the experiment, protein refolding was performed at low temperatures in order to reduce protein aggregation [6].

In this paper we present our study on unfolding and refolding pathway of GB1 protein. The unfolding simulation were performed at 300 K, 400 K and 500 K, while the refolding process were done at 200 K. Both folding and refolding simulation were run for 35 ns.

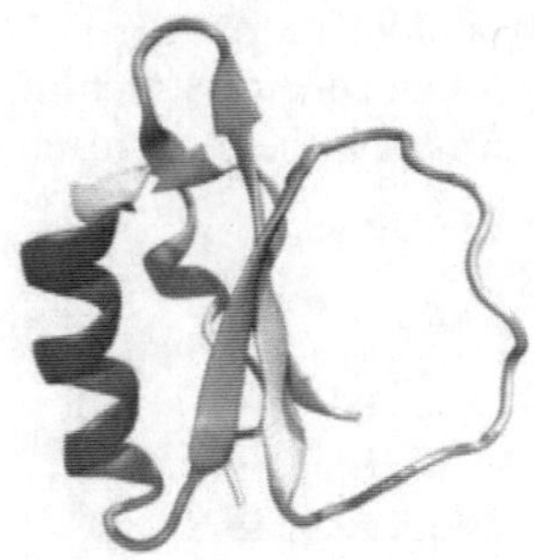

Figure 1. The initial structure of CI2 from Protein Data Bank (PDB).

2. Materials and Methods

The crystal structure of Chymotrypsin inhibitor 2 (CI2) (pdb code : 1YPC) [8] that used in this simulation was taken from the Protein Data Bank (PDB) with 1.7 Å resolution (figure 1). This original file still contains of water molecules and ligand (glycerol) which need to be removed. Protein was prepared in a simulation box with the size of 60 x 60 x 60 $Å^3$ filled with the solvent atoms, TIP3P [9]. To determine the potential energy of the crystal structure, CHARMM++ force field was used in this simulation. Neutralizing the total charge of the system was done by adding four Cl^{-1} ion to water box. All those preparations were done by using Virtual Molecular Dynamics program (VMD) [10].

The next step is MD simulation which consists of several steps i.e minimization, heating, equilibration and production run. All MD steps were performed using NAMD v.2.9 simulator [11]. Molecules that have been solvated then must be minimized. The purpose of minimization is to avoid the van der Waals contacts that do not match (bad contact) and to minimize the steric effects that favor a high energy. Each simulation was started with a minimization for 100 ps to put the protein in its lowest energy. The second stage is heating and equilibration. Protein was heated from the temperature of 0 K up to 300 K, 400 K, 500 K for the unfolding process and 200 K for the refolding process, for 40 ps. Equilibration was done using Langevin protocol for 10 ps. Production run is the final stage of MD simulations where the constraints that have been applied to the equilibration process is removed so that protein now free to move. Production run performed for 35 ns for 300 K, 400 K and 500 K simulation. For the refolding process, production run were performed at 200 K for 35 ns as well.

The time step for MD simulations was set at 2.0 femto seconds (fs). The simulation was done by using periodic boundary condition (PBC) method to eliminate the effect of surface tension and to achieve density and pressure conditions were more uniform. Electrostatic energy system thoroughly calculated using the particle mesh ewald (PME) method, while the van der Waals interaction was calculated using the Lennard–Jones potential with 12 Å cutoff. All simulations was performed in single CPU powered by 3.4 GHz Intel® core i7 processor with 12 GB of RAM and using Ubuntu 12.04 Linux platform.

To analyze the unfolding and refolding pathway, VMD program was used to produce the output such as secondary structure and Root Mean Square Deviation (RMSD).

3. Result and Discusions

3.1 Unfolding

3.1.1 The Stability of the Secondary Structure

The stability of secondary structure during simulation plays an important role in maintaining the folded/unfolded state. The loss or damage of α-helix and β-sheet can be used as an indicator of unfolding or refolding event. The sequence of structural changes (or damages) of protein during the simulation will reveal the mechanism and the most responsible interaction that keep the protein stable. [4].

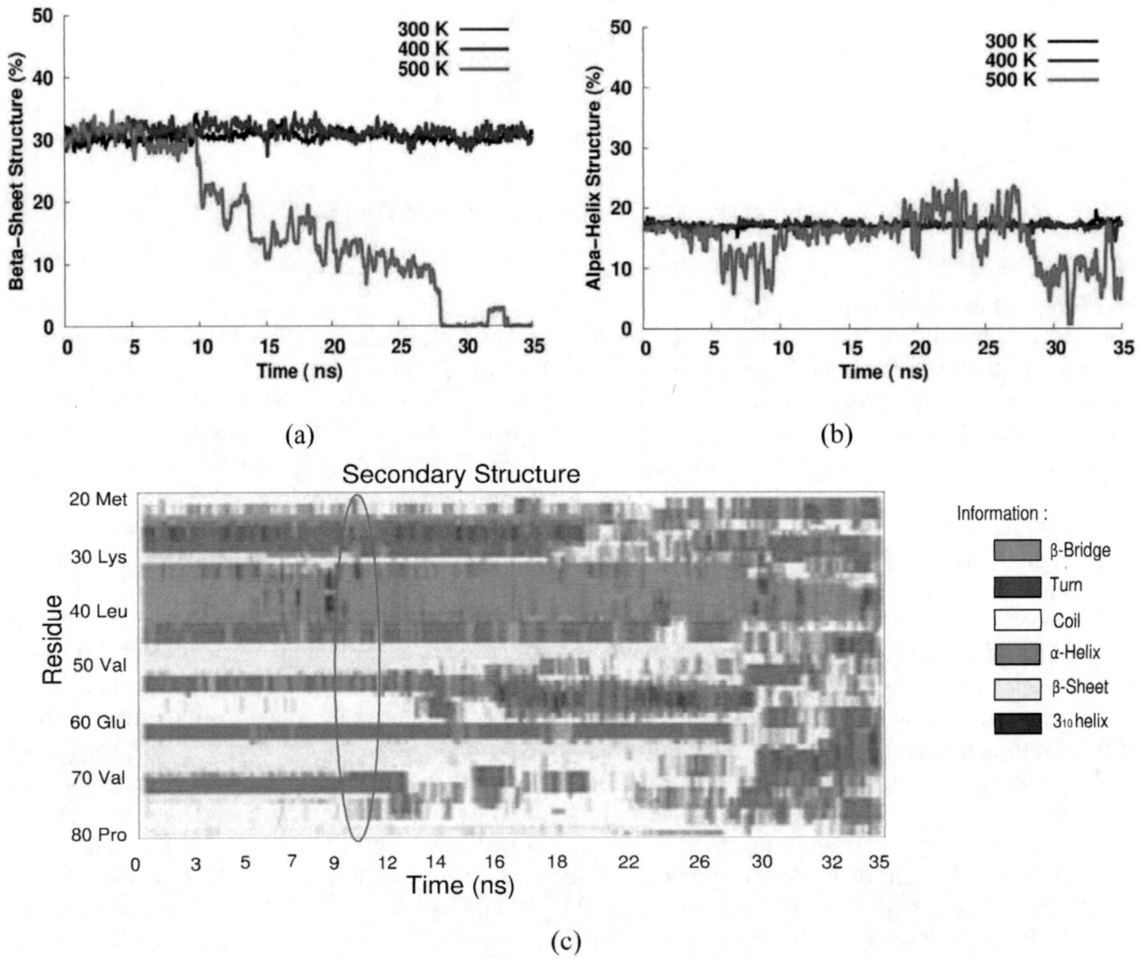

Figure 2. The percentage change in the composition of secondary structure during the simulation at 300 K (Black), 400 K (Blue) and 500 K (Red) for 35 ns for (a) beta-sheet structure. (b) α-helix structure. (c). Secondary structure changes in 500 K simulation for 50 ns; α-helix (Purple), β-sheet (Yellow) 3_{10} helix (Blue), Turn (Green) and Coil (White)

The composition of secondary structures, alpha-helix and beta-sheet, at temperature 300 K and 400 K has indicated a relatively stable and rigid structure. About 30% of the protein structure is in the form of beta-sheet and 18% α-helix. Whereas, at temperature of 500 K the composition of beta-sheet and α-helix were dramatically decreased. At this temperature beta-sheet is completely disappeared and

IOP Conf. Series: Earth and Environmental Science **31** (2016) 012010 doi:10.1088/1755-1315/31/1/012010

turned into coil and turn, while α-helix only slightly damaged. This found is similar with the results from other groups [1,2].

The composition of α-helix varies wildly at T=500 K indicates a very mobile and flexible structure. The α-helix structure at this temperature folds and refolds several times as also suggested by other [2]. The damage of secondary structure was initially occured in β_1 (beta-sheet number 1) at 10 ns, in β_2 at 14 ns and followed by α-helix at 30 ns. Finally, the structure of β-sheet and α-helix will completely disappear at 35 ns.

3.1.2 Root Mean Square Deviation (RMSD)

Root Mean Square Deviation analysis is used to describe the change in secondary structure during the simulation. The RMSD change above 5.0 Å is commonly used as an indicator of significant conformational changes in protein structure.

Figure 3 shows the fluctuations RMSD values at various temperatures. At simulation temperature of 300 K and 400 K, there is no significant change in RMSD values. At temperature 500 K the RMSD is relatively constant until 9.0 ns and jumps drastically from 10 ns until 35 ns with several plateaus. Those plateaus (12 to 14 ns, 14 to 17 ns, 24 to 27 ns and 27 to 30 ns) might correspond to the dynamics of the tertiary structure exhibited by fluctuation of α-helix and the disruption of beta-sheet as the simulation progressed.

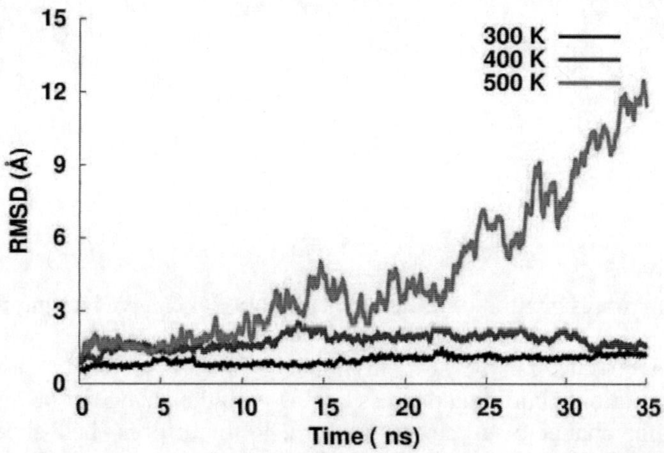

Figure 3. Root Mean Square Deviation (RMSD) as a function of simulation time (a) for temperature 300 K, 400 K, 500 K

3.2 Refolding

3.2.1 The Stability of the Secondary Structure

Studying the detailed pathway of refolding simulation from completely unfolded structure seems impossible and takes a very long time, but we can instead simulate the refolding event by starting from a transition state (TS) conformation [7]. Proteins were thought to fold not from a random coils or completely unfolded tertiary structure, but from a highly unfolded protein with some residual structures [12]. Based on that argument we choose CI2 unfolded structure at 15 ns as the initial structure for the refolding process. The low temperature of refolding was chosen to avoid aggregation of the protein as suggested by the experiment [6]. Similar to unfolding, the simulation of refolding will also be analysed from RMSD and secondary structure data.

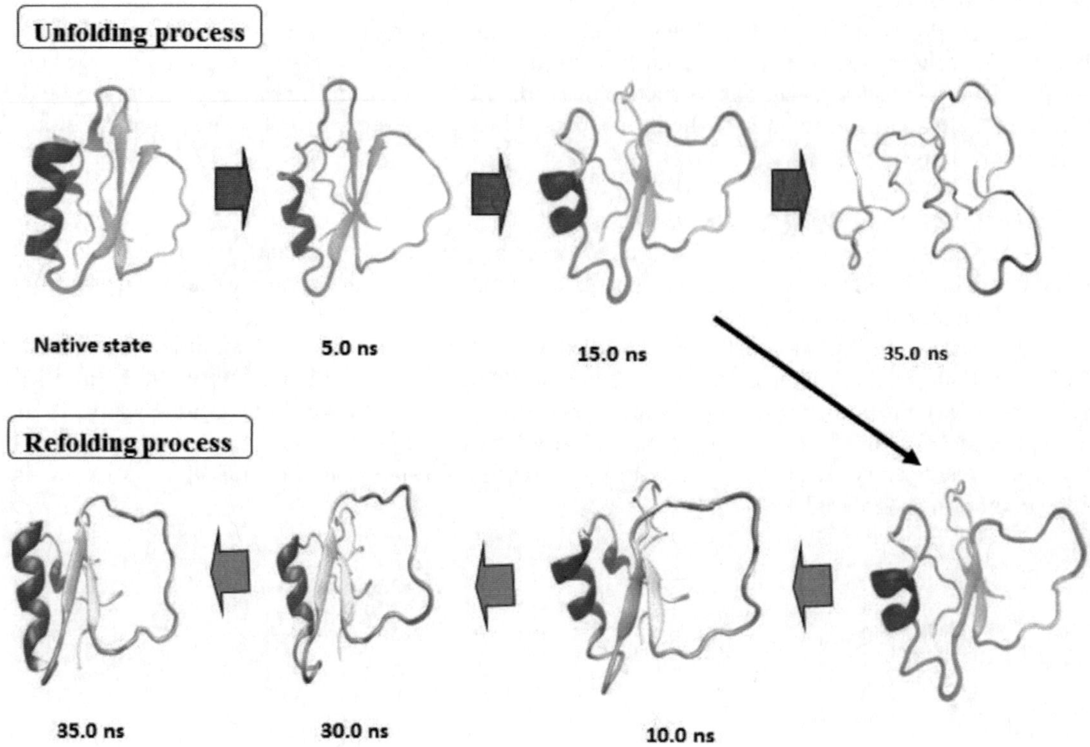

Figure 4. Cartoon representation of protein conformational changes during the simulations

Unfolding-refolding process as can be seen in figure 4 is represented by protein conformational changes during the simulation. Unfolding process begins with the change of beta-sheet structure into a coil and followed by the change of α-helix into a turn. In that process beta-sheet disrupted first and followed by α-helix. This unfolding simulation is in line with simulation from other groups [1,2]. After 35 ns all secondary structure was disappeared and transform into a completely random coil.

For refolding process, the rapid formation of α-helix precedes the formation of beta-sheet structure. At the end of refolding process, α-helix has been formed completely, while only some parts of beta-sheet structure reformed. Similar results have been reported by other researcher [2]. A Longer simulation time probably is needed here to obtain a completely folded structure of protein.

3.2.2 Root Mean Square Deviation (RMSD)

The initial structure for refolding process was taken from the unfolded structure at 15 ns with RMSD value about 5.1 Å. Figure 5 shows the fluctuation of RMSD values at various temperatures, at 300 K for the native state, 200 K for the refolding process, and 500 K for the unfolding process. For the native state, the average value of RMSD was 1.26 Å, which means that the protein was still in stable conditions. For the unfolding process, the RMSD value jumps drastically from 1.26 Å to 11.4 Å which indicates a transition from a folded to an unfolded structure. For the refolding process, the RMSD is slowly dropped from 5.1 Å to 3.06 Å, indicating a slow process of secondary structures reformation.

IOP Conf. Series: Earth and Environmental Science **31** (2016) 012010 doi:10.1088/1755-1315/31/1/012010

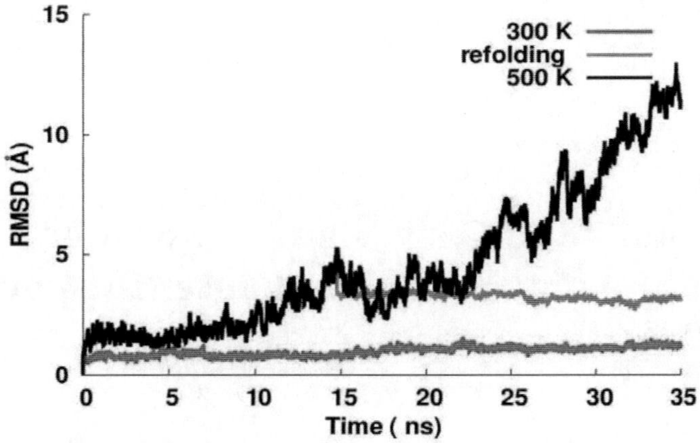

Figure 5. The RMSD values as a function of simulation time for temperature 300 K, (Native State), 200 K (refolding) and 500 K (unfolding).

4. Conclusion

Analysis of molecular dynamics simulation have shown that CI2 protein that has been unfolded at high temperature can be refolded back to its native state at low temperature simulation. The unfolding event in this simulation is characterized by a complete disruption of beta-sheet and a fluctuation of α-helix structure which also found by others [1,2]. Using a transition state conformation which consist of a small portion of residual beta-sheet and α-helix structures (obtained from the unfolding process), the refolding simulation is able to reform the whole α-helix and some part of beta-sheet structure. A longer refolding simulation time would probably allow a complete description of folding process of CI2 protein.

References

[1] Li A and Daggett V 1994 *Proc. Natl. Acad. Sci. USA*. **91**(22) 10430
[2] Kazmirski S L, Wong K, Freund S M V, Tan Y, Fersht A R and Daggett V 2001 *Proc. Natl. Acad. Sci. USA*. **98** (8) 4349
[3] Daggett V, Li A, Itzhaki L S, Otzen D E and Fersht AR 1996 *J. Mol. Biol* **257** (2) 430
[4] Day R, Bennion B J, Ham S and Dagget V 2002 *J. Mol. Biol*. **322** (1) 189
[5] Lindgren M, Westlund P O 2010 Biophysical Chemistry. **151** 46
[6] Yamaguchi S, Yamamoto E, Mannen T and Nagamune T 2013 *Biotechnol. J*. **8** (1) 18
[7] Pande V S and Rokhsar A 1999 *Proc. Natl. Acad. Sci. USA*. **96** (16) 9062
[8] Harpaz Y, Elmasry N, Fersht A R and Henrick K 1994 *Proc. Natl. Acad. Sci. USA*. **91** (1) 311
[9] Jorgensen W L, Chandrasekhar J, Madura J D, Impey R W and Klein M L 1983 *J. Chem.Phys*.**79** 926
[10] Humprey W, Dalke A and Schulten K 1996 *J.Molec.Graphics* **14** 33
[11] Phillips J C, Braun R, Wang W, Gumbart J, Tajkhorshid E, Villa E, Chipot C, SkeelR D, Kalé L and Schulten K 2005 *Journal of Computational Chemistry*, **26** (16) 1781
[12] Pappu R V, Srinivasan R and Rose G D 2000 *Proc. Natl. Acad. Sci. USA* **97** 12565

IOP Conf. Series: Earth and Environmental Science **31** (2016) 012011 doi:10.1088/1755-1315/31/1/012011

Multidimensional Analysis and Location Intelligence Application for Spatial Data Warehouse Hotspot in Indonesia using SpagoBI

Gamma Uswatun Hasanah[1], **Rina Trisminingsih**[2]
Department of Computer Science, Bogor Agricultural University
Jalan Meranti Wing 20 level 5 Kampus IPB, Bogor, West Java, Indonesia

E-mail: gammauswatun@gmail.com

Abstract. Spatial data warehouse refers to data warehouse which has a spatial component that represents the geographic location of the position or an object on the Earth's surface. Spatial data warehouse can be visualized in the form of a crosstab tables, graphs, and maps. Spatial data warehouse of hotspot in Indonesia has been constructed by researchers from FIRM NASA 2006-2015. This research develops multidimensional analysis module and location intelligence module using SpagoBI. The multidimensional analysis module is able to visualize online analytical processing (OLAP). The location intelligence module creates dynamic map visualization in map zone and map point. Map zone can display the different colors based on the number of hotspot in each region and map point can display different sizes of the point to represent the number of hotspots in each region. This research is expected to facilitate users in the presentation of hotspot data as needed.

1. Introduction

A spatial data warehouse is a collection of spatial data that is subject-oriented, integrated, time variant, non-volatile and used in decision-making process [1]. Spatial data is data that represents the geographic location of position or an object at the Earth's surface. The source of spatial data includes graphic data of analog map, satellite imagery, field survey results data, and measurements result in data using global positioning system [2].

Spatial data warehouse development by integrating thematic and geographical data to produce spatial data cube for multidimensional spatial data analysis [1]. Some of the research that utilizes spatial data warehouse include [3], [4], [5], and [6]. The studies did construction of spatial data warehouse for distribution in the hotspot in Indonesia that could perform a multidimensional analysis and show cartographic visualization which tabular charts features. The system was built using Geomondrian framework. The Geomondrian framework has OLAP visualization features, spatial query function with MultiDimensional eXpression (MDX) and visualization map supports using Geoserver as server folder. But in 2011, the Geomondrian framework stopped doing updates version so that the development of spatial data warehouse system to detected hotspot with Geomondrian difficult to proceed.

This research did spatial data warehouse visualization with multidimensional analysis module and location intelligence module using SpagoBI. SpagoBI is one of business intelligence tools that are complete, flexible, open source and has 16 features so that find the best many solutions who able to save time and economic resources [7]. According [8], benefit from SpagoBI is an addition facilitates integration system of new modules that appear to do the development and repair of deficient products. The multidimensional analysis module SpagoBI is a module that provides the OLAP engine is flexible and user friendly which can monitor data from different level – level through the drill down, drill up, slice, dice, and so on. On the location intelligence module, SpagoBI provides two geographical (GIS and GEO) that can make the relationship between geographic data and business data in the data warehouse.

In this paper, spatial data warehouse hotspot used are obtained from previous research [9]. The location intelligence machine feature that was selected is GIS that was able to display dynamic map. This research used SpagoBI application so that users expect to more quickly displayed hotspot summary data.

2. Data and Method

2.1. Data

The spatial data warehouse had been done in the previous research [9] which is sourced by Fire Information for Resource Management (FIRM) National Aeronautics and Space Administration (NASA) in year 2006 to 2015.

2.2. The analysis of spatial data warehouse hotspot

The early stages in this research were spatial data warehouse hotspot analysis from previous research [9]. The purpose spatial data warehouse hotspot analysis was to know the characteristics of the used data. The results is used to know the scheme, table dimensions, and table of fact. Therefore, the result is used to design of development system.

2.3. Design and implementation of multidimensional analysis module

OLAP visualization is aimed to spread hotspot data presentation. Design of multidimensional analayis module was make the multidimensional data cube using SpagoBI Studio. The steps in implementation of multidimensional analysis was made hierarchy of each dimension table, made server SpagoBI, made and loaded off the Mondrian template. The result is OLAP operations featuring analysis either a drill down, roll up, slice, dice, and pivot. The OLAP operations are displayed in tabular chart and crosstab table.

2.4. Implementation of dynamic map location intelligence

Spatial data warehouse hotspot from previous researched [9] visualized dynamic maps using SpagoBI applications. The steps in implementation of dynamic map location intelligence on SpagoBI i.e. saved Indonesia map in Geoserver, made SpagoBI Dataset, and made dynamic map SpagoBI. Indonesia map in Geoserver needed a worksheet (workspace) as a space to accommodate a map. Dataset SpagoBI can be CSV file, database queries, Web service Class, Java, scripts, QB, Custom, and Flat. Dynamic map SpagoBI was divide into 2 types map i.e. map zone (different color areas) and map point (point size). The result of this stage was the dynamic map showing the difference in the color and size of points based on the number of hotspots each year.

2.5. Testing multidimensional analysis module and location intelligence module
This stage was tested the main functions on the application using Blackbox method. The multidimensional analysis module has 4 main functions such as select data for the dimension and measure, display data in a graph, display data in a crosstab table, and display OLAP operations. The location intelligence module has 5 main functions such as map zone visualizations, map point, zoom in and zoom out, displays indicators map, and do filter map. Every function was tested to make sure the system could run and produce the expected output. The creation of OLAP visualization system and dynamic map using spatial data warehouse hotspot had been completed when the entire test was declared successful.

3. Result and Discussion

3.1. The analysis of spatial data warehouse hotspot
The data used was hotspot data of forest fire in year 2006 to 2015, and administrative map data of Indonesia by 2014 from Geospatial Information Agency (BIG). Spatial data warehouse design by [9] generated star schema with one table of fact as a center that contain a foreign key for each dimension and number of hotspot for measure. The results of the star schema can be seen in Figure 1. Analysis data was based on hotspot of the year, quarter, month, satellite name, island, provinces, and municipalities.

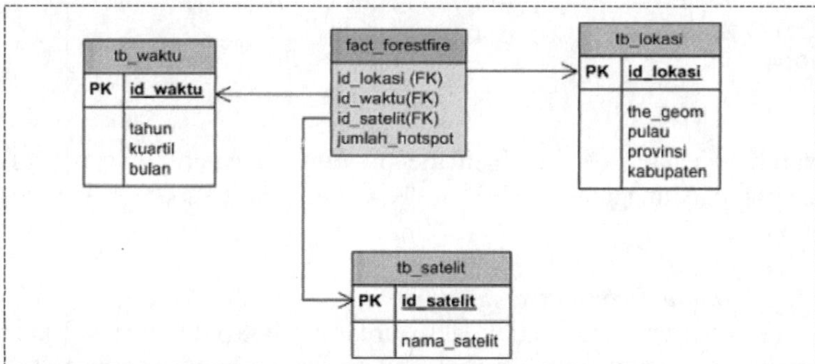

Figure 1. The result of star schema.

After created spatial data warehouse design, [9] did implementation ETL model using Geokettle. The results of the ETL was table of fact and dimension tables was loaded into DBMS PostgreSQL. Based on the ETL process performed [9], the time dimension table has four columns with the number of rows attribute as many as 120. Satellite dimension table contains two columns with the number of lines by two lines. Location dimension has five columns with a number of lines as many as 505 where there are 6 Islands, 34 provinces, and 505 districts. Table of Fact consists of four columns attribute with the number of lines as much as 32937.

3.2. Design and implementation of multidimensional analysis module
This stage was designed OLAP with build multidimensional data cube using SpagoBI Studio 5.1.0. The results of ETL implementation [9] in PostgreSQL was linked to SpagoBI Studio using PostgreSQL JDBC Driver 8. Type of fact table was changed to cube and jumlah_hotspot selected as a function of measure. Table type in the time dimension, satellite dimension and location dimension was converted into a type of dimension. Then every dimension was created a hierarchy.

OLAP implementation was made the server SpagoBI and made Mondrian OLAP templates with XML format. The result of the implementation spatial data warehouse OLAP was showed hotspot in tabular chart and crosstab table. There were OLAP operations features (roll-up/drill-up, drill down, slice and dice, pivot) featuring graphs, data storage, print files etc. Figure 2 is an example of a crosstab table results and graphs are formed.

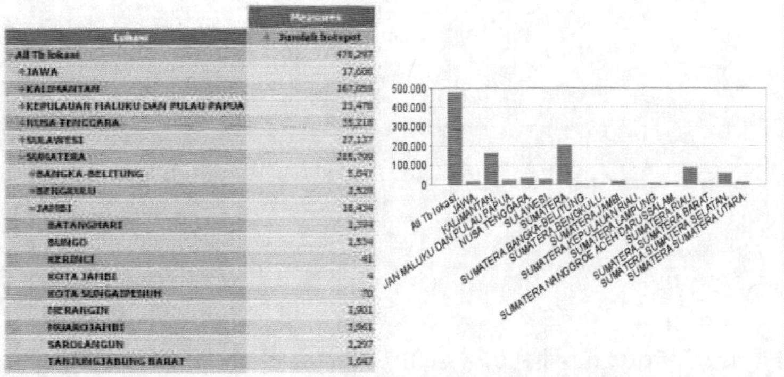

Figure 2. Representation hotspot data in the form of crosstab tables and graph.

3.3. Implementation of dynamic map location intelligence

Attribute on Indonesia map was modified by adding a new string type attribute columns using Quantum GIS application (QGIS). Indonesia map was modified because the existing of id_lokasi type was integer while id_lokasi can be shown on SpagoBI must of string type. The Indonesia map has been modified and then saved in Geoserver so that you can perform on SpagoBI. Geoserver can display large map with the URL to access it. Worksheet (workspace) and storage areas (stores) were created as a space to accommodate a map. Created workspace done in order the hotspot area only displays the region of Indonesia. At created workspace, configuration system did reference coordinates and bounding box. The coordinate reference system is a reference point to define the coordinates of a point vertically or horizontally. The bounding box represents the limits of the Earth's surface in the area that would later be shown. In this research, a coordinate reference system used WGS-84 (EPSG: 4326) and on the bounding box selected reproject native to declare because the value of native SRS unknown so it can't be matched with the EPSG.

Dataset serves as definition media of the number of hotspots data that appears on the map. This research used the dataset query in the form of a database showing the columns id_lokasi, district, province, and the number of hotspots each year. The addition of new string typed id_lokasi column at lokasi dimension table was done because SpagoBI read string typed column. Any program code requires union to combine a number of hotspots from 2006 – 2015.

Map visualization type used in this research is a dynamic map that was able to display a map in the shape of polygon on map zone and in the form of point on map point. The selection of module for visualization of map on SpagoBI was location intelligence. JSON formatted dynamic map template was needed to making dynamic map.

Dynamic map SpagoBI has 4 sections. In figure 3, the biggest section was the Indonesia map that displayed the different of colors or the sizes of the number of hotspots. The top right section was the part to select map zone appearance which displayed the map in the form of polygon and map point that displayed the map in the form of point. Indicator section displayed selection of hotspot each year (2006 – 2015) in Indonesia. The bottom right section is the filter menu based on districts, provinces and the number of hotspots. The number of filter shown based on datasets that have been previously entered.

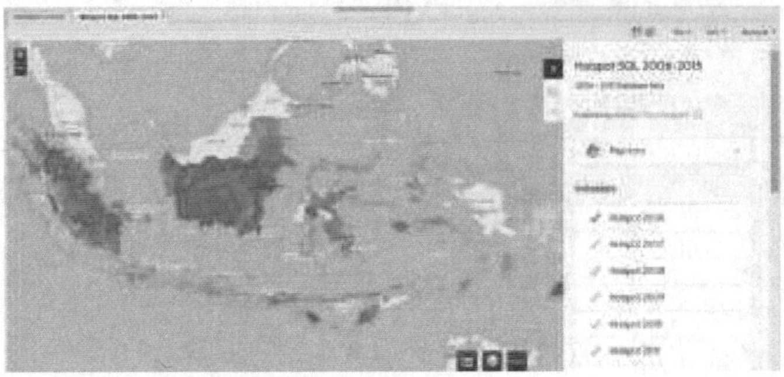

Figure 3. Polygon-shaped visualization of dynamic map SpagoBI.

Dynamic maps visualization displayed a map of the hotspot in 2006 in the form of polygons by default. The differences colors from light blue to red indicated changes the number of hotspots. The more red color was displayed, then the larger the number of hotspots in the district. The number of hotspot analysis was represented by number of hotspots in 2006 with the largest number of hotspot dominated by Sumatra and Borneo. The largest number of hotspot in Sumatra were located in the province of Riau and South Sumatra Province. The largest number of hotspots in Borneo were in the province of West Kalimantan, Central Kalimantan, and some territory in East Kalimantan.

Figure 4. Point-shaped visualization of dynamic map SpagoBI.

Figure 4 is point-shaped visualization of dynamic map SpagoBI. The sizes of each point represents the frequency of the number of hotspots in each district. The point location that appears to represent hotspot for one district which are not necessarily representative of regions with specific longitude latitude has a particular hotspot. When zoom in map, the map did not show the visible area to the level of districts because the maps are stored on the Geoserver is an administrative district map and the scope of the data used [9] only up to the level of the district.

Each point describes the location of the hotspot in the area that contains the code of the district, the name of district, the code of province, and the name of the province. If seen in use terms of map visualization, map zone used when users want to know the overall area that has one of the largest known hotspots of red on the map. Map point used when the user wants to see a more detailed areas that have the largest number of hotspots. That is because the size of the point represents the frequency of the number of hotspots so that users know which areas that became the top priority in handling the emergence of hotspots.

3.4. Testing multidimensional analysis module and location intelligence module

Testing system was performed with scenario test in functions for each module. The black box testing method is used in this research. Testing was conducted to find the function of multidimensional analysis module and location intelligence module was already running. Table 1 shows the scenario of multidimensional analysis module and table 2 shows the scenario for location intelligence module.

Table 1. Scenario of multidimensional analysis module.

No	Testing	Scenario	Expected Result	Test Result
1	Select data for a dimension and measure	Choose one of the dimensions (time, satellite, location)	The data is displayed according to the selected dimensions	Successful
2	Display the data in a graph	Select the dimension, and then select the 'graph'.	The data represented in a graph	Successful
3	Display data in a crosstab table	Select the dimension, then by default the crosstab table appears.	The data represented in the crosstab table	Successful
4	Featuring OLAP operations	Select the dimension, the OLAP crosstab will display operation, select the desired OLAP operations	The data appear according to the selected OLAP operations	Successful

Table 2. Scenario of location intelligence module.

No	Testing	Scenario	Expected Result	Test Result
1	Display visualization map zone	Choosing the zone map	Differences in color maps for each district	Successful
2	Display visualization map point	Choosing the point map	Appears spheres of different sizes corresponding number of hotspots for each district	Successful
3	Displays zoom in and zoom out map	Selecting enlarge or reduce the map view	A map is displayed to be larger or smaller	Successful

4. Summary

This work is able to make multidimensional analysis module and the location intelligence module for spatial data warehouse of hotspots in Indonesia from the years 2006 to 2015 using SpagoBI. Multidimensional analysis module is able to display OLAP visualization in crosstab table and graph, therefore it can help to present the hotspot summary. Location intelligence module uses GIS engine that is able to display a dynamic map in the form of map zone and map point that capable of running its functions properly. Visualization of the map zone can display different colors based on the number of hotspots in each district and visualization of map point can display different sizes of point that represent the number of hotspots in each district.

References

[1] Han J, Kamber M, Pei J. 2012. *Data Mining: Concept and Techniques. Ed ke-3.* Amsterdam, Belanda: Elsevier, Morgan Kaufmann.

[2] Ekadinata A, Dewi S, Hadi DP, Nugroho DK, Johana F. 2008. *Sistem Informasi Geografis untuk Pengelolaan Bentang Lahan Bebasis Sumber Daya Alam.* Bogor, Indonesia: World Agoforestry Centre.

[3] Trisminingsih R. 2010. *Pembangunan spatial data warehouse berbasis web untuk persebaran hotspot di wilayah Indonesia*, Bogor, Indonesia: Institut Pertanian Bogor.

[4] Fadli MH. 2011. *Data warehouse spatio-temporal kebakaran hutan menggunakan Geomondrian dan Geoserver*. Bogor, Indonesia: Institut Pertanian Bogor.

[5] Imaduddin A. 2012. *Sinkronisasi antara visualisasi peta dan query olap pada spatial data warehouse kebakaran hutan di Indonesia*. Bogor, Indonesia: Institut Pertanian Bogor.

[6] Kusumah B. 2014. *Pengembangan modul update data pada sistem spatial data warehouse hotspot*. Bogor, Indonesia: Institut Pertanian Bogor.

[7] Cazzin G. 2012. *Business Intelligence with SpagoBI.* Padua, Italy: SpagoBI Competency

Center.
[8] Golfarelli M. 2009. "Open source BI platforms: a functional and architectural comparison" in *11th International Conference DaWak.* Bologna, Italy, pp. 287-297
[9] Astriani W. 2015. *Modul extract, transform, dan load untuk spatial data warehouse titik panas menggunakan Geokettle*. Bogor, Indonesia: Institut Pertanian Bogor.

Determination of Optimal Epsilon (Eps) Value on DBSCAN Algorithm to Clustering Data on Peatland Hotspots in Sumatra

Nadia Rahmah[1]*, Imas Sukaesih Sitanggang[1]

[1]Computer Science Departement, FMIPA, Bogor Agricultural University, Bogor 16680

E-mail: nadiarahmahh@gmail.com

Abstract. In this work we determine the optimal epsilon value on peatland on DBSCAN Algorithm to clustering data on peatland hotspots in sumatera. DBSCAN is a base algorithm for density based data clustering which contain noise and outliers. We found using this method that the area which has the highest density of hotspots in Sumatra in 2013 peatland is contained in cluster 1 of Riau Province that is equal to 2112 hotspots.

1. Introduction

Monitoring of hotspots area could be performed by knowing if the hotspots are randomly distributed or clustered. DBSCAN is a base algorithm for density based clustering containing large amount of data which has noise and outliers. DBSCAN has 2 parameters namely Eps and MinPts. However, conventional DBSCAN cannot produce optimal Eps value. DBSCAN modifications is required to determine the optimal Eps value automatically. Determination of the optimal Eps value is aquired using DMDBSCAN algorithm on a single density level corresponding to the k-dist plot. DMDBSCAN is one method of DBSCAN algorithm modification on Eps value to obtain optimal value automatically with different densities.

The parameter of values are used in the clustering data on hotspots peatlands in Sumatra. The data set of peatlands hotspots in Sumatera in the period 2013. This research modifies DBSCAN algorithm using the R programming language to get the optimal Eps value automatically to clustering data on peatland hotspots in Sumatera. Implementation of DBSCAN modifications in determining the value of parameter Eps is expected to be a reference to the DBSCAN algorithm, when searching the Eps value and determining the spread of hotspots that could be known clusters of the region that have the potential occurrence of peatlands fire.

IOP Conf. Series: Earth and Environmental Science **31** (2016) 012012 doi:10.1088/1755-1315/31/1/012012

2. Results and Discussions

Determination of Optimal Eps Value Using R Programming Languange

This phase is done by modifying the algorithm DBSCAN in the programming language R. Modifications done by inserting a search algorithm to the algorithm DBSCAN Eps value on R found in accordance with the package fpc. DMDBSCAN pseudocode is presented in Figure 1.

Algorithm 1 The pseudo code of the proposed technique DMDBSCAN to find suitable Epsi for each level of density in data set	
Purpose	*To find suitable values of Eps*
Input	*Data set of size n*
Output	*Eps for each varied density*
Procedure	*for i* *for j = 1 to n* *d(i,j) ← find distance (x_i, x_j)* *find minimum values of distances to nearest 3* *end for* *end for* *sort distances ascending and plot to find each value* *Eps corresponds to critical change in curves*

Figure 1 Pseudocode DMDBSCAN Algorithm (Elbatta 2012)

The algorithm begins with the calculation euclidean distance on each pair of data latitude and longitude by dist function. Furthermore dist transform data into a matrix using as.matrix function in R. A function of matrix is used because the results dist upper triangular form that needs to be normalized into the matrix intact. Normalization is used to facilitate the search for the *k* nearest neighbors on each line of the distance calculation.

After calculating the distance and the matrix, the initialization is done to count the number of rows in the data by using the function nrow. Further searches carried out three nearest neighbors of each matrix line spacing for sorting is done in ascending for each result the closest distance to the neighbors. Sorting the results of each line of the nearest neighbor made a plot with the x-axis and y-axis is the object and the distance k nearest neighbors. Plots that have formed in ascending calculated the difference in slope of the line to get the value of Eps (Gaonkar and Sawant 2013). A point which has a slope changes or changes in the slope of the plot will be a significant Eps optimal value (Elbatta 2012).

In this study, an algorithm DMDBSCAN was adopted is to use single-level density to determine the optimal value Eps so that only produce one value Eps alone. Eps rate determination for single-level density is obtained by calculating the slope between points with equation (y2-y1) / (x2-x1) using a threshold value of 1% so that the slope has a slope of 1% difference is an optimal Eps value. Figure 2 depicts the result of a plot that has been sorted in ascending Using data hotspot 2013. Searches Eps value was done by calculating the slope of the line from any point and sought-after pair of points that have the greatest slope to locate the point. The slope of the line is located at the point of 0.1118034, a point which is the optimal value Eps (Elbatta 2012).

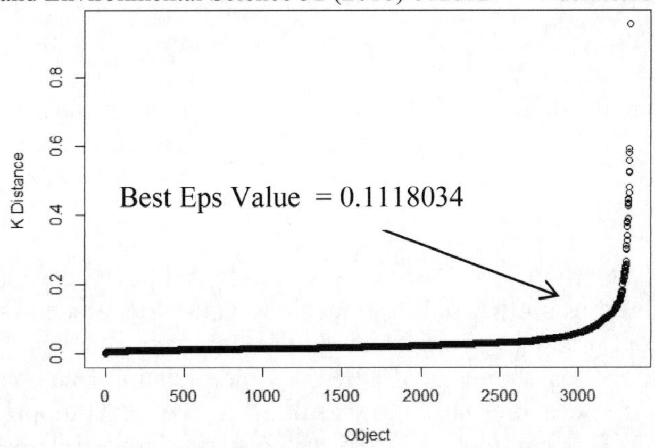

Figure 2 Points sorted by distance to the 3rd nearest neighbor

Clustering of Data Hotspots Using Optimal Eps Value on DBSCAN Algorithm
Clustering process is successfully executed from the search results DBSCAN modification Eps optimal value automatically with the idea of k-dist plot to generate the parameter Eps = 0.1118034 and MinPts = 3. This parameter is used for clustering processes that generate 43 clusters with outliers at 54 points. Then the area has the highest density of hotspots in the area of peatland in Sumatra in 2013 contained in cluster 1 (Riau Province) that is equal to 2112 hotspots.

Figure 3 shows the area in Riau province were included in cluster 1. Districts included in cluster 1 is Bengkalis, Pelalawan, Rokan Hilir, Siak, Rokan Hulu and Indragiri Hulu. Subdistrict contained in the districts included in cluster 1 can be seen in Figure 3.

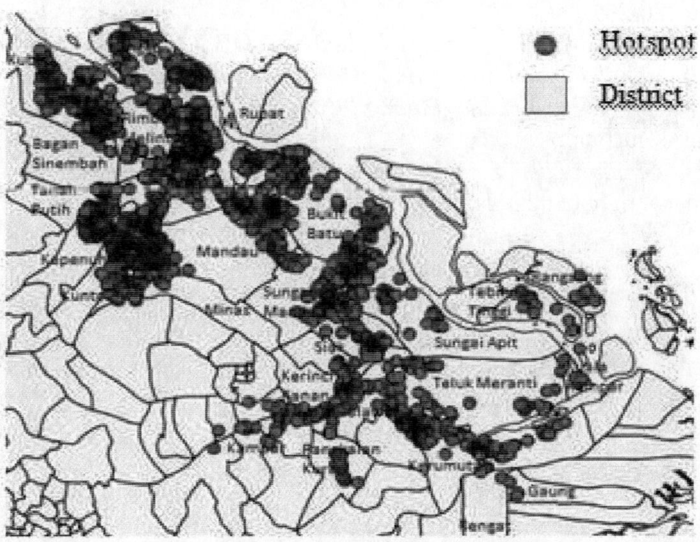

Figure 3 Area/Subdistrict was included in *cluster* 1

Cluster Analyst

The analysis process performed analysis of running time to the result of determining the optimal Eps automatically on DBSCAN algorithms for clustering data using R. This analysis includes a comparison generated in this studied with research conducted Usman (2014).

This research resulted in a runtime of 1.72 seconds to 32.29 seconds for the system and user time use 2 GB of RAM. The system time is the speed of the system running the job before the user enter data. User time is the time used after users enter data into the system. Therefore, this research is more efficient in classifying the hotspots data because it can determine the optimal Eps value automatically.

3. Conclusion

In this study, an algorithm DMDBSCAN was adopted is to use single-level density to determine the optimal value Eps so that only produce one value Eps alone. The slope of the line is located at the point of 0.1118034, a point which is the optimal value Eps. Clustering process is successfully executed from the search results of DBSCAN modification. We obtain the an Eps optimal value automatically with the idea of k-dist plot resulting in the generatiopn of parameters Eps = 0.1118034 and MinPts = 3. This parameter is used for clustering processes that generate 43 clusters with outliers at 54 points. It follows that the area which has the highest density of hotspots in Sumatra in 2013 peatland is contained in cluster 1 (Riau Province) that is equal to 2112 hotspots. The advantage of this method is that the optimal Eps value can be determined automatically.

References

[1] Elbatta MNT. 2012. An improvement for DBSCAN algorithm for best results in varied densities [disertasi]. Gaza (PS): Islamic University of Gaza.
[2] Ester M, Kriegel HP, Sander J, Xu X. 1996. A density-based algorithm for discovering clusters in large spatial databases with noise. Di dalam: Simoudis E, editor. *Proceedings of 2nd International Conference on Knowledge Discovery and Data Mining (KDD-96)*; 1996 Agu 4-6. hlm 226-231.
[3] Gaonkar MN, Sawant K. 2013. AutoEps DBSCAN: DBSCAN with Eps automatic for large dataset. *International Journal on Advanced Computer Theory and Engineering*. 2(2): 11-16.
[4] Han J, Kamber M, Pei J. 2012. *Data Mining: Concepts and Techniques*. 3rd ed. San Francisco (US): Morgan-Kauffman.
[5] Usman M. 2014. *Spatial clustering* berbasis densitas untuk penyebaran titik panas sebagai indikator kebakaran hutan dan lahan gambut di Sumatera [tesis]. Bogor (ID): Institut Pertanian Bogor.

Simulation of Heat Transfer in Cylinder Husks Furnace with Finite Difference Method

I Noor[1*], Irzaman[2], H Syafutra[2], F Ahmad[3]

[1] Department of Physics, Bogor Agricultural University, Jalan Meranti Kampus Dramaga IPB, 16880, INDONESIA
[2] Material Physics Division, Department of Physics, Bogor Agricultural University, Jalan Meranti Kampus Dramaga IPB, 16880, INDONESIA
[3] Theoretical Physics Division, Department of Physics, Bogor Agricultural University, Jalan Meranti Kampus Dramaga IPB, 16880, INDONESIA

E-mail: iman_noor09@apps.ipb.ac.id

Abstract. Simulation of heat transfer on cylinder husk furnace with FDM (Finite Difference Method) has been performed. This simulation aimed to study the heat transfer based on conduction and convection mechanism on the cylinder husk furnace. Fluid dynamics inside the cylinder of husk furnace was obtained by solving the Navier-Stokes equations with laminar flow approach. The initial temperature in all parts of the cylinder is room temperature, except at the bottom of the cylinder is 491 °C. Through numerical calculation of heat conduction and convection equation by FDM method, we got that the velocity of fluid flow is ranged between 3:32 - 13.75 m/s and 2:00 - 13.75 m/s for fixed and unfixed temperature of the cylinder blanket, respectively.

Introduction

Husk furnace is a stove which uses fuel from rice husk. The husk furnace components consist of cone part and cylinder part (see Fig 1). The cylinder part of husk furnace is the important part of the furnace that will ensure the heat is delivered well. Heat distribution in the cylinder husk furnace can be determined theoretically by studying heat transfer based on conduction and convection process. As we knew that the Graetz problem is the heat transfer problem of laminar fluid flow in ducts [1]. Hsu [2] studied the Graetz problem with axial diffusion in circular tube, using a semi- infinite domain formulation with a specified inlet condition, while Michelsen and Villadsen [3] analyzed the effects of axial diffusion in a infinite domain formed by an insulated preparation region followed by an isothermal wall. Both studies used numerical schemes to complete the solution of the problem. Barros and Sphaier [4] proposed analytical approximations for solving an extended version of the Graetz problem with axial diffusion in an infinite domain. The Graetz problem for non-Newtonian fluids with constant wall temperature and heat flux was studied numerically by Shih and Tsou [5]. Fluid dynamics inside the cylinder husk furnace determines how convection works, that can be obtained by solving the Navier-Stokes equations [6]. The pressure gradient of the fluid inside the cylinder can be derived from temperature gradient or fluid density gradient. The aim of this research is to study the heat transfer on

[1] To whom any correspondence should be addressed.

the cylinder husk furnace based on conduction and convection mechanism with laminar fluid flow approach.

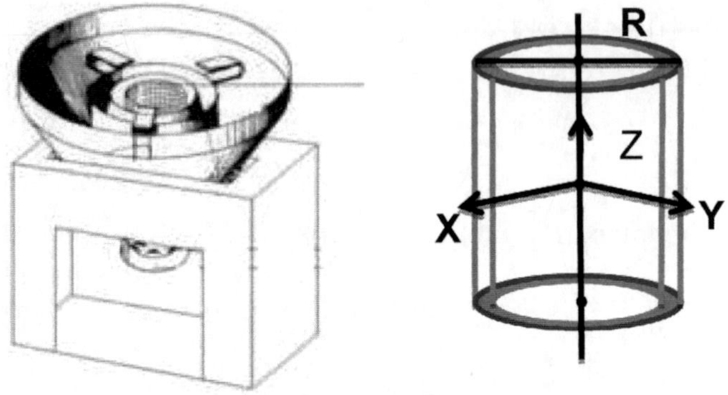

Figure 1. Husk Furnace and Cylinder

Result and Discussion

2.1 Mathematical Modelling of Heat Transfer Cylinder

Heat transfer on the cylinder husk furnace is associated with the phenomenon of conduction and convection. The governing equation for the heat transfer can be expressed as

$$\rho C_p \frac{\partial T}{\partial t} + \vec{U} \cdot \nabla T = k \nabla^2 T \tag{1}$$

With $\vec{U}$ is fluid flow velocity, ρ is air density, C_p is constant of material specific heat, k is constant of conductivity thermal matter, α is constant of diffusivity thermal matter, and T is temperature on cylinder [7].

The term on the right side of Equation (1) corresponds to the heat conduction. The Laplacian on the equation in cylindrical coordinate is written as :

$$\nabla^2 T = \frac{1}{r}\frac{\partial T}{\partial r} + \frac{\partial^2 T}{\partial r^2} + \frac{1}{r^2}\frac{\partial^2 T}{\partial \theta^2} + \frac{\partial^2 T}{\partial z^2} \tag{2}$$

While, the second term on left side of the equation corresponds to the heat convection and in cylindrical coordinate the term can be written as:

$$\vec{U} \cdot \nabla T = \vec{U}_r \frac{\partial T}{\partial r} + \vec{U}_z \frac{\partial T}{\partial z} + \vec{U}_\emptyset \frac{\partial T}{\partial \theta} \tag{3}$$

With $\vec{U}_r$ is fluid flow velocity for r radius direction and $\vec{U}_z$ is fluid flow velocity for z height direction. $\vec{U}_\emptyset$ is fluid flow velocity for ∅ degrees direction. From equation (2) and (3), we combine it to equation (1), so the equation be :

$$\frac{\partial T}{\partial t} = \alpha \left(\frac{1}{r}\frac{\partial T}{\partial r} + \frac{\partial^2 T}{\partial r^2} + \frac{1}{r^2}\frac{\partial^2 T}{\partial \theta^2} + \frac{\partial^2 T}{\partial z^2} \right) - \frac{1}{\rho C_p} \left(\vec{U}_r \frac{\partial T}{\partial r} + \vec{U}_z \frac{\partial T}{\partial z} + \vec{U}_\emptyset \frac{\partial T}{\partial \theta} \right) \tag{4}$$

Because of cylindrical is symmetry, there is no gradient temperature respect to Θ angles.

$$\frac{\partial T}{\partial \theta}=0 \text{ and } \frac{\partial^2 T}{\partial^2 \theta}=0$$

So, the heat transfer equation on cylinder husk furnace can be simplified to:

$$\frac{\partial T}{\partial t}=\alpha\left(\frac{1}{r}\frac{\partial T}{\partial r}+\frac{\partial^2 T}{\partial r^2}+\frac{\partial^2 T}{\partial z^2}\right)-\frac{1}{\rho C_p}\left(\vec{U}_r\frac{\partial T}{\partial r}+\vec{U}_z\frac{\partial T}{\partial z}\right) \tag{5}$$

2.2 Finite Difference Methods (FDM)

Finite difference method is one of numerical technique to solved differential equation problem [8]. Here we used FDM to solve equation (7) based on forward and center formulas that are given by [9] :

$$\frac{df(x_i)}{dx}\approx\frac{f_{i+1}-f_i}{dx}$$
$$\frac{d^2f(x_i)}{dx^2}\approx\frac{f_{i+1}-2f_i+f_{i-1}}{dx^2}$$

Based on the relations, we transform the continuous differential equation problem for heat distribution on cylinder to discrete differential equation problem:

$$\frac{\partial T}{\partial t}=\frac{T(z_j,r_i,t_k+\Delta t)-T(z_j,r_i,t_k)}{\Delta t} \tag{6}$$
$$\frac{\partial T}{\partial r}=\frac{T(z_j,r_i+\Delta r,t_k)-T(z_j,r_i-\Delta r,t_k)}{\Delta r} \tag{7}$$
$$\frac{\partial^2 T}{\partial r^2}=\frac{T(z_j,r_i+\Delta r,t_k)-2T(z_j,r_i,t_k)+T(z_j,r_i-\Delta r,t_k)}{\Delta r^2} \tag{8}$$
$$\frac{\partial^2 T}{\partial z^2}=\frac{T(z_j+\Delta z,r_i,t_k)-2T(z_j,r_i,t_k)+T(z_j-\Delta z,r_i,t_k)}{\Delta z^2} \tag{9}$$

To simplify those equations, we introduce new index notation as follows:

$$T(z_j,r_i,t_k+\Delta t)=T_{j,\,i,\,k+1} \tag{10}$$
$$T(z_j,r_i+\Delta r,t_k)=T_{j,\,i+1,\,k} \tag{11}$$
$$T(z_j,r_i-\Delta r,t_k)= T_{j,\,i-1,\,k} \tag{12}$$
$$T(z_j+\Delta z,r_i,t_k)=T_{j+1,\,i,\,k} \tag{13}$$
$$T(z_j-\Delta r,r_i,t_k)= T_{j-1,\,i,\,k} \tag{14}$$

Where j index for height cylinder, i index for radius cylinder, k index for time. So, the governing numerical equation for heat distribution conduction and convection on cylinder is:

$$T_{j,\,i,\,k+1}= T_{j,\,i,\,k}+\alpha\left(\frac{1}{r}\frac{(T_{j,\,i+1,\,k})-(T_{j,\,i-1,\,k})}{\Delta r}+\frac{(T_{j,\,i+1,\,k}-2T_{j,\,i,\,k}+T_{j,\,i-1,\,k})}{\Delta r^2}+\frac{(T_{j+1,\,i,\,k}-2T_{j,\,i,\,k}+T_{j-1,\,i,\,k})}{\Delta z^2}\right)-\frac{1}{\rho C_p}\left(\vec{U}_r\frac{T_{j,\,i+1,\,k}-T_{j,\,i-1,\,k}}{\Delta r}+\vec{U}_z\frac{T_{j+1,\,i,\,k}-T_{j-1,\,i,\,k}}{\Delta z}\right) \tag{15}$$

2.3 Fluid Dynamics

Fluid dynamics inside the cylinder husk furnace was obtained by solving the Navier-Stokes equations with laminar flow approach. Analytical solution of the equation for laminar flow case is given by [10]:

$$u(x) = \frac{1}{2}\frac{h^2}{\mu}\frac{dP}{dz}\left[1 - \left(\frac{x}{h}\right)^2\right] \tag{16}$$

Where boundary condition at the wall is :

$$u = 0 \text{ at } x = \pm R$$

The pressure gradient on equation (27) as function of temperature can be written as follows:

$$\frac{dP}{dz} = \left(\frac{\rho}{1+\beta(T-T_\infty)}\right) gh \tag{17}$$

2.4 Numerical Calculation of Heat Conduction and Convection Equation

The initial temperature in all parts of the cylinder is at room temperature, except the bottom of the cylinder is at 491 °C. There are two simulation case namely fixed and unfixed cylinder blanket temperature. For fixed cylinder blanket temperature assumed that the cylinder walls in contact with heat reservoir and unfixed temperature assumed that the cylinder walls in contact with low heat conductivity medium.

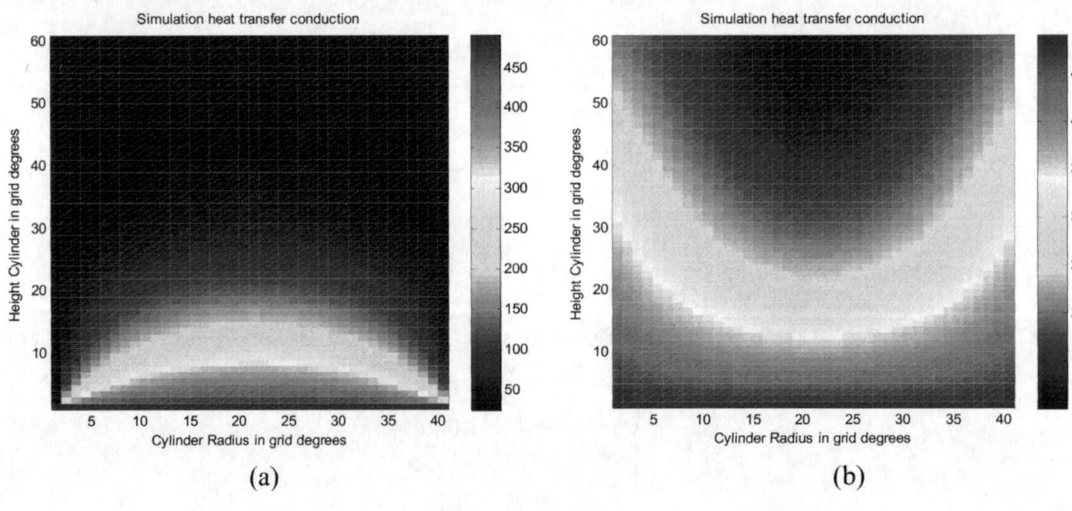

Figure 2. Heat distribution in the cylinder by conduction for fixed (a) and unfixed (b) blanket temperature

Figure 2 show the simulation of heat distribution on the cross section of cylinder furnace caused by conduction process for one minute. It is appears that the rate of heat transfer in the fixed temperature case is lower than unfixed temperature case. In fixed temperature case, the heat transfer only occurs in the fluid, there is no contribution from the wall heat conduction. While in unfixed temperature case, there is a contribution from the wall heat conduction with higher conductivity than fluid, so the temperature of cylinder edge is higher than its center.

IOP Conf. Series: Earth and Environmental Science **31** (2016) 012013 doi:10.1088/1755-1315/31/1/012013

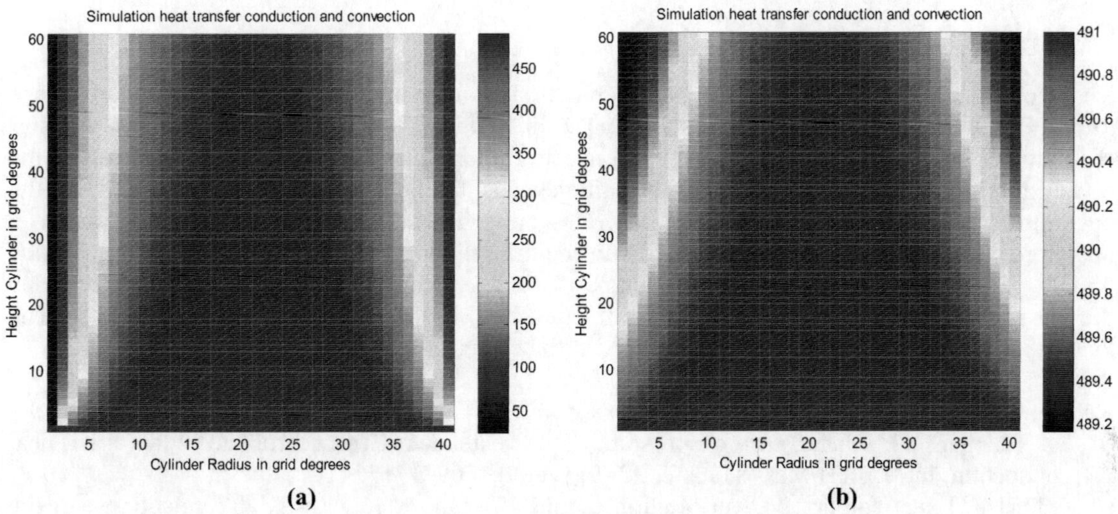

Figure 3. Heat distribution in the cylinder by conduction and convection for fixed (a) and unfixed (b) blanket temperature

Figure 3 show the simulation of heat distribution on the cross section of cylinder furnace caused by conduction and convection process simultaneously, for one minute. By comparing Fig 2 and Fig 3, it can be concluded that the heat transfer in the cylinder husk furnace is dominated by convection mechanism. On the mechanism of convection, the heat transfers through the movement of air particles due to the pressure gradient caused by the temperature gradient. While in conduction, there is only vibration between atoms of air due to a temperature gradient. As with the previous simulation, the rate of heat transfer in the fixed temperature case is lower than unfixed temperature case.

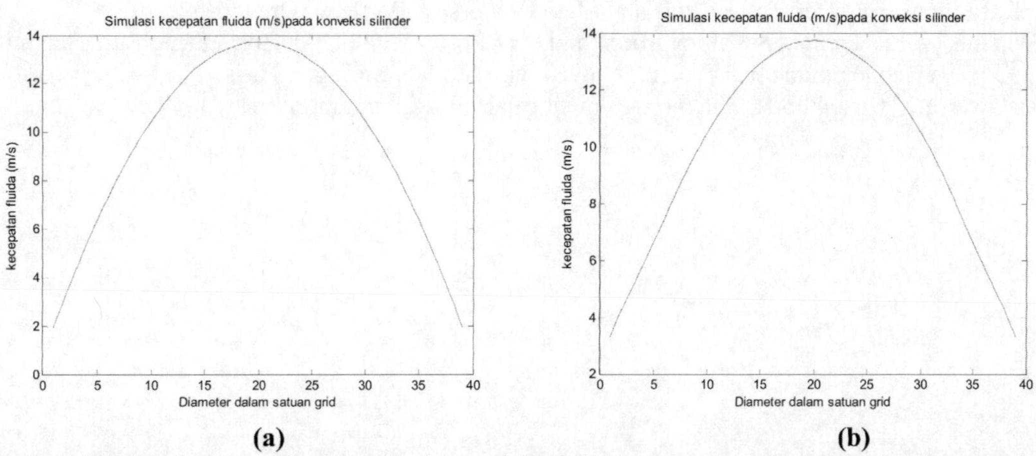

Figure 4. Fluid velocity profile for unfixed (a) and fixed (b) temperature of the cylinder blanket, respectively (m/s).

Figure 4 above shows the fluid velocity profile in the middle of cylinder for unfixed and fixed temperature case. The velocity profile is parabolic with maximum speed at the center of the cylinder. The velocity of fluid flow inside the cylinder is ranged between 2:00 - 13.75 m/s and 3:32 - 13.75 m/s

IOP Conf. Series: Earth and Environmental Science **31** (2016) 012013 doi:10.1088/1755-1315/31/1/012013

for unfixed and fixed temperature case, respectively. The different result is due to the pressure gradient near the wall in fixed temperature is bigger than unfixed temperature case.

3. Conclusion

We have performed simulation of heat transfer on cylinder husk furnace with FDM (Finite Difference Method). Based on numerical calculation with FDM for heat conduction and convection, heat transfer in the cylinder husk furnace is dominated by convection mechanism. On convection mechanism, the heat transfer through the movement of air particles due to the pressure gradient caused by the temperature gradient. While in conduction, there is only vibration between atoms of air due to a temperature gradient. The velocity of fluid flow inside the cylinder is ranged between 2:00 – 13.75 m/s and 3:32 – 13.75 m/s for unfixed and fixed temperature case, respectively. The different result is due to the pressure gradient near the wall in fixed temperature is bigger than unfixed temperature case. The velocity profile is parabolic with maximum speed at the center of the cylinder.

4. References

[1] M. Ebadian, H. Zhang, An exact solution of extended Graetz problem with axial heat conduction, Int. J. Heat Mass Transfer 32 (9) (1989) 1709–1717.

[2] C.J. Hsu, Exact solution to entry-region laminar heat transfer with axial conduction and the boundary condition of the third kind, Chem. Eng. Sci. 23 (5) (1968) 457–468.

[3] M.L. Michelsen, J. Villadsen, The Graetz problem with axial heat conduction, Int. J. Heat Mass Transfer 17 (11) (1974) 1391–1402.

[4] L.S. De Barros, L. A. Sphaier, Enhanced lumped formulations for the graetz problem in infinite domains, In: Proceedings of the 22nd International Congress of Mechanical Engineering (COBEM 2013), November 3-7, Ribeirão Preto, SP, Brazil, 2013, pp. 3299–3306.

[5] Y.P. Shih, J. D. Tsous, Extended Leveque solutions for heat transfer to power fluids in laminart flow in a pipe, Chem. Eng. J. 15 (1978) 55–62.

[6] Lienhard, John H. 2005. *A Heat Transfer Textbook* .Third edition. Phlogiston Pressridge, Massachusetts, U.S.A.

[7] Williams, J. 2000. *Engineering Heat Transfer*. Second edition. CRC Press, New York, U.S.A

[8] Rectenwald, Gerald W. 2011. *Finite-Difference Approximations to The Heat Equation*. Mechanical Engineering Department Portland State University, Portland, Oregon.

[9] Danaila, I., J. Pascal, S. M. Karber, and M. Postel, 2006, An Introduction to Scientific Computing, Twelve Computational Projects Solved with Matlab. Springer. Paris

[10] E. Ower, R.C. Ppankhurst, The measurement of air Flow, Pergamon Press, 1974.

IOP Conf. Series: Earth and Environmental Science **31** (2016) 012014 doi:10.1088/1755-1315/31/1/012014

A Mathematical Model of Protectant and Curative Fungicide Application and its stability analysis

N. Anggriani*[1], N. Istifadah[2], M. Hanifah[1] and A.K. Supriatna[1]
[1]Department of Mathematics, Universitas Padjadjaran, INDONESIA.
[2]Department of Agriculture, Universitas Padjadjaran, INDONESIA.

E-mail: nursanti.anggriani@gmail.com

Abstract. In this paper we introduce a mathematical model for fungicide application with effect of protectant and curatives factor. We show the value of the Basic Reproduction Number (R_0) of the fungal disease, which is computed from the largest eigen value of the next generation matrix of the model. The result show that in the region where R_0 greater than one there is only one single stable endemic equilibrium. However, in region where R_0 less than one some parameters affect the number of posibble equilibria. Some numerical simulation are also given to illustrate our analytical results.

1. Introduction

Recently Castle et al. (2012) have developed a mathematical model of fungicide application. They argued that most models of fungicides application have been done by simply modifying the parameters of the underlying epidemiological models, for example by reducing the infection rates and/or increasing the host recovery rates. Many work have been done to investigate the effect of the application of fungicide for several underlying mathematical but the dynamics of the fungicides themselves are not well explored. Castle et al. were pioneering this work by investigating the dynamics of fungicide application to infected plant by considering the curative factor effect of the fungicide and showed how the timing of initial infection relative to host population growth affects invasion and persistence in a chemically controlled system. It followed by the work of Anggriani et al. (2015) to consider the optimal control of the fungicide application.

Since Castle et al (2012) have argued that generally the application of fungicide can be modeled by one or more of the following assumptions that "there is complete coverage by the fungicide for either the entire host population or a fixed subset of the host population, that a generic, multipurpose, fungicide has been applied to the hosts (i.e. they have not readily separated out the effects of different fungicide types such as protectants, curatives or eradicants); or that the fungicides are permanent (i.e. that the chemicals do not decay or that their effects do not change over time)", then we develop a mathematical model to combine the effect of curative and protective function of the fungicide in this paper. Specifically the paper discusses a model of fungicide application in a fungal disease epidemic, where explicitly take fungicide dynamics into the model together with the curative and protective effect of the fungicide. We show the value of the Basic Reproduction Number (R_0) where it is computed from the largest eigen value of the next generation matrix. In the region where R_0 greater than one there is only one single stable endemic equilibrium. However, in region where R_0 less than one some parameters affect the number of posibble equilibria (Chavez et al 2002). Some numerical simulation are also given to ilustrate our results.

2. Mathematical Model

Following Castle et al (2012) in this model we assume that :

1. The fungal disease spread through one population;
2. There is there is a simple density dependent growth of the host up to a carrying capacity;
3. Infected host tissue consumes resources but does not contribute to new host growth;
4. The hosts cannot be re-infected after they cease to be infectious;
5. A protectant fungicide affects susceptible hosts, reducing their capacity to become infected;
6. All susceptible hosts are protected;
7. The population is divided into two classes, susceptible $(\hat{S}_t)$, infected $(\hat{I}_t)$ and protected $(\hat{P}_t)$.

Table 1. Factor that influence the model

Symbol	Explanation
ρ	All the factor that influence effectiveness of fungicide $(0 \le \rho \le 1)$
t	Time $(\hat{\beta}\hat{i}^{-1})$
ϑ	Growth rate of host $(\hat{\vartheta}\hat{\beta}^{-1})$
μ	Removal rate $(\hat{\mu}\hat{\beta}^{-1})$
π	Fungicide application rate $(\hat{\pi}\hat{\beta}^{-1})$
δ	Fungicide decay rate $(\hat{\delta}\hat{\beta}^{-1})$
α	Curatives rate of fungicide $(\hat{\alpha}\hat{\beta}^{-1})$
ε	Fungicide effectiveness
$\hat{k}$	Disease free carrying capacity

Table 1 shows the symbols used in the subseqment model.Using the symbols in Table 1 and from the assumption, we can build a transition diagram

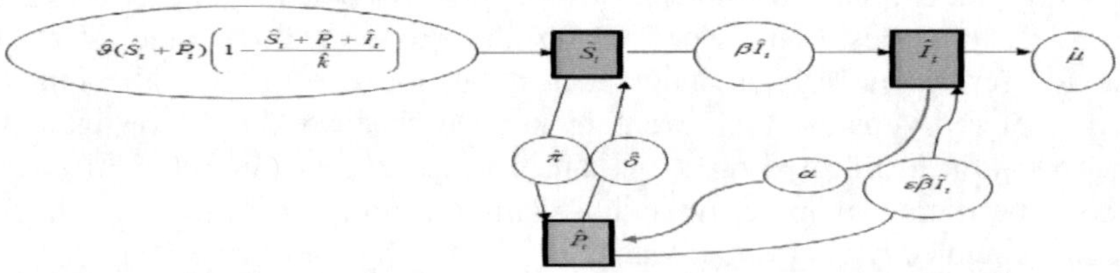

Figure 1 Scheme of susceptible ($\hat{S}_t$), infected ($\hat{I}_t$) and protected ($\hat{P}_t$) population (modified from Castle)

Based on the assumption and by referring to the work of Anggriani et al 2015, the differential equations model for the protective and curative explicit model is given by :

$$\frac{d\hat{S}_t}{d\hat{t}} = \hat{\vartheta}(\hat{S}_t + \hat{P}_t)\left(1 - \frac{\hat{S}_t + \hat{P}_t + \hat{I}_t}{\hat{k}}\right) - \hat{\pi}\hat{S}_t + \hat{\delta}\hat{P}_t - \frac{\hat{\beta}\hat{S}_t\hat{I}_t}{\hat{k}}$$
$$\frac{d\hat{P}_t}{d\hat{t}} = \hat{\pi}\hat{S}_t - \hat{\delta}\hat{P}_t - \frac{\varepsilon\hat{\beta}\hat{P}_t\hat{I}_t}{\hat{k}} + \alpha I_t \qquad (1)$$
$$\frac{d\hat{I}_t}{d\hat{t}} = \frac{\hat{\beta}\hat{I}_t(\hat{S}_t + \varepsilon\hat{P}_t)}{\hat{k}} - \hat{\mu}\hat{I}_t - \alpha I_t$$

After the reduction and normalization by changing the compartment from density to proportion (substituting $S_t = \hat{S}_t\hat{k}^{-1}, I_t = \hat{I}_t\hat{k}^{-1}, P_t = \hat{P}_t\hat{k}$ and $\vartheta = \hat{\vartheta}\hat{\beta}^{-1}, \pi = \hat{\pi}\hat{\beta}^{-1}, \delta = \hat{\delta}\hat{\beta}^{-1}, \mu = \hat{\mu}\hat{\beta}^{-1}, \alpha = \hat{\alpha}\hat{\beta}^{-1}, t = \hat{\beta}\hat{t}$) we get the non-dimensional system:

$$\frac{dS_t}{dt} = \vartheta(S_t + P_t)(1 - (S_t + P_t + I_t)) - \pi S_t + \delta P_t - S_t I_t$$
$$\frac{dI_t}{dt} = I_t(S_t + \varepsilon P_t) - \mu I_t + \alpha I_t \qquad (2)$$
$$\frac{dP_t}{dt} = \pi S_t - \delta P_t - \varepsilon P_t I_t + \alpha I_t$$

Anggraini et all (2015) showed the optimal control of the curvature intervention but did not present the stability analysis of the equilibrium points. Here we present the stability analysis of the resulting equilibrium points.

3. Mathematical Analysis

The non-endemic fixed point of the equation (2) is

$$(S_{t0}, P_{t0}, I_{t0}) = \left(\frac{\delta}{\delta + \pi}, \frac{\pi}{\delta + \pi}, 0\right)$$

and the endemic fixed point is

$$S_t = \mu\alpha - \varepsilon P_t$$
$$I_t = 2\vartheta P_t^2\varepsilon + 2\vartheta\varepsilon P_t\mu + 2\vartheta\varepsilon P_t\alpha + \pi\varepsilon P_t + \delta P_t + \vartheta\mu(1-\mu) - +\alpha\vartheta(1-2P_t) + \vartheta P_t(1-P_t)$$
$$-\vartheta\varepsilon^2 P_t^2 - \vartheta\varepsilon P_t - \alpha^2\vartheta - \pi\mu - \pi\alpha - 2\vartheta\mu\alpha - 2\vartheta P_t\mu) / (\vartheta\alpha + \vartheta P_t + \vartheta\mu + \mu + \alpha - \vartheta\varepsilon P_t - \varepsilon P_t)$$

By substituting S_t and P_t before to the equation (2) the third line, then P_t is :

$$((\vartheta\varepsilon - 2\vartheta\varepsilon^2 + \varepsilon^3\vartheta)P_t^3 +$$
$$(-3\varepsilon^2\vartheta\alpha + 2\vartheta\varepsilon\mu + \vartheta\varepsilon^2 + 4\vartheta\varepsilon\alpha - 2\varepsilon^2\vartheta\mu + \delta\vartheta\varepsilon - \vartheta\alpha - \pi\vartheta\varepsilon - \alpha\varepsilon - \delta\vartheta + \pi\vartheta\varepsilon^2)P_t^2 +$$
$$(-\delta\vartheta\mu + \pi\vartheta\alpha - 2\vartheta\varepsilon\alpha - \pi\varepsilon\mu - 2\vartheta\alpha^2 - \vartheta\varepsilon\mu - \delta\vartheta\alpha + \pi\vartheta\mu + 4\varepsilon\vartheta\mu\alpha + \vartheta\alpha$$
$$-2\pi\vartheta\varepsilon\mu + 3\varepsilon\vartheta\alpha^2 - 2\pi\vartheta\varepsilon\alpha + \varepsilon\vartheta\mu^2 - \delta\mu - 2\vartheta\mu\alpha)P_t +$$
$$\vartheta\mu\alpha + \pi\mu^2 + \pi\vartheta\mu^2 - \alpha\vartheta\mu^2 + 2\pi\vartheta\mu\alpha - 2\vartheta\mu\alpha^2 - \vartheta\alpha^3 + \pi\mu\alpha + \pi\vartheta\alpha^2 + \vartheta\alpha^2) = 0$$

By using Routh Hurwitz criteria, we can conclude that the polynomial has the real positive root provided that condition $0 < \varepsilon < \frac{1}{4}$, $0 < \mu < \frac{3\varepsilon\alpha}{2}$, $\pi\mu + \alpha + \pi\alpha < \delta\mu + 2\varepsilon\alpha + \delta\alpha + 2\pi\varepsilon\mu + 2\pi\varepsilon\alpha$ is satisfied, with the Basic Reproduction Ratio given by $R_{EPC} = \frac{1}{\mu}\left(\left(\frac{\delta + \varepsilon\pi}{\delta + \pi}\right) - \alpha\right) = R_0 - \frac{\alpha}{\mu}$

Stability Analysis

The jacobian matrix for the equation (2) is

$$MJ_E = \begin{bmatrix} \vartheta(1 - S_t - P_t - I_t) - \vartheta(S_t + P_t) - \pi - I_t & -\vartheta(S_t + P_t) - S_t & \vartheta(1 - S_t - P_t - I_t) - \vartheta(S_t + P_t) + \delta \\ I_t & S_t + \varepsilon P_t - \mu - \alpha & I_t\varepsilon \\ \pi & -\varepsilon P_t + \alpha & -\delta - I_t\varepsilon \end{bmatrix}$$

The stability testing at non- endemic fixed point s given as follows.

Theorem 1

The non-endemic fixed point (S_{t0}, P_{t0}, I_{t0}) *locally asymptotically stable if* $R_{EP} < 1$ *and not stable if* $R_{EP} > 1$.

Proof :

Following (Diekmann dan Heesterbeek, 2000) substitution of the non-endemic fixed point into the jacobian matrix results in :

$$J_E(E_0) = \begin{bmatrix} \vartheta\left(1 - \frac{\delta}{\delta+\pi} - P_t\right) - \vartheta\left(\frac{\delta}{\delta+\pi} + P_t\right) - \pi & -\vartheta\left(\frac{\delta}{\delta+\pi} + P_t\right) - \frac{\delta}{\delta+\pi} & \vartheta\left(1 - \frac{\delta}{\delta+\pi} - P_t\right) - \vartheta\left(\frac{\delta}{\delta+\pi} + P_t\right) + \delta \\ 0 & \frac{\delta}{\delta+\pi} + \varepsilon P_t - \mu - \alpha & 0 \\ \pi & -\varepsilon P_t + \alpha & -\delta \end{bmatrix}$$

From the jacobian matrix, we can get three eigen values:

$$\lambda_1 = \frac{\delta + \varepsilon P_t\delta + \varepsilon P_t\pi - \mu\delta - \mu\pi - \delta\alpha - \alpha\pi}{\delta + \pi} = \mu(R_{EC} - 1) < 0,$$

$$\lambda_2 = -(\delta + \pi) < 0,$$

$$\lambda_3 = -\frac{\vartheta(-\pi + 2P_t\pi + 2\delta P_t + \delta)}{\delta + \pi} = -\frac{\pi\vartheta + \delta\vartheta}{\delta + \pi} < 0$$

It is clear that if $R_{EPC} < 1$, all of the eigen values are negative, and hence the equilibrium point of (S_{t0}, P_{t0}, I_{t0}) is locally asymptotically stable.

IOP Conf. Series: Earth and Environmental Science **31** (2016) 012014 doi:10.1088/1755-1315/31/1/012014

4. Numerical Simulation

In this simulation suppose that growth rate of host $\vartheta = 0,5$, the removal rate $\mu = 0,2$,. The initial condition to the class of *susceptible*, *infected*, and *protected* respectively are (0.7; 0.3;0). And the other parameters divided into two condition, without fungicide application has a data set $\{\pi = \delta = \varepsilon = \alpha = 0$ and with fungicide has a data set $\pi=0.1$; $\varepsilon=0.1$; $\alpha=0.05$, $\delta=0.001$. Matlab solution for the system is given in figure 2 to 4.

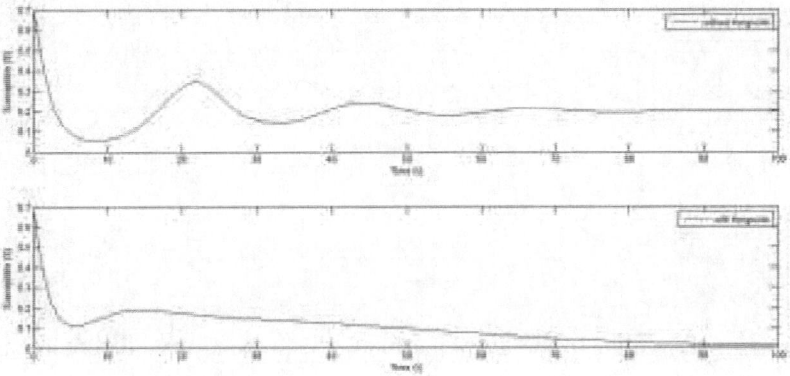

Figure 2. The graph of the influence of the fungicide to the susceptible host (S_t)

In Figure 2, when the fungicide is applied, the susceptible population will decrease towards zero.

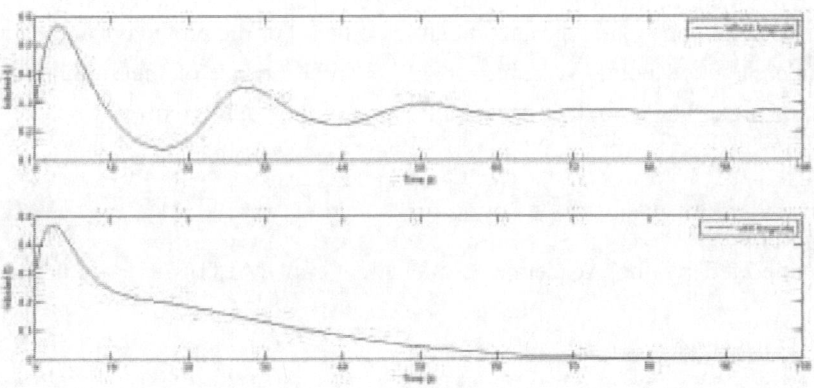

Figure 3. The graph of the influence of the fungicide to the infected population (I_t)

IOP Conf. Series: Earth and Environmental Science **31** (2016) 012014 doi:10.1088/1755-1315/31/1/012014

In Figure 3, when the fungicide is applied, the infected population will decrease towards zero.

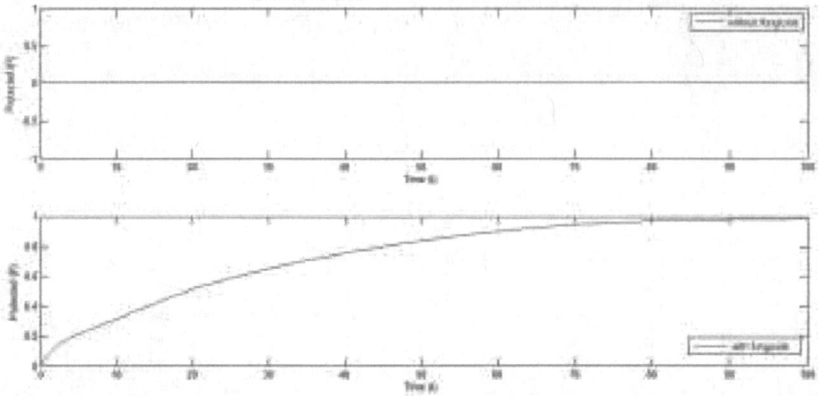

Figure 4. The graph of the influence of the fungicide to the protected population (P_t).

In Figure 4, when the fungicide is applied, the protected population increase toward one.

5. Conclucion

We have discussed a model of the protective and curative of fungicide. It is shown that the model has two equilibrium points, i.e non – endemic and endemic points. For the protective and curative explicit model the non-endemic fixed point is stable at $R_{EPC} < 1$. The result of the simulation shown that fungicide application are effective to decrease the infection of the population.

6. Acknowledgment

This work was supported by the Academic Leadership Grant (ALG) 2015 from the Universitas Padjadjaran.

IOP Conf. Series: Earth and Environmental Science **31** (2016) 012014 doi:10.1088/1755-1315/31/1/012014

7. References

[1] Castle, Matthew D., Giligan, Christopher A. 2012. *An Epidemiological Framework for Modelling Fungicide Dynamics and Control.* Department of Plant Sciences, University of Cambridge, Cambridge, United Kingdom.

[2} Castillo-Chavez, C., Feng, Z., dan Huang, W. 2002. "On the Computation R_0 and its role on global stability", Mathematical Approaches for Emerging and Reemerging Infectious Disease : An Introduction. IMA 125 : 229 – 250.

[3] Diekmann, O., dan Heesterbeek, J. A. P. 2000. *Mathematical Epidemiology of Infectious Disease,* Model Building, Analysis and Interpretation. John Willey & Son.

[4] N. Anggriani., L. Nurul Putri., A. Supriatna. 2015." Stability and Opti al Control of Plant Fungal Epidemic : An Explicit Model with Curative Factor. AIP Conf. Proc. 1651, 40-47; doi : 10.1063/1.4914430.

SNPs Selection using Gravitational Search Algorithm and Exhaustive Search for Association Mapping

W A Kusuma[1], L S Hasibuan, M A Istiadi

Department of Computer Science, Bogor Agricultural University, INDONESIA
Jalan Meranti Wing 20 level 5 Kampus IPB, Bogor, West Java, Indonesia

Email: ananta@apps.ipb.ac.id

Abstract. Single Nucleotide Polymorphisms (SNPs) are known having association to phenotipic variations. The study of linking SNPs to interest phenotype is refer to Association Mapping (AM), which is classified as a combinatorial problem. Exhaustive Search (ES) approach is able to be implemented to select targeted SNPs exactly since it evaluate all possible combinations of SNPs, but it is not efficient in terms of computer resources and computation time. Heuristic Search (HS) approach is an alternative to improve the performance of ES in those terms, but it still suffers high false positive SNPs in each combinations. Gravitational Search Algorithm (GSA) is a new HS algorithm that yields better performance than other nature inspired HS. This paper proposed a new method which combined GSA and ES to identify the most appropriate combination of SNPs linked to interest phenotype. Testing was conducted using dataset without epistasis and dataset with epistasis. Using dataset without epistasis with 7 targeted SNPs, the proposed method identified 7 SNPs – 6 True Positive (TP) SNPs and 1 False Positive (FP) SNP- with association value of 0.83. In addition, the proposed method could identified 3 SNPs- 2 TP SNP and 1 FP SNP with association value of 0.87 by using dataset with epistases and 5 targeted SNPs. The results showed that the method is robust in reducing redundant SNPs and identifying main markers.

1. Introduction

Association mapping is a study aiming to detect linkage between genetic polymorphisms and phenotypic variations in existing germplasm. One of the most popular marker to identify genetic polymorphisms is Single Nucleotide Polymorphisms (SNPs), since it allows generation of abundant information on genetic variability at DNA level. The analysis of AM has increased human understanding of phenotipic variations and heritability in human, animals and plants [1][2].

A number of methods for analyzing AM have been proposed before. In the conventional methods, AM is conducted based on investigation of single locus and the interest phenotype [3]. Nevertheless, the methods are not suitable for complex phenotypes with epistases interaction among genes [3][4]. The search for multi-locus association is potential to explain the phenomenon of complex phenotypes and genetic polymorphisms. Multi-locus association is classified as a combinatorial problem, it seems simple to be solved but hard to be implemented because the search space would increase exponentially according to the growth of data.

Multi-locus AM with 1000 SNPs require computer with high performance to evaluate 8.25×10^{12} combinations for five SNPs. Reducing the redundant and uncorrelated SNPs in the search space is the important task. One of the reducing techniques could be applied is using heuristic search and followed by conducting exhaustive search to the remaining SNPs to find the most appropriate combination of SNPs. This approach was implemented using Ant Colony Optimization (ACO) and exhaustive search in the case of case-control phenotype [3].

Gravitational Search Algorithm (GSA) is one of the latest heuristic search that developed based on the law of gravity and mass interaction. It was first introduced by Rashedi et al in 2009 [5]. It becomes to continue growing and being popular in researchers. Many modified and hybrid GSA have already proposed by researchers. They showed that GSA has better performance in terms of computation time and convergence than the previous heuristic search [6]. In this study, we proposed a method combining GSA and exhaustive search to find the most appropriate combination of SNPs that has linkage to quantitative phenotype.

2. Material

The proposed method was tested using two types of simulated dataset which were used by Oliveira et al [7]. The first dataset only has main effects without interaction among SNPs, while the second one with epistasis among SNPs. The dataset were generated by the function of *simulateSNPglm* of the *scrime* package in the R software. No filters HWE and *call rate* in simulated dataset were used, since there were not any missing values and no markers in Hardy-Weinberg disequilibrium.

2.1. Simulated dataset without epistasis 1

The SNP data were given in form of 1, 2 or 3. Notation 1 for homozygous reference sequence, 2 for heterozygous genotype and 3 for homozygous variant sequence. A thousand SNPs were simulated for 250 subjects, with a minor allelic frequency (MAF), simulated for each SNP, based on a uniform distribution with minimum and maximum limits were 0.10 and 0.40 respectively. The phenotype is continous and generated by Equation 1:

$$Y = \beta_0 + \beta_1 L_1 + \beta_2 L_2 + \beta_3 L_3 + \beta_4 L_4 + \beta_5 L_5 + \beta_6 L_6 + \beta_7 L_7 + error \quad (1)$$

where *error* was a normal random variable with mean was 0 and standard deviation was 5, L_1 = (SNP1 == 2), L_2 = (SNP10 == 1), L_3 = (SNP20 == 3), L_4 = (SNP30 == 3), L_5 = (SNP40 == 3), L_6 = (SNP50 == 2), L_7 = (SNP60 == 2). The beta coefficients were set as $\beta_0 = 0$, $\beta_1 = \beta_2 = \beta_3 = 200$, $\beta_4 = 900$, $\beta_5 = \beta_6 = \beta_7 = 200$.

2.2. Simulated dataset with epistasis 2

Epistasis could be defined as a form of functional interaction among genes that caused not common phenomenon in phenotypic variations [4]. Illustration about epistatis is showed in Figure 1.

The SNP data were given in form as used before in simulated dataset 1, likewise MAF adjusment. Ten thousand SNPs were simulated for 600 subjects with phenotype as defined in Equation 2:

$$Y = \beta_0 + \beta_1 L_1 + \beta_2 L_2 + \beta_3 L_3 + error \quad (2)$$

where error was a normal random variable with mean was 0 and standard deviation was 1, L_1 = (SNP4 != 2)[3] & (SNP3 != 1), L_2 = (SNP5 == 3), L_3 = (SNP12 != 1) & (SNP9 == 3). The beta coefficients were set as $\beta_0 = 0$, $\beta_1 = \beta_2 = 150$ and $\beta_3 = 40$. [3]The SNP4 != 2 symbol means that the SNP4 does not have the largest allele frequency represented by 2. Figure 2 shows the histogram and boxplot for the datasets.

3. Method

The SNPs selection method proposed in this study consist of two stages, selection using heuristic search in the first stage and exhaustive search in the second one. As mention in introduction, finding the most appropriate combination of SNPs that has linkage to quantitave phenotype is classified as a combinatorial problem, for which the worst case time requirement grows exponentially with the number of SNPs. To address this problem, heuristic search was conducted to select a number of possible SNPs that have linkage to phenotype. In other words, the first stage was aimed to reduce the number of redundant SNPs, this stage was conducted two times. However, solution provided by heuristic seach is not necessary as the best solution for the given problem [6]. Therefore, exhaustive search was needed in the final stage to select targeted SNPs. The association between SNPs and phenotype was conducted using SVR [8] first, SVR generated predict phenotype based on the given combination of SNPs and the predict phenotype was associated to the phenotype in simulated dataset. The correlation between predicted and simulated phenotype was being the evaluation of the fitness function.

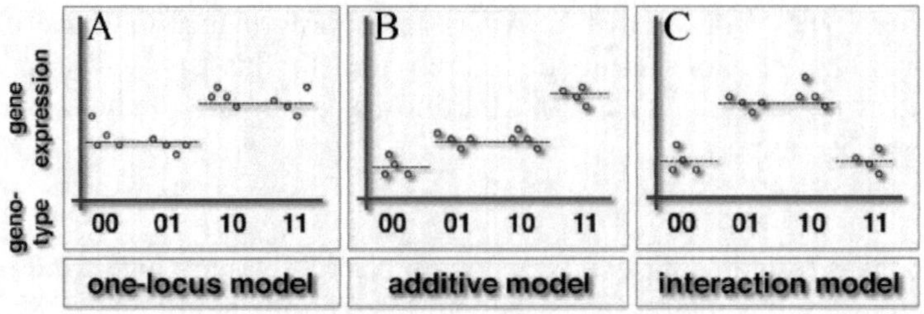

Figure 1. Illustration of the one-locus, additive and epistatic effects models in haploid organisms. The x-axis represent the combined genotype of the pair of loci while the y-axis represent expression level of the underlying gene. **(A)** In the one-locus model the genotype of one locus (here the first in the pair) drives the given phenotype. **(B)** In the additive model both loci contribute to the phenotype in an additive way. **(C)** When an epistatic interaction occurs, the effect of the two loci on the trait is non-additive [4].

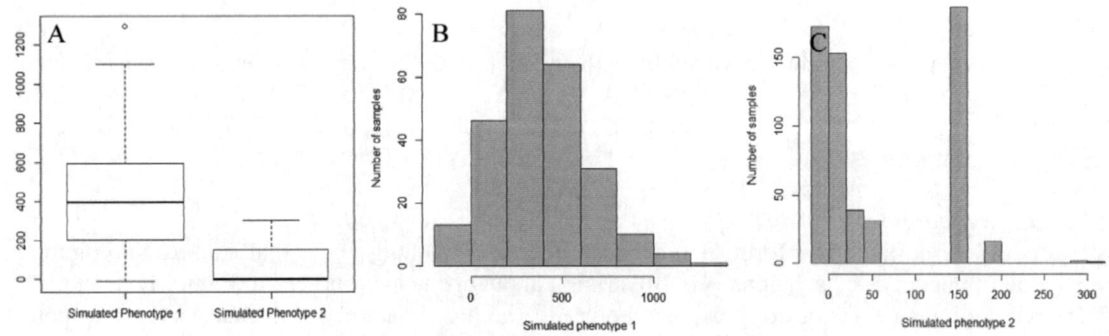

Figure 2. Boxplot and histogram of the datasets

3.1. *Gravitational Search Algorithm (GSA)*

One of the latest heuristic search proposed by Rashedi et. al. in 2009 is GSA. It was developed based on the law of gravity and mass interactions [5]. As the common heuristic search, GSA is comprised of parallel artificial agents and each agent represents a solution of the problem. Consider an artificial population consist of N agents, then solution of the ith agent is defined by

$$X_i = \left(x_i^1, \ldots, x_i^d, \ldots, x_i^n\right) \; for\, i = 1, 2, \ldots, N \tag{3}$$

The solution of the ith agent represents a combination of n SNPs, where x_i^d represents the dth SNP in a combination and n is the dimension, represent the desired number of SNPs for a combination.

All agents attract each other by a gravity force, and this force causes a movement of all agents globally towards the agents with heavier mass. Mass of an agent correspond to the evaluation of the fitness function for the given solution. The attraction and the movement are iterative procedures which would be stopped at a predefined number of iterations, T .

At a specific time t , the force acting on agen i from agen j is defined as the following:

$$F_{ij}^{d} = G(t)\frac{M_{pi}(t)\times M_{aj}(t)}{R_{ij}(t)+\varepsilon}\left(X_{j}^{d}(t)-X_{i}^{d}(t)\right) \tag{4}$$

where M_{aj} is the active gravitational mass related to agent j, M_{pi} is the passive gravitational mass related to agent i, $G(t)$ is the gravitational constant at time t, ε is a small constant, and $R_{ij}(t)$ is the Euclidian distance between two agents i and j. The effects of distance between SNPs is obviated, so that $X_j^d - X_i^d$ is set to 1 and R_{ij} is set to 0. The total force that acts on agen i in a dimension d is randomly weighted sum of dth component of the forces exerted from *Kbest* agents:

$$F_{i}^{d}(t) = \sum_{j\in Kbest,\, j\neq i} rand_{j} F_{ij}^{d}(t) \tag{5}$$

where $rand_j$ is a random number in the interval [0,1] and *Kbest* is the set of first K agents with the best fitness value and biggest mass. *Kbest* is a function of time, initialized to K_0 at the beginning and decreasing with time.

According to the law of motion, the acceleration of the agent i, at time t, in the dth dimension, is given as follows:

$$a_{i}^{d}(t) = \frac{F_{i}^{d}(t)}{M_{ii}(t)} \tag{6}$$

where M_{ii} is the is the inertial mass of the ith agent. The next velocity of an agent is considered as a fraction of its current velocity added to its acceleration. Therefore, its velocity and its position could be calculated as follows:

$$v_{i}^{d}(t+1) = rand_{i}\times v_{i}^{d}(t)+a_{i}^{d}(t) \tag{7}$$

$$x_{i}^{d}(t+1) = x_{i}+v_{i}^{d}(t+1) \tag{8}$$

where $rand_i$ is an uniform random number in the interval [0,1].

The gravitational constant, G, is initialized at the begining and will be reduced with time to control the search accuracy. In other words, G is a function of the initial value (G_0) and time (G):

$$G(t) = G(G_0, t) \tag{9}$$

$$G(t) = G_0 e^{-\frac{\alpha t}{T}} \tag{10}$$

The masses of the agents are calculated using fitness function. A heavier mass means a more efficient agent to solve the problem and moves more slowly. Supposing the equality of the gravitational and inertia mass, the values of masses is calculated using the map of fitness. The gravitational and inertial masses are updated by the following equations:

$$M_{ai} = M_{pi} = M_{ii} = M_i, i = 1, 2, \ldots, N \tag{11}$$

$$q_i(t) = \frac{fit_i(t) - worst(t)}{best(t) - worst(t)} \tag{12}$$

$$M_i(t) = \frac{q_i(t)}{\sum_{j=1}^{N} q_j(t)} \tag{13}$$

where $fit_i(t)$ represents the fitness value of the agent i at time t, and the $best(t)$ and $worst(t)$ in the population indicate the strongest and the weakest agent according to their fitness evaluation. For a maximization problem:

$$best(t) = max_{j\in\{1,...,N\}} fit_j(t) \tag{14}$$

$$worst(t) = min_{j\in\{1,...,N\}} fit_j(t) \tag{15}$$

The principle of GSA is shown in Figure 3.

SNPs of the best agent, $x_j^1,\ldots,x_j^d,\ldots,x_j^n$, in every iteration were collected and then added into detected SNPs set, L_d, at the end of iteration. The reduction of redundant SNP is conducted two times to get L_d with less redundant SNPs at the end of process.

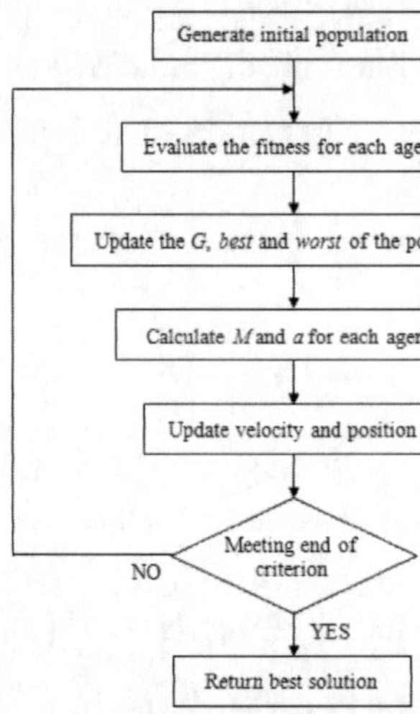

Figure 3. General principle of GSA [5]

3.2. Exhaustive Search

The number of outcome SNPs of the previous stage were less than the number of outset SNPs, $L_d < L_{all}$. Exhaustive search was conducted to select the most appropriate combination of SNPs that has linkage to the interest phenotype. The first step, exhaustive search determined the best combination of two SNPs. The two selected SNP were excluded from L_d and added into solution set L_s. Each remaining SNP in L_d were attempted to be added into L_s iteratively. The addition was allowed if it can improve association value between SNPs in L_s and phenotype. The addition will be stopped when no more SNP in L_d will improve association value if it was added into L_s.

4. Results and Discussions

The performance of the proposed method applied to both simulated dataset was compared to the method proposed by Oliveira et al [7] and generic GSA [5] itself. The comparison was conducted using two simulated dataset, without epistasis and with epistasis.

4.1. Simulated dataset without epistasis 1

Combination of SNPs generated by the proposed method and the other two methods using simulated dataset 1 were showed in Table 1. From the table, it was seen that the generic GSA was able to identify all of targeted SNPs but suffer of many false positive SNPs. The false positive SNPs may lead models' performance to predict phenotype become low, even though all of targeted SNPs consist in the generated combination. As could be seen in Tabel 1, correlation of the generic GSA was the lowest, 0.262, compared to the other methods.

Table 1. Combination of SNPs and correlation of methods in simulated dataset 1

Method	Identified SNPs	Correlation
Oliveira method	**1**, **10**, 15, **20**, **30**, **60**, 158, 177, 269, 274, 391, 446, 516, 673, 686, 693, 717, 725, 739, 825, 930**	0.750
Generic GSA*	**1**, **10**, **20**, **30**, **40**, **50**, **60** and 170 redundant SNPs**	0.262
Proposed Method	**1**, **10**, **20**, **30**, **50**, **60**, 72**	0.830

*)SNPs collection of best agen in every iteration
**)The bold SNPs is the targeted SNPs

The method proposed by Oliveira et al [7] was able to reduce 977 of 993 redundant SNPs, but the given combination still suffer of many false positive SNPs, 16 false positive SNPs. Besides that, there were two other targeted SNPs, the SNP40 and the SNP50, unidentified by the method. The correlation was higher than the generic GSA but lower than the proposed method.

As mention before, the generic GSA [5] was able to identify all of targeted SNPs. Besides that, it was also able to reduce redundant SNPs, 823 of 993 redundant SNPs, although the result was not better than that of the Oliveira method. Because all of targeted SNPs were consisting in the combination generated by generic GSA and the number of redundant SNPs was lower than those of in the outset SNPs, it was possible to conduct exhaustive search to find the most appropriate combination of SNPs. From Table 1, it could be seen that the performance of the proposed method was better than the Oliveira method in term of identifying targeted SNPs. Besides that, the number of false positive SNPs in the proposed method was lowest than the two other methods. Moreover, the correlation yielded by the proposed method was the highest among others, it was 0.83.

There was one SNP, the SNP40, which was not able to be identified by the Oliveira method and the proposed method. Figure 4 showed the relationship between the SNP40 and the simulated phenotype 1. From the Equation 1, it was known that SNP40 would affect to the phenotype when its value was 3. The Figure 4 showed that the number of sample with the value of SNP40 is 3 was less than the number of sample that the value of SNP40 is not 3. Figure 5 showed the relationship between the SNP30 and the simulated phenotype 1, the same thing also happening to SNP30. From the Equation 1, it was known that SNP30 will affect to phenotype when its value is 3. Figure 5 showed that the number of sample that the value of SNP30 is 3 was less than the number of sample that the value of SNP30 is not 3 but the models were able to identified it. From the Equation 1, it could be seen that the beta coeficient of SNP30 was greater almost five folds than the other beta coeficients. It means that, SNP30 is the main marker for the given phenotype. Meanwhile, the SNP40 is not the main marker and only a few samples show effect to phenotype because of this marker. Therefore, the models could not identified the SNP40 except generic GSA.

IOP Conf. Series: Earth and Environmental Science **31** (2016) 012015 doi:10.1088/1755-1315/31/1/012015

The comparison of models performance based on true positive rate, false positive rate and correlation was showed in Figure 6. Generally, the proposed method and the Oliveira method [7] were able to reduce redundant SNPs than the generic GSA [5]. The proposed method was better in identifying targeted SNPs than the Oliveira method, it could be seen from TPR and correlation metrics were higher than the Oliveira methods.

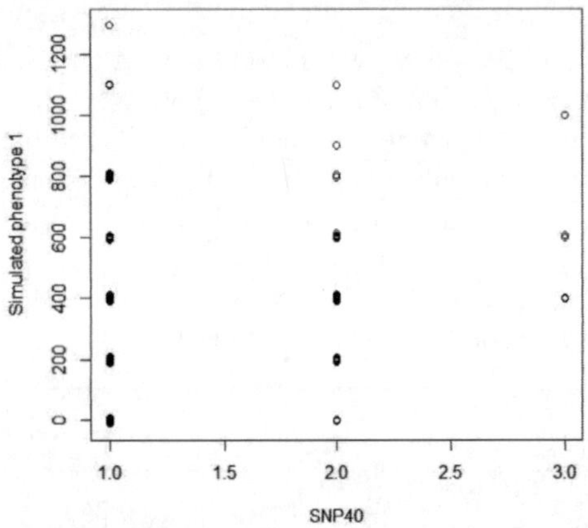

Figure 4. Scatter plot between the SNP40 and simulated phenotype 1

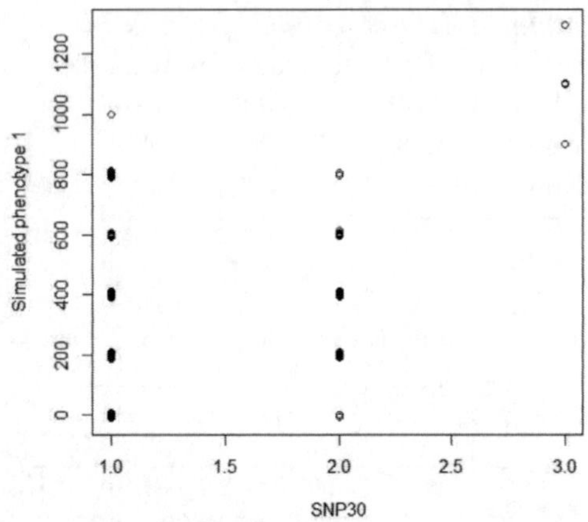

Figure 5. Scatter plot between the SNP30 and simulated phenotype 1

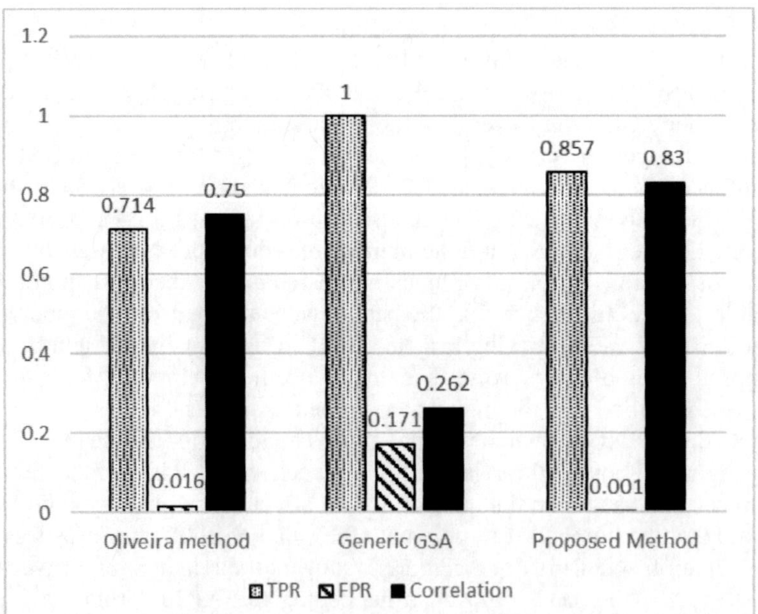

Figure 6. Methods performance comparison in simulated dataset 1

4.2. Simulated dataset with epistasis 2

The performance of the proposed method using simulated dataset 2 was showed in Table 2. From the table, it could be seen that all methods were able to reduce redundant SNPs except the method by generic GSA [5]. Nevertheles, the ability of all methods in identifying targeted SNPs was lower than before.

The generic GSA was able to identify two targeted SNPs, SNP3 and SNP4, while it suffer with 62 false positive SNPs and its correlation was 0.23. The Oliveira method [7] was able to identify one targeted SNP, the SNP3, with no false positive SNP and the correlation was 0.95. The proposed method was able to identify two targeted SNPs, SNP3 and SNP4, with one false positive SNP. The correlation was 0.87. From Table 2, it was seen that only SNP3 could be identified by the proposed methods.

In the simulated dataset 2, there was only one sample with condition (SNP4 != 2) and (SNP3 != 1) and there was no sample with condition (SNP12 != 1) and (SNP9 == 3). According to [7], the simulated dataset 2 was generated with high deviation and asymmetry data. It was done to examine the robustness of the method.

The comparison of models performance based on true positive rate, false positive rate and correlation using simulated dataset 2 was showed in Figure 7. From the figure, it could be seen that TPR of all methods were low and the correlations were high, except the generic GSA method [5]. It is mean that the proposed method and the Oliveira method [7] are robust in reducing the redundant SNPs and identifying main marker affected the simulated phenotype dataset 2.

Table 2. Combination of SNPs and correlation of methods in simulated dataset 2

Method	Identified SNPs	Correlation
Oliveira method	**3**	0.95
Generic GSA	**3**, **4** and 62 redundant SNPs*	0.23
Proposed Method	**3**, **4**, 8029	0.87

*)SNPs collection of the best agen in every iteration
**)The bold SNPs is the targeted SNPs,

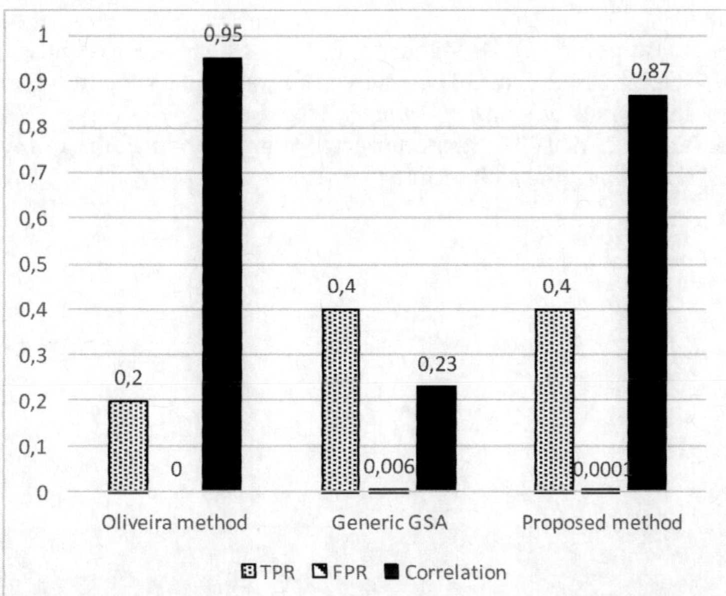

Figure 7. Methods performance comparison in simulated dataset 1

5. Conclusion

The proposed method successfully combined GSA and exhaustive search to solve the association mapping problem. The method was tested using two types of simulated datasets, without epistasis and with epistasis. In both simulated datasets, the proposed method shows the robussnest in reducing the redundant SNPs and identifying the main marker SNP which affect the simulated phenotype. It is concluded that the proposed method has good potentiality to be applied in association mapping with real dataset.

Acknowledgment
The authors would like to thank to Indonesia Ministry of Agriculture for funding this research through the KKP3N 2015 program.

References

[1] Cobb, J. N., DeClerck, G., Greenberg, A., Clark, R., & McCouch, S. 2013 Next-generation phenotyping: requirements and strategies for enhancing our understanding of genotype–phenotype relationships and its relevance to crop improvement *Theoretical and Applied Genetics* **126(4)** 867-887

[2] Minzhu, X., Jing, L., & Tao, J. Detecting genome-wide epistases based on the clustering of relatively frequent items *Bioinformatics* **28.1 (2012)** 5-12

[3] Wang, Y., Liu, X., Robbins, K., & Rekaya, R 2010 AntEpiSeeker: detecting epistatic interactions for case-control studies using a two-stage ant colony optimization algorithm *BMC research notes* **3(1)** 117

[4] Huang, Y., Wuchty, S., & Przytycka, T. M. 2013 eQTL epistasis–challenges and computational approaches *Frontiers in genetics* **4**

[5] Rashedi, E., Nezamabadi-Pour, H., & Saryazdi, S 2010 BGSA: binary gravitational search algorithm *Natural Computing* **9(3)** 727-745A

[6] Sahoo, G. 2014 A Review on gravitational search algorithm and its applications to data clustering & classification *International Journal of Intelligent Systems and Applications (IJISA)* **6(6)** 79

[7] de Oliveira, F. C., Borges, C. C. H., Almeida, F. N., e Silva, F. F., da Silva Verneque, R., da Silva, M. V. G., & Arbex, W 2014 SNPs selection using support vector regression and genetic algorithms in GWAS *BMC genomics*, **15**

[8] Smola, A., & Vapnik, V 1997 Support vector regression machines *Advances in neural information processing systems* **9** 155-161.

Model Averaging Method for Supersaturated Experimental Design

Deiby T. Salaki[1] , Anang Kurnia[2], Bagus Sartono[3]
[1]Department of Mathematics, Sam Ratulangi University, Manado, Indonesia
Graduate School, Bogor Agricultural University, Indonesia
[2,3]Department of Statistics, Bogor Agricultural University, Indonesia

E-mail: deibyts17@gmail.com

Abstract. In this paper, a new modified model averaging method was proposed. The candidate model construction was performed by distinguishing the covariates into focus variables and auxiliary variables whereas the weights selection was implemented using Mallows criterion. In addition, the illustration result shows that the applied model averaging method could be considered as a new alternative method for supersaturated experimental design as a typical form of high dimensional data. A supersaturated factorial design is an experimental series in which the number of factors exceeds the number of runs, so its size is not enough to estimate all the main effect. By using the model averaging method, the estimation or prediction power is significantly enhanced. In our illustration, the main factors are regarded as focus variables in order to give more attention to them whereas the lesser factors are regarded as auxiliary variables, which is along with the hierarchical ordering principle in experimental research. The limited empirical study shows that this method produces good prediction.

1. Introduction

Nowadays, the necessity of high dimensional data is increasing in all fields of study and has wide applications. Fortunately, the rapid growth in information technology supports the development of high dimensional data analysis. According to [1], high dimensional statistics refer to statistical inference when the number of unknown parameter is much larger order the sample size. Nevertheless, analysis of such data is impossible to be performed without additional assumption or penalization.

A typical form of high dimensional data stemmed from experimental data is supersaturated factorial design. A supersaturated factorial design is an experimental series in which the number of factors exceeds the number of runs. The supersaturated design (SSD) is recommended in high cost experiments and in situation where the number of effects is large, to screen few of a huge number of potentially active factors might be believed to be significant. There were some designs initiated to SSD. According to [2], Satterthwaite (1959), the initiator of the SSD, considered SSD to be useful in the initial investigation of large experiments and also proposed a random balance design for constructing SSD. Another construction method suggested by [3], namely $E(s^2)$ criterion to measure the efficiency of SSD. Hadamard matrix was a well-known design proposed by [2] in constructing SSD. Several authors have proposed various strategies in solving the challenges in SSD data analysis using some regression procedures, such as forward selection by [4], stepwise and all-subsets regression by [5] and partial least squares by [6]. There are also shrinkage regression approaches such as Smoothly Clipped Absolute Deviation procedure [7] and the Dantzig selector [8].

Many physical experiments can be high costly and time consuming to be studied using conventional method. In computer experiments or computer-based simulation, for example, the experiments can only be simulated by a computer modelling program in which the number of factor can be large as possible. Consequently, the running time is the mostly thing to be considered instead and the goal is turning to how to produce a best prediction model at any combination of values of the inputs that fitted the data [9]. Such concerns are in accordance with the motivation of model averaging as an alternative to variable selection or screening in SSD point of view.

Model averaging (MA) is another approach to improve prediction accuracy in analysing high dimensional data. Instead of relying on only one best model, the method refers both inference and prediction to the average over a set potential models in particular manner. The key issues of the methods are for example, weight selection criterion and the way of construction the candidate model [10]. Getting the best prediction model is preferred goal instead of estimation. Application of MA in SSD, have not received many attention yet. So far, the major focus of SSD analysis has been on screening or variable selection. MA approach in SSD, based on literature study, had been conducted by [11]. The paper proposed model averaging method by using BIC weight. Another weight used in model averaging namely mallows averaging which is refer to least square model averaging proposed by Hansen [12]. In his paper, Hansen performed different approach in preparing candidate models that is by constructing nested models while the weights is selected by minimizing Mallows criterion.

The current paper introduced a new typical MA approach which is tailored to candidate models preparation. Here modification, the main factors of matrix model are categorized as focus variables that must be included in the model and their interactions that are less certain as auxiliary variables. The approach went along with effect hierarchical principle to be considered to address the relative importance and relationship among effects. The principle set the lower order effects as more important factors than the higher level effects while the same order consider to be equally likely important [4]. Henceforth, the candidate model is constructed by augmenting the matrix of focus variable with the different combination of one or several auxiliary variables. Besides reducing the number of candidate model, the reason of this differentiation is to improve the prediction performance of the model [13]. In weights selection, Mallows criterion is applied which is proven asymptotically optimal in the sense of achieving the lowest possible squared error [12].

This paper is organized as follows. In Section 2, the supersaturated experimental design is described. Section 3 reviews the intended model averaging. In the Section 4, an illustration of MA approach in SSD is performed to show how well the prediction matched the actual model. Section 6 gives some concluding remarks

2. Supersaturated Design

Supersaturated design is fractional factorial design in which the number of actual effects exceeded the number of runs. The SSD is divided into two classes, multi-level designs with *s* level on each factor and mixed-level in which each factor has different level. The current paper was concerned with multi levels of linear main effects model. Supposed the model intended is

$$\boldsymbol{y} = \mathbf{X}\,\boldsymbol{\theta} + \boldsymbol{\varepsilon} \tag{2.1}$$

where $\boldsymbol{y}$ is the $n \times 1$ response vector, $\mathbf{X}$ is an $n \times (m+1)$ model matrix, vector of $\boldsymbol{\theta} = (\theta_0, \dots, \theta_m)^T$ constitutes the parameter and $\boldsymbol{\varepsilon}$ is a vector of independent normally distributed random errors with mean 0 and variance σ^2. The matrix $\mathbf{X}$ can be a union of main effects and their interactions. It is assumed that each of the m factors has two levels, contrast between high and low denoted by 0 and 1.

If the model matrix $\mathbf{X}$ is non-singular then the matrix $(\mathbf{X}'\mathbf{X})^{-1}$ exists and the least squares estimator of parameter vector $\boldsymbol{\theta}$ is denoted by $\widehat{\boldsymbol{\theta}} = (\mathbf{X}'\mathbf{X})^{-1}\mathbf{X}'\boldsymbol{y}$. According to [14], if the matrix model is away from orthogonality and presence of multicollinearity then detection of the true active factors is hardly to be performed.

3. Model Averaging

Model averaging is considered the alternative to model selection, especially in estimation of model with high dimensional data. Model selection will choose one of a wide set of competed models based on certain selection criterion with possibly contradictory conclusion such as R^2, BIC and AIC. On the other hand, model averaging treats each potential model as an information contributor about the parameters of interest. The contribution is expressed in a weight assigned to the candidate model which is determined based on certain criterion. The goal of model averaging is not to find the best possible model but rather to get the best possible prediction model.

The first step in model averaging is to construct the model candidate. After getting estimates of each model, model averaging assigning weight to each of them and finally, takes the weighted averaging of all parameter vectors to obtain the model averaging estimator.

3.1. Framework of MA

Consider the n observations $\{(y_i, x_i); i = 1, \ldots, n\}$, where y_i is response variable and $\boldsymbol{x}_i$ is a vector of m-dimensional explanatory variables where the number of predictor is allowed to increase with the sample size n. The relationship between the response variable and the explanatory variables will be considered from observed data. The relationship is explained in a multiple linear regression model, can be expressed as

$$\boldsymbol{y} = \sum_{j=1}^{m} \theta_j \boldsymbol{x}_j + \boldsymbol{\varepsilon} \tag{2.2}$$

where $E(\varepsilon) = 0$ and $Var(\varepsilon) = \sigma^2$. The intercept term is dropped by assuming that the means of the response variable and the input variable are already subtracted out.

Claimed M candidate models and the ith model can be written as follow

$$M_i\colon\ \boldsymbol{y}_i = \sum_{j \epsilon T_i} \theta_j \boldsymbol{x}_j + \boldsymbol{\varepsilon} \tag{2.3}$$

Where T_i was subset of explanatory variables contained in ith model. In matrix form, equation (2.2) can be written as equation (2.1) and equation (2.3) as below

$$M_i\colon\ \boldsymbol{y}_i = \mathbf{X}_i \boldsymbol{\theta}^i + \boldsymbol{\varepsilon} \tag{2.4}$$

Where $\boldsymbol{\theta}^i = [\theta_j]_{j \in T_i}$ is the vector of parameter $n \times s; s \leq m$

In model selection methods, according to certain criterion, it will be chosen a single best model:

$$\widehat{\boldsymbol{y}} = \widehat{\boldsymbol{\mu}} = \textstyle\sum_{j \in T_i} \widehat{\theta}_j \boldsymbol{x}_j.$$

Recall the model M_i in equation (2.4), its prediction can be written as

$$\widehat{\boldsymbol{\mu}}_i = \mathbf{X}_i \widehat{\boldsymbol{\theta}}^i$$

Supposed w_i is the weight associated with ith model and let $\boldsymbol{w} = (w_1, \ldots, w_M)'$ be the vector in unit simplex in $\mathbb{R}^M$

$$H_n = \{\boldsymbol{w} \in [0,1]^M\}$$

then *model averaging* estimator for equation (2.1) takes the form

$$\widehat{\boldsymbol{\mu}}_{MA} = \sum_{i=1}^{M} w_i \widehat{\boldsymbol{\mu}}_i \tag{2.5}$$

Based on the previously framework, the steps of MA analysis is itemized below:

1. Construct several candidate models M_i; $i = 1, \ldots, M$, as described in equation (2.4).
2. Estimate each model M_i
3. Determine weights based on certain criterion
4. Average out weighted prediction of those models, as in equation (2.5)

3.2. Model and Estimation

Supposed the linear regression model of model in equation (2.1) is mentioned in the framework of [13] as below

$$\boldsymbol{y} = \mathbf{X}\boldsymbol{\theta} + \boldsymbol{\varepsilon} = \mathbf{X}_1 \boldsymbol{\beta}_1 + \mathbf{X}_2 \boldsymbol{\beta}_2 + \boldsymbol{\varepsilon} \tag{2.6}$$

The matrix of explanatory variables is $\mathbf{X} = (\mathbf{X}_1\,|\mathbf{X}_2)$; $\mathbf{X}_1$ $(n \times q)$ is matrix of focus variables refer to intercept and main factor; $\mathbf{X}_2$ $(n \times r)$ is matrix of auxiliary variables refer to their interaction. Thus, vector of parameter consists of $\boldsymbol{\beta_1} = (\theta_0, \theta_1, \theta_2, \ldots, \theta_q)'$ and $\boldsymbol{\beta_2} = (\theta_{q+1}, \theta_{q+2}, \ldots, \theta_r)'$. It is also assumed that $q \geq 1\,; r \geq 0,\ +r \leq n-1$; that is $\mathbf{X} = (\mathbf{X}_1\,|\mathbf{X}_2)$ has full column-rank. According to Magnus [13] the model in equation (2.6) is more general than the conventional case where typically all variables are auxiliary except the constant term. Such form also allows some variables to be included in the model even when diagnostic tests suggest it to be removed or permitted ones as must contained in the model. Moreover, such classification could reduce the number of candidate model.

The least squares estimator of $\boldsymbol{\theta}$ for a full model where all the auxiliary variables are included in model is

$$\widehat{\boldsymbol{\theta}} = \begin{pmatrix} \widehat{\boldsymbol{\beta}}_1 \\ \widehat{\boldsymbol{\beta}}_2 \end{pmatrix} = (\mathbf{X}'\mathbf{X})^{-1}\mathbf{X}'\boldsymbol{y}$$

The set of candidate models consist of different models arise when a different subset of the $\boldsymbol{\beta}_2$, is set to zero in other words each model matrix of candidate models consist of focus variables incorporate with combination of auxiliary variables so there are totally M=2^r candidate models to be consider.

Consider $S_i = (\,r_i \times r)$ a selection matrix to choose the included auxiliary variables, where r_i is the number of auxiliary variables contained in M_i. So, the estimators for the candidate model M_i, is

$$\widehat{\boldsymbol{\theta}}_i = \left(A_{M_i}{}'A_{M_i}\right)^{-1} A'_{M_i}\,\boldsymbol{y} \tag{2.7}$$

where $A_{M_i} = (\,\mathbf{X}_1|\mathbf{X}_2 S_i'\,)$. For a full model least square estimator, $\widehat{\boldsymbol{\theta}}_i = \widehat{\boldsymbol{\theta}}$, with $S_i = I_r$, an identity matrix with r dimension. On the contrary, $S_i = \mathbf{0}$ is for the least square estimator of the smallest model among all possible candidate models, that was the model with $\mathbf{X}_1$ as the design matrix. General form of the selection matrix is $S_i = (I_{r_i}|0_{r-ri})$ and its column combination.

Suppose $\boldsymbol{w} = (w_1, \ldots, w_M)'$ is a weight vector with $w_i \geq 0$ and satisfied $\sum_1^M w_i = 1$, that is the weight vector belongs to the unit simplex in $\mathbb{R}^M$:

$$\mathcal{H}_M = \left\{\boldsymbol{w} \in [0,1]^M\,;\ \sum_1^M w_i = 1\right\}$$

then the averaging estimator of $\boldsymbol{\theta}$ is

$$\widehat{\boldsymbol{\theta}}_{MA} = \textstyle\sum_1^M w_i \widehat{\boldsymbol{\theta}}_i$$

If the corresponding prediction candidate model is

$$\widehat{\boldsymbol{\mu}}_i = A_{M_i}\widehat{\boldsymbol{\theta}}_i = A_{M_i}\left(A_{M_i}{}'A_{M_i}\right)^{-1} A'_{M_i}\,\boldsymbol{y} = P_{M_i}\boldsymbol{y}$$

where

$$P_{M_i} = A_{M_i}\left(A_{M_i}{}'A_{M_i}\right)^{-1} A'_{M_i}$$

then model averaging prediction of equation (2.6) can be written as

$$\widehat{\boldsymbol{\mu}}_{MA} = \textstyle\sum_1^M w_i\widehat{\boldsymbol{\mu}}_i = \sum_1^M w_i P_{M_i}\boldsymbol{y} = P(\boldsymbol{w})\boldsymbol{y} \tag{2.8}$$

3.3. Mallow Weight Criterion

There are various methods in determining the weights of model averaging. One of them was Mallow averaging proposed by Hansen [12] in which the selection weights is performed by minimizing a Mallows criterion. Such criterion for the averaging estimator is

$$C(\boldsymbol{w}) = \hat{\varepsilon}(\boldsymbol{w})'\hat{\varepsilon}(\boldsymbol{w}) + 2\sigma^2 tr(P(\boldsymbol{w}))$$

where

$$\hat{\varepsilon}(\boldsymbol{w}) = \boldsymbol{y} - A\,\widehat{\boldsymbol{\theta}}_{MA},$$

If $tr(P(\boldsymbol{w})$ is rewritten as $\boldsymbol{K} = (k_1, \ldots, k_M)'$; k_i is the number of parameters in ith model, then

$$\boldsymbol{C(w)} = \boldsymbol{w}'\hat{\boldsymbol{\varepsilon}}'\boldsymbol{w} + 2\sigma^2\boldsymbol{w}'\boldsymbol{K} \tag{2.9}$$

which is a quadratic form in the vector $\boldsymbol{w}$.

By minimizing the criterion in (2.9), the weight vector is

$$\widehat{\boldsymbol{w}} = \underset{\boldsymbol{w} \in \mathcal{H}_M}{\operatorname{argmin}} C(\boldsymbol{w}) \tag{2.10}$$

4. Illustration of MA Application in Supersaturated Designs

In order to apply model averaging method in SSD, an illustration will be presented based on [6] on pg. 443. Its goal was to develop an epoxide adhesive system for bonding a polyester cord. The experiment involved 23 factors with 14 runs or experiments. The response is strip adhesion, which needs to be maximized. After a screening analysis using stepwise regression, there are 6 significant main effects to be considered. Those effects are 4,10,12,15 and 19. A further analysis was planned by involving also two-factor interaction effects. As there are 5, hereafter denoted by A,B,C,D and E respectively, the analysis would consist of 5 main effects and 10 interaction effects with an additional parameter of intercept. Thus, there were as many as 16 parameters to be estimated with only 14 observations that was a 14×16 design matrix. It was a typical problem of high dimensional data, typically, supersaturated design experiment that reasonable to be solved using model averaging analysis.

Based on those resulted factors, the further analysis conducted by using model averaging with the follow steps.

1. Define the model by setting the main effects A, B, C, D and E as focus variables and the interaction effects, namely AB, AC,…, DE as auxiliary variables as was displayed in Table 1. Supposed the intercept and all factors are denoted by $\boldsymbol{x}_0, \boldsymbol{x}_1, \dots, \boldsymbol{x}_{15}$ respectively, the model could be written as

$$\boldsymbol{y} = \beta_0 + \sum_{i=1}^{5} \beta_i \boldsymbol{x}_i + \sum_{i=6}^{15} \beta_i \boldsymbol{x}_i + \boldsymbol{\varepsilon} \tag{2.11}$$

2. Prepare the candidate models based on the factors in step 1, each of which is composed of intercept and focus variables appended with combination of auxiliary variables sequentially. Thus, the design matrix of the first group of candidate model consists of model with focus variables only because $A_{M_0} = (\mathbf{X}_1 | \mathbf{X}_2 S_0') = X_1$ where $S_0' = \mathbf{0}$. The second group consists of model with focus variables and one of auxiliary variables. In this case, the form of selection matrix was $S_i = (I_1 | 0_{10-1}) = (I_1 | \mathbf{0}_9)$ and its column combination, those are $(1\,0\,0\,0\,0\,0\,0\,0\,0\,0), (0\,1\,0\,0\,0\,0\,0\,0\,0), \dots, (0\,0\,0\,0\,0\,0\,0\,0\,0\,1)$ respectively. The next group is contained model with focus variables and two auxiliary with the form of selection matrix was $S_i = (I_2 | \mathbf{0}_8) = \left(\begin{array}{cc|cccccccc} 1 & 0 & 0 & 0 & 0 & 0 & 0 & 0 & 0 & 0 \\ 0 & 1 & 0 & 0 & 0 & 0 & 0 & 0 & 0 & 0 \end{array}\right)$ and its column combination, for example $\begin{pmatrix} 1 & 0 & 0 & 0 & 0 & 0 & 0 & 0 & 0 & 0 \\ 0 & 0 & 1 & 0 & 0 & 0 & 0 & 0 & 0 & 0 \end{pmatrix}$ and $\begin{pmatrix} 1 & 0 & 0 & 0 & 0 & 0 & 0 & 0 & 0 & 0 \\ 0 & 0 & 0 & 0 & 1 & 0 & 0 & 0 & 0 & 0 \end{pmatrix}$. So do the next groups.

3. This step was performed using R program. The package `optimx` used to solve quadratic programming in order to get the vector weight proposed in equation (2.10). To estimate the candidate models, least squares method is applied. The predictive ability of model averaging prediction is validated using leave-one out-cross validation (LOOCV). According to [15], LOOCV is almost unbiased but has high variance, however, recommended to be used where the sample data is limited. It is a typical case of *k*-fold CV with *k* = 1 in which each of 14 iteration nearly all the data except for a single observation are used for training and the model is tested on the single observation. The MAPE used to measure mean of the absolute value of the percent errors, that is $\text{MAPE} = \frac{1}{n}\left(\sum_1^n \frac{actual - predicted}{actual}\right) 100\%$

Table 1. Supersaturated design matrix with 5 main factors and their interactions

Run	Adhesion	A	B	C	D	E	AB	AC	AD	AE	BC	BD	BE	CD	CE	DE
1	133	0	1	0	0	1	1	0	0	1	1	1	0	0	1	1
2	62	0	0	0	1	0	0	0	1	0	0	1	0	1	0	1
3	45	1	1	1	1	0	0	0	0	1	0	0	1	0	1	1
4	52	1	1	0	1	1	0	1	0	0	1	0	0	1	1	0
5	56	1	0	0	1	1	1	1	0	0	0	1	1	1	1	0
6	47	1	1	0	1	1	0	1	0	0	1	0	0	1	1	0
7	88	0	1	1	0	1	1	1	0	1	0	1	0	1	0	1
8	230	0	1	0	0	0	1	0	0	0	1	1	1	0	0	0
9	32	0	0	1	1	0	0	1	1	0	1	1	0	0	1	1
10	53	1	0	1	1	0	1	0	0	1	1	1	0	0	1	1
11	276	1	0	0	0	0	1	1	1	1	0	0	0	0	0	0
12	145	0	1	1	0	0	1	1	0	0	0	1	1	1	1	0
13	130	1	0	1	0	1	1	0	1	0	1	0	1	1	0	1
14	127	0	0	1	0	1	0	1	0	1	1	0	1	1	0	1

The value of actual and prediction are described in Figure 1 and Figure2

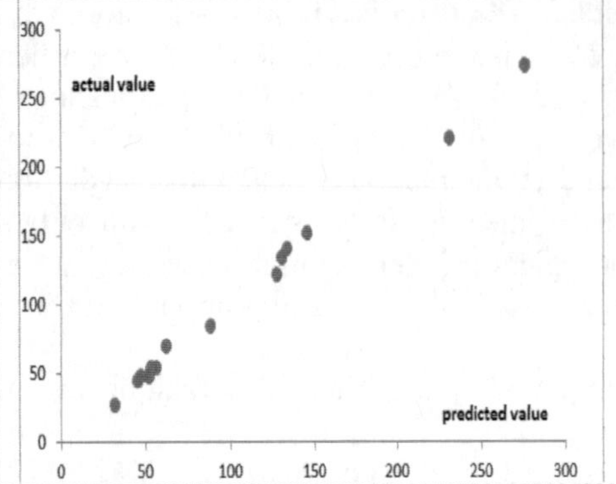

Figure 1. Plot of predicted against actual values

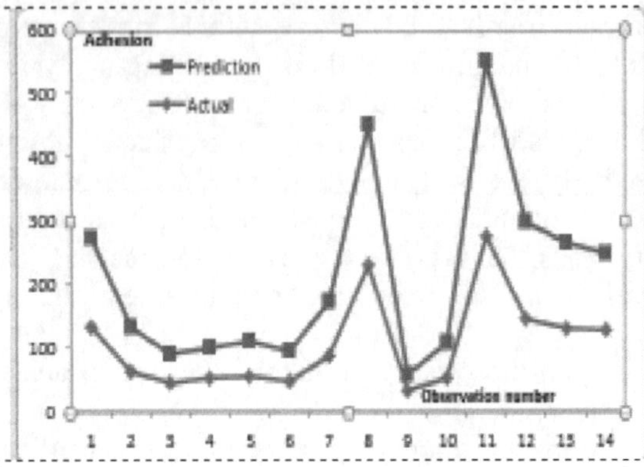

Figure 2. Plot of actual and predicted values

Refer to Figure 1, the value of actual against prediction is scattered closely to the absolute line. The value of both actual and prediction are shown in Figure 2. Their value is relatively having the same trend. As one of best prediction performance criterion, the mean absolute percent error (MAPE) is equal to 91%. The value of R^2 is equal to 0.99. It designates a high performance result of fitting because performed in collaboration with cross validation that is able to avoid overfitting. Thus, the value of R^2 could indicate a good prediction.

5. Conclusion

This study concludes that model averaging method could be applied in supersaturated experimental design as a typical of high dimensional data. That method was a modification of model averaging in which the candidate model construction is performed by distinguishing the matrix model into focus variables and auxiliary variables whereas the weighted selection is using Mallows criterion. In the case of experimental data set, main effects are regarded as focus variables, while the interactions as auxiliary variables. This differentiation follows the hierarchical ordering principle in experimental researches to address the relative importance among effects.

Based on our limited empirical study as was illustrated above, this method results a prediction model with good performance according to the value of MAPE and R^2. However, an advance study is needed to assess its performance in handling multilevel and mixed level experiments. Comparison with other type of model averaging is also highly recommended to be done in order to evaluate whether the modified model averaging method is better than others. Moreover, the further study in concern with criterions of estimator such as unbiasedness, efficiency and consistency would be a worth complement task.

References

[1] Buhlmann P and Van de Geer S 2011 *Statistics for High Dimensional Data* (Berlin: Spinger Verlag)

[2] Wu C F J and Hamada M S 2009 *Experiments Planning, Analysis, and Optimization* 2nd ed (NJ: John Wiley & Sons)

[3] Booth K H V and Cox D R 1962 Some Systematic Supersaturated Experimental Designs. *Technometrics* 4: 489-495

[4] Wu C F J 1993 Constructing of Supersaturated Designs Through Partially Aliased Interactions *Biometrika* 80:661-669

[5] Abraham B, Chipman H and Vijayan K 1999 Some Risks in the Construction and Analysis of Supersaturated Designs *Technometrics* 41(2):135-141

[6] Zhang Q Z, Zhang RC and Liu M Q 2007 A method for Screening Acrive Effect in Supersaturated Designs. *Journal of Statistical Planning and Inference* 137(9):235-245

[7] Li R and Lin D K J 2002 Data Analysis in Supersaturated Designs *Statistics & Probability*

Letters 59: 135-144
[8] Phoa F K H, Pan YH and Xu H. 2009 Analysis of Supersaturated Designs via Dantzig Selector. *Journal of Statistical Planning and Inference* 139(7):2362-2372
[9] Minqian L, Kaitai F 2006 A Case Study in The Application of Supersaturated Designs to Computer Experiment. *Acta Mathematica Scientia* 26B(4): 595-602
[10] Ando T and Li K C 2014 A Model Averaging Approach for High Dimensional Regression. *Journal of the American Statistical Association* 505(109): 254-265
[11] Marley, C J and Woods, D C 2010 A Comparison of Design and Model Selection Methods for Supersaturated Experiments *Journal of Computational Statistics & Data Analysis* 54(12): 3158-3167
[12] Hansen B E 2007 Least Squares Model Averaging. *Econometrica* 75:1175-1189
[13] Magnus J R, Powell O R and Schneck P 2010 A Comparison of Two Averaging Techniques with an Application to Growth Empirics. *Journal of Econometrics* 154: 139-153
[14] Georgiou S D 2014 Supersaturated designs: A review of their construction and analysis. *Journal of Statistical Planning and Inference* 144: 92-109
[15] Efron B 1983 Estimating the error rate of a prediction rule: improvement on cross-validation. *Journal American Statistics Association* 78:316-331

IOP Conf. Series: Earth and Environmental Science **31** (2016) 012017 doi:10.1088/1755-1315/31/1/012017

How Nonlinear Optics can be Simplified to Study Molecular Deposition and Surface Vicinality of a Si(001) Interface

Hendradi Hardhienata

Theoretical Physics Division, Department of Physics, Bogor Agricultural University, Jl. Meranti, Gedung Wing S, Kampus IPB Darmaga, Bogor 16680, Jawa Barat, Indonesia

E-mail: hendradi@ipb.ac.id

Abstract. Second-harmonic-generation (SHG) has long been considered as a powerful optical surface diagnostic tool due to its ability to sense optical surface properties in centrosymmetric material. In a series of papers Aspnes et. al. showed that azimuthal measurements of rotated samples can be described by the so called simplified bond-hyperpolarizabilty model (SBHM) [G. D. Powell, J. F. Wang, and D. E. Aspnes, Phys. Rev. B 65, 205320 (2002)]. In this work, their model is applied to show how small changes in the step hyperpolarizability and vicinal angle of a Silicon (001) interface affects the second harmonic generation (SHG) intensity profile significantly, suggesting that real time surface deposition and interface vicinal angle monitoring using SHG is possible.

1. Introduction

The application of nonlinear optics is not limited in the fields of physics but also in chemistry, material science, and biology [1]. One of the possible and widespread method in studying surface layers using nonlinear optics is second harmonic generation (SHG) in non centrosymmetric crystals [2]. In a recent work, SHG microscopy combined with interferometry can also be used as a nonlinear imaging device to study corrosion effects at material surfaces [3]. All the mentioned technological applications and benefit of nonlinearity cannot be achieved without a firm theoretical understanding of the basic physical laws that can be constructed from a physical model.

Although higher order multipole effects such a quadrupole or magnetic dipole can still contribute to SHG from within the bulk, its effect are usually minor in vicinal centrosymmetric semiconductor due to the large surface anisotropy. Notable work was performed to explain the Rotational Anisotropy (RA) SHG data from a vicinal Si surface using phenomenological model [4] but is still to complex because it requires up to 10 fitting parameters. The model was later modified by Powell et. al. [5] who showed using an anharmonic bond model, assuming that the SHG comes from charges polarized along the atomic bonds that it require just three independent fitting parameters. Later it was shown that the SHG susceptibility in the previous phenomenological model (group theory) and bond model (bond products) are connected [6] and even developed to explain thid harmonic (THG) and electric field induced second harmonic (EFISH) in metal oxide semiconductors [7]. Most recent works that applies the hyperpolarizability model is the investigation of SHG from bulk dipolar contribution in

IOP Conf. Series: Earth and Environmental Science **31** (2016) 012017 doi:10.1088/1755-1315/31/1/012017

zincblende structure [8] and interface dipole versus bulk quadrupolar SHG contribution in diamond semiconductor using arbitrary input polarization angle [9]. It will be shown here that such model can also be applied to interpret RASHG intensity during molecular deposition by assigning a change in the effective hyperpolarizability, which will influence the peak profile due to the modification of the interface e.g. filled terrace-steps.

2. Simplified Bond Hyperpolarizability Model

SBHM is a phenomenological model that is based on a classical electrodynamics. As a phenomenology theory, SBHM is still based on the fact that one requires experimental fitting to determine the nonlinear hyperpolarizability. A previous phenomenological model describes the nonlinear polarization source at the surface/interface through a polarization sheet and apply standard transmission and reflection analysis at the boundary of this sheet, deriving the nonlinear susceptibility from group theory (GT) [10]. However it can be shown that GT and SBHM are closely related [6, 11]. In addition, the bond model in [5] handle this source term in a straightforward fashion by viewing it as dipole sources driven by the incomming electric field. Governed by an anharmonic Lorentz model, the charge inside the material experience anharmonic oscillation and producing nonlinear radiation that can be detected in the far field.

Originally, SBHM takes the following assumptions: The nonlinear radiations are only produced by anharmonic dipoles that are oscillating along the bonds. It has to be stated though that other sources can still generate SHG inside the material such as bulk quadrupole, spatial dispersion, and EFISH. Nevertheless, SHG from vicinal diamond semiconductor seems to be very well modelled by assuming only surface dipoles since bulk dipole radiation inside the bulk is forbidden due to parity [12] whereas for a flat (non vicinal) Si surface/interface, bulk quadrupole contribution should be considered too as the p-in p-out sixfold SHG intensity profile from Rotational Anisotropy SHG can only be fitted if bulk effects such as spatial dispersion and bulk quadrupole effects are applied [13].

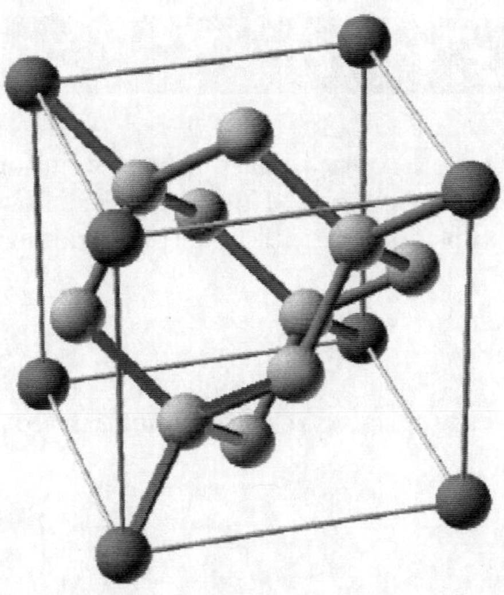

Figure 1. IUPAC unit cell for Si(001) facet and definition of the up bond unit vectos (blue) and down bond unit vectors (red) [14].

It can be seen from Fig. 1 which depict the unit cell according to the International Union of Pure and Applied Chemistry (IUPAC) that a Si(001) surface/interface has the following bond unit vectors:

$$\hat{\mathbf{b}}_1 = \begin{pmatrix} -\frac{1}{\sqrt{2}}\sin\frac{\beta}{2} \\ -\frac{1}{\sqrt{2}}\sin\frac{\beta}{2} \\ -\cos\frac{\beta}{2} \end{pmatrix} \quad \hat{\mathbf{b}}_2 = \begin{pmatrix} \frac{1}{\sqrt{2}}\sin\frac{\beta}{2} \\ \frac{1}{\sqrt{2}}\sin\frac{\beta}{2} \\ -\cos\frac{\beta}{2} \end{pmatrix} \quad \hat{\mathbf{b}}_4 = \begin{pmatrix} -\frac{1}{\sqrt{2}}\sin\frac{\beta}{2} \\ \frac{1}{\sqrt{2}}\sin\frac{\beta}{2} \\ \cos\frac{\beta}{2} \end{pmatrix} \quad \hat{\mathbf{b}}_5 = \begin{pmatrix} \frac{1}{\sqrt{2}}\sin\frac{\beta}{2} \\ -\frac{1}{\sqrt{2}}\sin\frac{\beta}{2} \\ \cos\frac{\beta}{2} \end{pmatrix}$$

where $\beta = 109.47°$ is the angle between two bonds in a Si lattice. In reality the outermost surface layer in Si(001) can differ due to cut roughness by one or two steps of one atomic layer which will change the bond orientation by a shift of $\frac{\pi}{2}$. Therefore we should have to define two times four bond vectors in respect to the double domain nature of the surface:

$$\hat{\mathbf{b}}_5 = -\hat{\mathbf{b}}_1 \quad \hat{\mathbf{b}}_6 = -\hat{\mathbf{b}}_2 \quad \hat{\mathbf{b}}_7 = -\hat{\mathbf{b}}_3 \quad \hat{\mathbf{b}}_8 = -\hat{\mathbf{b}}_4$$

The ratio between the two domains can differ from $50\% - 50\%$ to $0\% - 100\%$ depending on the puritiness of the cut which can be considered in the calculations by multiplying both groups of four bond vectors with the surface ratio respectively. For the sake of simplicity, the ideal case is assumed thus only one domain consisting of the previous four bond vectors is considered in the simulation.

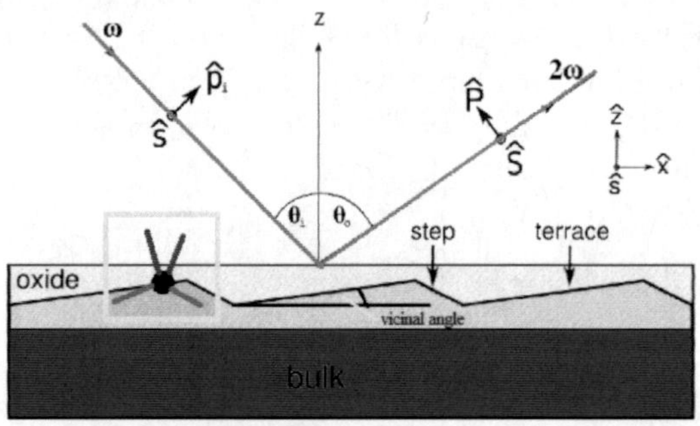

Figure 2. Vicinal Si(001) interface and optical axis of the system for a p polarized fundamental input and p polarized second harmonic output. The yellow box shows that the two down bonds are affected by the terrace and steps and thus its hyperpolarizability must be modified. Image extended from Ref. [15].

To account for the vicinality of the surface the rotation matrix around the y-axis is introduced:

$$R_y(\gamma) = \begin{pmatrix} \cos\gamma & -\sin\gamma & 0 \\ 0 & 1 & 0 \\ -\sin\gamma & 0 & \cos\gamma \end{pmatrix} \tag{1}$$

where γ is the vicinal angle. Considering that the sample can be rotated about the z axis, rotation of the bond direction along this axis for an angle ϕ gives:

$$\hat{\mathbf{b}}_j(\phi) = \mathbf{R_z}(\phi) \cdot \hat{\mathbf{b}}_j \tag{2}$$

where $R_z(\phi)$ is the common rotation matrix:

$$R_z(\phi) = \begin{pmatrix} \cos\phi & -\sin\phi & 0 \\ \sin\phi & \cos\phi & 0 \\ 0 & 0 & 1 \end{pmatrix} \tag{3}$$

The SHG polarization produced by dipolar interfacial sources can be modelled in the SBHM via [5, 13]:

$$\begin{aligned} \mathbf{P}_D^{(2),\text{interface}} &= \chi_D^{(2)} \cdot\cdot \mathbf{E}_{\text{int}}\mathbf{E}_{\text{int}} \\ &= \frac{1}{V}\sum_{j=1}^{n}\left(\alpha_{2j}\hat{\mathbf{b}}_j\hat{\mathbf{b}}_j\hat{\mathbf{b}}_j\right) \cdot\cdot \mathbf{E}_{\text{int}}\mathbf{E}_{\text{int}} \end{aligned} \tag{4}$$

where $\chi_D^{(2)}$ is the third rank susceptibility tensor; $\mathbf{E}_{\text{int}}$, is the field on the interface, α_{2j} is the interface dipolar SHG hyperpolarizability, V is the volume, n is the total number of bonds inside the considered conventional cell, and $\hat{\mathbf{b}}_\mathbf{j}$ is the bond unit vector. For a non vicinal surface there are two hyperpolarizabilities namely *up* hyperpolarizability labelled as α_u for the two *up* bond $(\hat{\mathbf{b_1}}, \hat{\mathbf{b_2}})$ and a hyperpolarizability α_d for the two *down* bonds $(\hat{\mathbf{b_3}}, \hat{\mathbf{b_4}})$, whereas we only use one bulk hyperpolarizability for all of the bulk bonds because of symmetry. All hyperpolarizabilities can generally have a real and imaginary term. For a vicinal surface the two down bond hyperpolarizability are clearly affected by the cut as can be seen in Fig. 2. Therefore the two down hyperpolarizability should be changed accordingly to [16]:

$$\alpha_{eff,int} = (1 - |\tan\gamma|)\alpha_{terrace} + |\tan\gamma|\,\alpha_{step} \tag{5}$$

which reduces immediately to the initial hyperpolarizability if the surface is flat (e.g. $\gamma = 0$)

With the assumption before that the bond charges oscillate only along the bonds, the far-field radiation E_{ff} at the interface is [13, 5]:

$$\mathbf{E}_{ff,\text{int}}(\mathbf{r}) \propto \left[(\mathbf{I} \quad \hat{\mathbf{k}}_{\mathbf{o2}\omega,\text{int}}\hat{\mathbf{k}}_{\mathbf{o2}\omega,\text{int}}) \cdot \mathbf{P}\right] \tag{6}$$

where $\hat{\mathbf{k}}_{\mathbf{o2}\omega,\text{int}}$ is the outgoing SHG vector from the interface.

3. Results and Discussions

Using Eq. (6) the far field can be obtained for a *p*-in *p*-out polarization because *pp* gives the most accurate changes in the SHG intensity profile compared to the other three polarization (e.g. *ps*, *sp*, and *ss*).

Fig. 3a depicts the SHG intensity for variation in the vicinality of the surface for a fix hyperpolarizability which we take for simplicity to be one. It can be seen that the vicinality will produce a change in the middle peak for an azhimuthal angle of $\frac{\pi}{2}$ which is sensitive even to a 1° of change. For $\gamma = 0°$ it is obvious that the simulation produces four equal heighted peaks corresponding to a C_{2v} surface symmetry whereas the bulk has T_d symmetry thus producing no even harmonic dipolar contribution. Therefore, using Rotational Anisotropy Spectroscopy (RA) SHG one can determine the vicinality of an unknown diamond sample by looking at the SHG intensity. However after the vicinality reaches 8° the middle peak vanishes and the simulation cannot distinguish further inclinations. Results for $\gamma = 9°$ looks quite similar to experimental result given in Fig. 3 of Ref. [5] where there is a small asymmetry between the two peaks arising from a small difference in the terrace and step hyperpolarizability in the effective hyperpolarizability of the down bonds and in addition the middle peak is higher than in the simulation but can in principle be fitted. The point is nevertheless, that even small changes in vicinality can be observed from the peak profile height variation at the middle.

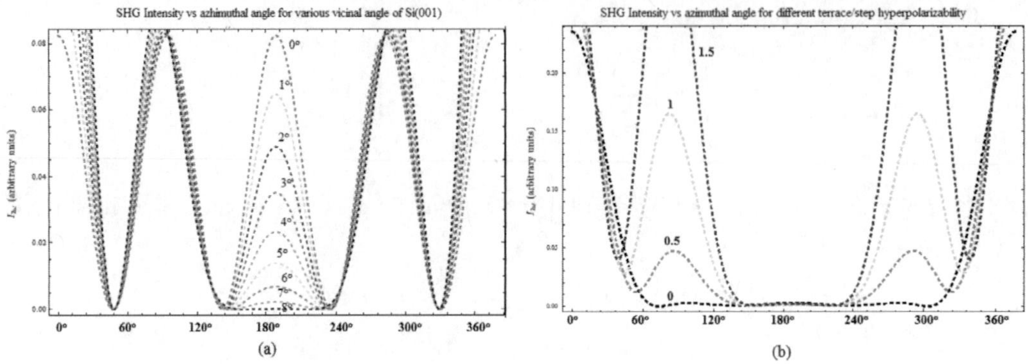

Figure 3. SBHM simulation for a Si(001) vicinal interface with (a) variations of the vicinal angle and (b) effective hyperpolarizability.

To investigate the effect of changes in the step and terrace hyperpolarizability, for example due to depositions of molecules on the surface of a vicinal diamond semiconductor, the effective hyperpolarizability is changed gradually by setting the $\alpha_{terrace} = \alpha_{step}$ respectively from $0, 0.5, 1$ and 1.5. It can be seen from Fig. 3b that the peaks between $120° - 160°$ and $240° - 300°$ varies accordingly in a symmetric manner if we assume that the effective hyperpolarizability of the two down bonds are the same. An asymetric peak will result if the hyperpolarizabilities of the two down bonds are nor similar. Therefore one can use SBHM to determine possible surface contamination in vicinal structures. Finally, this method can also be applied for other different Si interface orientation such as Si(111) or Si(011) with different polarizations. Further work will be performed to investigate whether bulk effects such as spatial dispersion or quadrupole effects also play a significant role in vicinal centrosymmetric semiconductors.

4. Conclusions

It is demonstrated that for a Si(001) vicinal semiconductor that the *pp* polarization RASHG data can be combined with SBHM simulation to determine the surface vicinal angle and possible molecule deposition by monitoring the change in the SHG intensity peak profile.

5. References

[1] S.Yamada, B. Song, S.Jeon J. Upham, Y. Tanaka, T. Asano, and S. Noda. , Optics Letters, Vol. 39, No. 7 (2014).M. P. Brown and K. Austin, Appl. Phys. Letters 85, 25032504 (2004).

[2] Y. R. Shen. The Principles of Nonlinear Optics, (Wiley-Interscience, 2003). Rev. 174, 813 (1968).

[3] A. Prylepa, J. Duchoslav, C. Reitböck, K. Hingerl, and D. Stifter, Lasers and Electro-Optics Europe (CLEO EUROPE/IQEC), Munich (2013)

[4] G. Lüpke, D. J. Bottomley, and H. M. van Driel,"Second- and third-harmonic generation from cubic centrosymmetric crystals with vicinal faces: phenomenological theory and experiment," J. Opt. Soc. Am. B **11**, 33 - 44 (1994).

[5] G. D. Powell, J.-F. Wang, and D. E. Aspnes, Simplified bond-hyperpolarizability model of second harmonic generation, Phys. Rev. B **65**, 205320 (2002).

[6] A. Alejo-Molina, H. Hardhienata, and K. Hingerl, " Simplified bond-hyperpolarizability model of second harmonic generation, group theory, and neumann's principle," J. Opt. Soc. Am. B, Vol. **31**, 526-533 (2014)

[7] A. Alejo-Molina, K. Hingerl, and H. Hardhienata,"Model of third harmonic generation and electric field induced optical second harmonic using the simplified bond-hyperpolarizability model," J. Opt. Soc. Am. B., Vol. **32**, 562 - 570 (2015).

[8] H. Hardhienata, A. Alejo-Molina, C. Reitböck, A. Prylepa, D. Stifter, and K. Hingerl,"Bulk dipolar

contribution to second-harmonic generation in zincblende," J. Opt. Soc. Am. B., Vol. **33**, No. 2, 195 - 201 (2016).
[9] C. Reitböck, D. Stifter, A. Alejo-Molina, K. Hingerl, and H. Hardhienata ,"Bulk quadrupole and interface dipole contribution for second harmonic generation in Si(111)," J. Opt., **18**, 035501 (2016).
[10] V. Mizrahi and J. E. Sipe, *J. Opt. Soc. Am. B*, **5**, 233402 (1988).
[11] E.S. Jatirian-Foltides, J.J. Escobedo-Alatorre, P.A. Márquez-Aguilar, H. Hardhienata, K. Hingerl, and A. Alejo-Molina, "About the calculation of the second-order $\chi^{(2)}$ susceptibility tensorial elements for crystals using group theory " Rev. Mex. Fis. **E 62**,5-13 (2016) .
[12] G. New. Introduction to Nonlinear Optics. (Cambridge University Press, 2011).
[13] H. Hardhienata, A. Prylepa, D. Stifter, and K. Hingerl,"Simplified bond-hyperpolarizability model of second-harmonic-generation in Si(111): theory and experiment," J. Phys. **423**, 012046 (2013).
[14] Emminger, C. Transformation of the Third Rank Tensor. (2013) JKU Bachelor Thesis.
[15] Noebauer K., Modelling Second-Harmonic Generation on a Vicinal Surfaces using a Simplified Bond-Hyperpolarizability Model. Dissertation. JKU. 2010
[16] J. F. McGilp, Using steps at the $Si-SiO_2$ interface to test simple bond models of the optical second-harmonic response, J. Phys. 19, 016006 (2007)

IOP Conf. Series: Earth and Environmental Science **31** (2016) 012018 doi:10.1088/1755-1315/31/1/012018

Landmark Analysis Of Leaf Shape Using Polygonal Approximation

Zakhi Firmansyah[1], Yeni Herdiyeni[1], Bib Paruhum Silalahi[2], Stephane Douady[3]
[1]*Department of Computer Science, Faculty of Mathematics and Natural Sciences. Bogor Agricultural University, West Java Indonesia*
[2]*Department of Mathematic, Faculty of Mathematics and Natural Sciences. Bogor Agricultural University, West Java Indonesia*
[3]*Laboratoire: Matiere et System Complex, Université Paris Diderot, France*

E-mail: zakhi99@gmail.com, yeni.herdiyeni@ipb.ac.id, bibparuhum@gmail.com

Abstract. This research proposes a method to extract landmark of leaf shape using static threshold of polygonal approximation. Leaf shape analysis has played a central role in many problems in vision and perception. Landmark-based shape analysis is the core of geometric morphometric and has been used as a quantitative tool in evolutionary and developmental biology. In this research, the polygonal approximation is used to select the best points that can represent the leaf shape variability. We used a static threshold as the control parameter of fitting a series of line segment over a digital curve of leaf shape. This research focuses on seven leaf shape, i.e., eliptic, obovate, ovate, oblong and special. Experimental results show static polygonal approximation shows can be used to find the important points of leaf shape.

1. Introduction

Botanist observe various features in plants using taxonomic key to identify the species of a plant. One feature that is observed to determine the species of plants are leaf shape. Each class has a unique shape. In addition, the general class of leaf shape is also determined by the ratio of length to width leaf and the widest position of the leaf. Previous research related to plant identification such as Gwo *et al.* (2013) use key point along leaf *boundary*, *centroid* dan fuzzy histogram to representation leaf shape. There are so many works related shape identification using landmark analysis. Takemura *et al.* (2002) introduces a new shape analysis approach using the well known wavelet transform and exploring fish shape representation by landmarks. In this paper, a new method using static polygonal approximation to analyze landmark of leaf shape is proposed.

1.1. Landmark

Landmark is a biologically definable point on an organism, that can be sensibly compared between related organisms. It typically requires domain-specific expert knowledge to choose a suitable set of landmarks.

Landmark-based shape analysis has been the core of morphometrics geometry and has been used as a quantitative tool in evolutionary and developmental biology (Klingenberg, 2010). However, it exhibits three main fundamental limitations (Laga 2014) :
First, it assumes that ~~the~~ landmarks are already incorrespondence across shapes. Shapes of plant leaves, however exhibit large non-rigid variations, even with in a single species. This makes finding the correspondences across shapes even more challenging.
Second, landmark-based methods are not invariant to all the shape preserving transformations. They can be made, in a preprocessing step, invariant to translation, scale and rotation of the shapes. However, they are not invariant to there- parameterization of the shapes.

Finally, landmark-based methods assume that the shapes to be analyzed are points in a high-dimensional Euclidean space and are distributed following a Gaussian model. This limits the type of statistical analysis that can be performed to Gaussian distribution only.

1.2. Leaf Shape

Leaf shape is one of the most important features in describing the plant. Humans can easily identify the different types of leaves and classify them into different species based on information found on the leaves. Based on the shape, the leaves are grouped into 16 classes (Harlow dan Harrar, 1969). There are classes to distinguish the width and length ratio and their unique shape. ~~In~~ Fig. 1 shows the 16 classes of shape i.e. there are: (1) Accicular, (2) Scalelike, (3) Linear, (4) Oblong, (5) Lanceolate, (6) Oblanceolate (7) Ovate, (8) Obovate, (9) Elliptical, (10) Oval, (11) Orbicular, (12) Reniform, (13) Cordate, (14) Deltoid, (15) Rhomboid, (16) Spatulate. Each class has a characteristic shape of leaves typical.

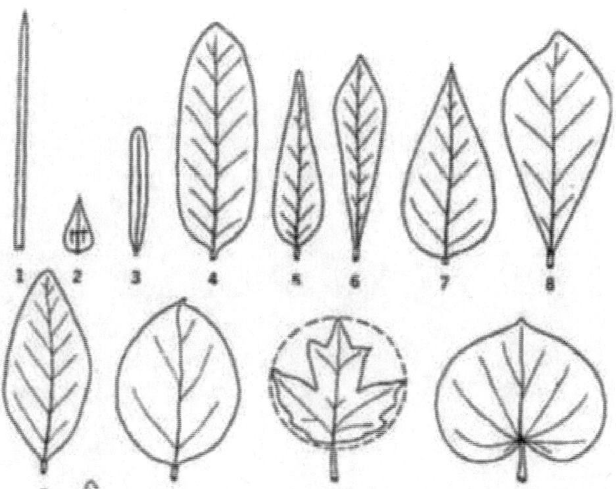

Fig. 1.Leaf Shapes Of Harlow and Harrar.

IOP Conf. Series: Earth and Environmental Science **31** (2016) 012018 doi:10.1088/1755-1315/31/1/012018

1.3. Laminar shape

The simplest way to describe the overall shape of the lamina is to locate the axis or, in some cases, the zone of greatest width that lies perpendicular to the axis of greatest length or long axis (Wing et al. 1999):

a. **Elliptic**
The widest part of the leaf is on an axis in the middle of the long axis of the leaf.

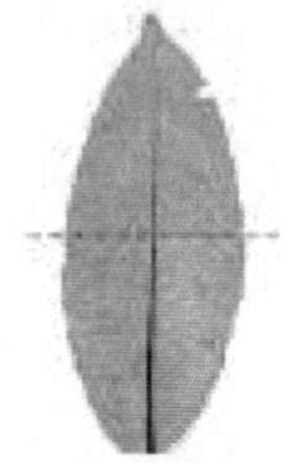

Fig. 2. Elliptic

b. **Obovate**
Widest part of the leaf is on an axis in the apical 2/5 of the leaf

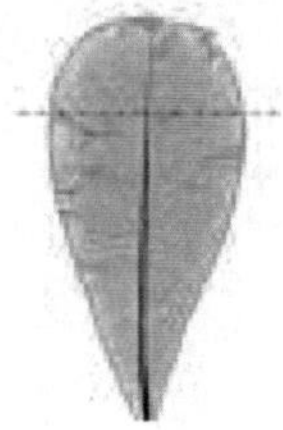

Fig. 3. Obovate

c. **Ovate**
Widest part of the leaf is on an axis in the basal 2/5 of the leaf

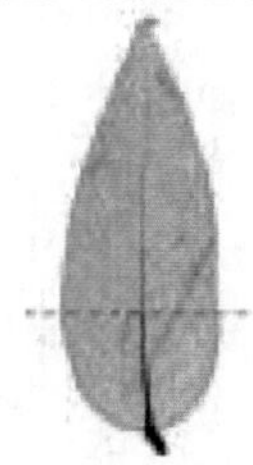

Fig. 4. Ovate

d. **Oblong**
Widest part of the leaf is a zone in the middle 1/3 of the long axis where the opposite margins are roughly parallel.

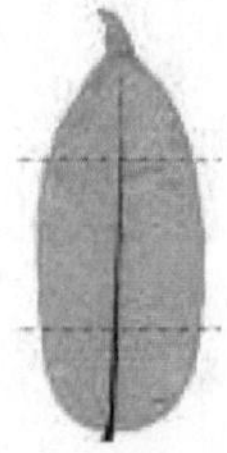

Fig. 5. Oblong

2. Methods

2.1. Polygonal Approximation

Polygonal Approximation is a simplification of the method of representation form the curve (Prasad *et al.* 2012). For example, the set $\{P_1, P_2, P_3, \dots, P_n\}$ as shown in Figure 6, is a series of dots representing a sequential digital curve, then given treatment so that the set of points is reduced to $\{Q_1, Q_2, Q_3, \dots, Q_m\}$ with m < n

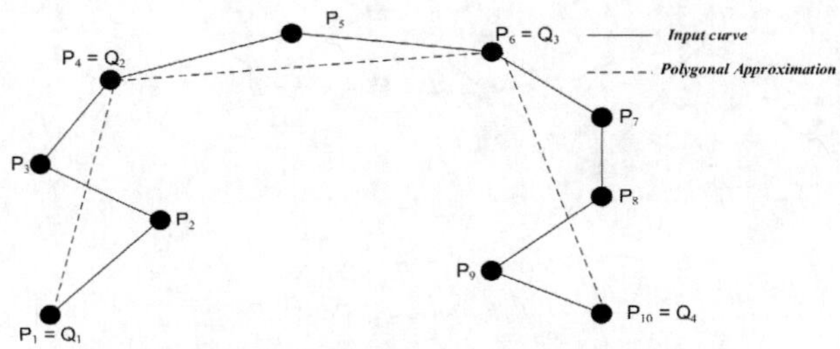

Figure 6. *Polygonal Approximation*

There are two main ways to reduce the polygonal approximation point (Grigore O et al. 2003):

- Min – ε : is the approach to determine the number of points so that predictions can represent essentially the shape of the curve as shown in Figure 7.

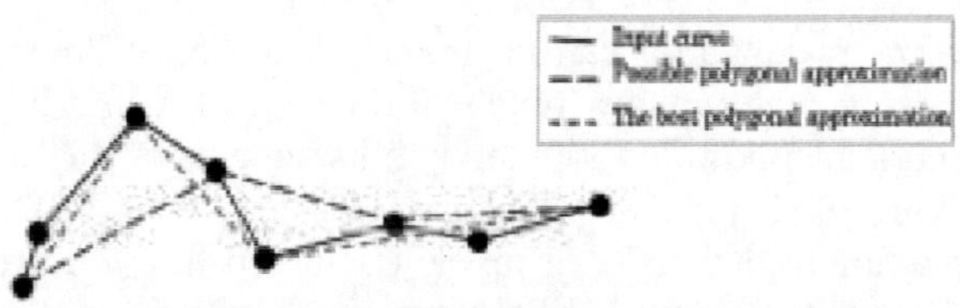

Figure 7. Min - ε *Polygonal Approximation* (Grigore O et al. 2003)

- Min - #. is an approach that determines the margin of error which is a deviation limits in representing the final results as shown in Figure 8.

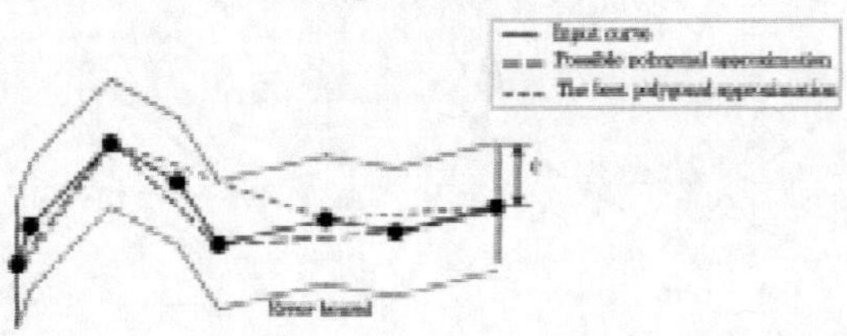

Figure 8. Min - # *Polygonal Approximation* (Grigore O et al. 2003)

Prasad et al. (2012) optimize the polygonal approximation of digital curves that had previously been done by Lowe (1987), Ramer (1972), Douglas & Peucker (1973) or L-RDP, digital curve e = {P1 P2 ... PN}, with Pi is i the number of points the edge of the pixel in the digital curve e. Line through a pair of pixels Pa (xa, ya) and Pb (xb, yb) is given by Equation 1.

$$x(y_a - y_b) + y(x_a - x_b) + y_b x_a - y_a x_b = 0 \tag{1}$$

and deviation d_i of the pixels Pi (xi, yi) ε e of the line passing through the pair {P1, PN} is given by Equation 2.

$$d_i=|x_i(y_1-y_N)+y_i(x_N-x_1)+y_Nx_1-y_1x_N| \quad (2)$$

From the level difference d_i can be seen how far the deviations that occur, as shown in Figure 9.

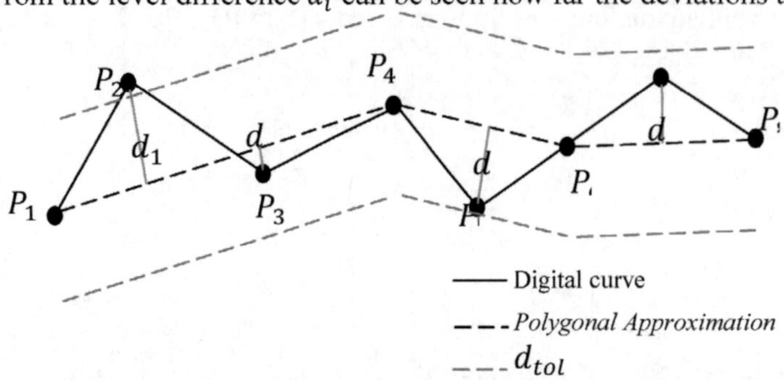

Figure 9. The degree of difference d_i

Based on the above picture, d_i is the degree of difference of P_2 against the pair of points P_1 and P_4, and d_1 have the biggest difference level. By this means that the greater the degree of similarity between the d_i the digital curve and the results are becoming increasingly different polygonal approximation.

Thus, pixels with a maximum deviation can be found, denoted by Pmax, then given the pair {P1, Pmax} and {Pmax, PN}, so we get two new pixels of the curve E using Equation 1 and 2 above. This process is repeated until certain conditions are met by all segments of the line.

The conditions used by L-RDP [Douglas & Peucker (1973), Lowe (1987), Ramer (1972)] is that for each line segment, the maximum deviation of pixels contained in the corresponding edge segment is less than a certain tolerance value as Equation 3 below.

$$\max(d_i) < d_{tol} \quad (3)$$

With d_{tol} is the threshold or margin selected. As shown in Figure 10.

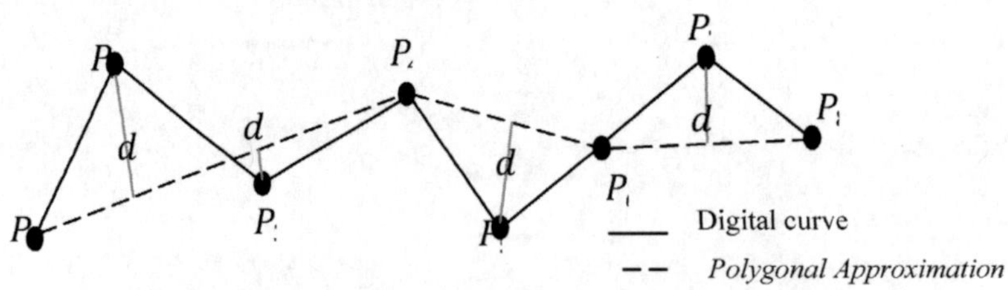

Figure 10. Illustration d_{tol} on Polygonal Approximation

IOP Conf. Series: Earth and Environmental Science **31** (2016) 012018 doi:10.1088/1755-1315/31/1/012018

3. Results and discussions

In this paper used image dataset from Machine Learning Repository UCI [21]. The Leaves were collected in the Royal Botanic Gardens, Kew, UK. All there are 100 species of leaf and every species there are 16 image data of a binary image. In this paper have been classified some image data according to Benson leaf shape. There are five leaf shape that can be analyzed is Eliptic, obovate, ovate, oblong and special

3.1. Static Threshold Polygonal Approximation in Leaf Shape
We know that edge shape of leaf consists of successive points. That meant the points can choose the best points which could indicate landmarks of leaf shape (Fig. 11). Static value d_{tol} 0.5, 1.0, 1.5, and 2.0 Polygonal Approximation is used in this research

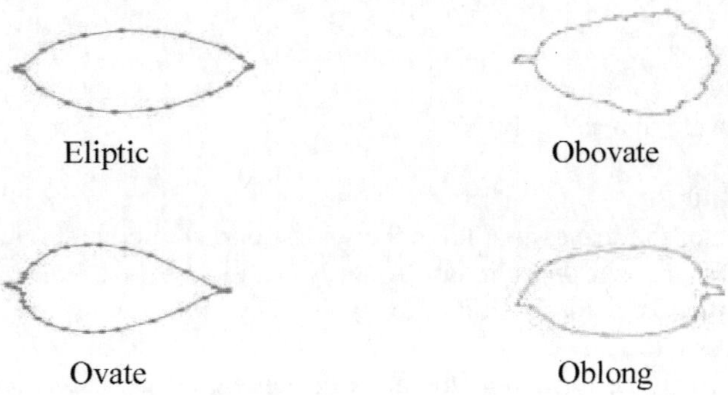

Figure 11. Sample Leaf shape with static value d_{tol} 2.0

3.2. Eliptic Shape Analysis
Eliptic is the shape of the leaves that have the widest part at the center of the leaf. Elliptical leaves have the broadest width in the middle and then taper off at the ends As we see in the Fig. 12 a and b, the elliptic shape is determine in point 4, 5, 6 which have more point in the middle of the leaf for threshold 2.0.

In point 1, 2, 3 respectively be formed as the base of the leaf shape and 7, 8 be formed as the apex of the leaf shape which have less broader than the middle.

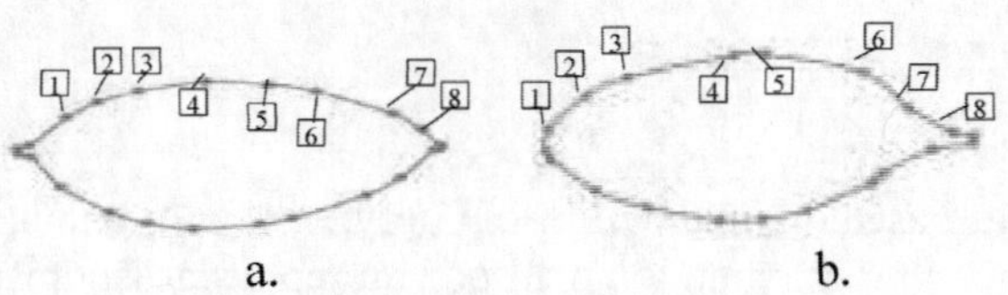

Figure 12. Eliptic Shape Analysis

3.3. Obovate Shape Analysis
Obovate is the shape of the leaves that have the widest part of the leaf is on an axis in the apical 2/5 of the leaf. As we see in the Fig. 13 a and b is two species from obovate shape, the obovate shape is determine in point 7, 8, 9, 10, 11 which have more point in the apex of the leaf for threshold 1.0.

In point 1, 2, 3 respectively be formed as the base of the leaf shape and 4, 5, 6 be formed as the middle of the leaf shape which have less broader than the apex

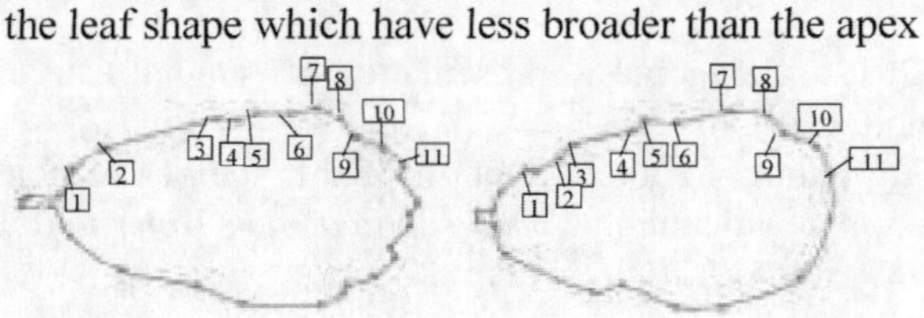

a. b.

Figure 13. Obovate Shape Analysis

3.4. Ovate Shape Analysis

Obovate is the shape of the leaves that have widest part of the leaf is on an axis in the basal 2/5 of the leaf. As we see in the Fig. 14 a and b is two species from ovate shape. the ovate shape is determine in point 1, 2, 3, 4 which have more point in the basal of the leaf for threshold 2.0.

In point 5, 6 and 7 respectively be formed as the middle of the leaf shape and 8, 9 be formed as the apex of the leaf shape which have less broader than the middle.

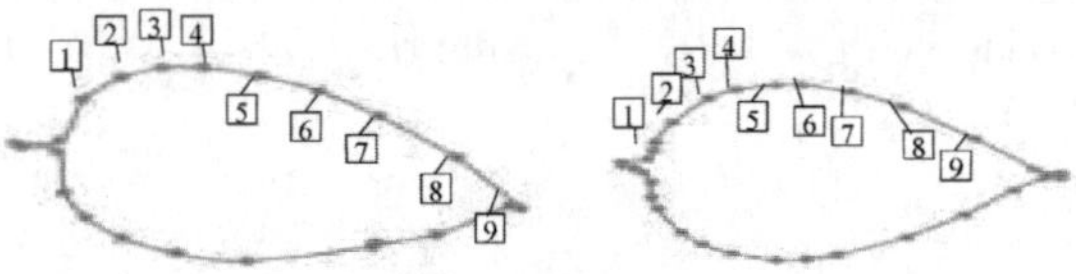

Figure 14. Ovate Shape Analysis

3.5. Oblong Shape Analysis

Oblong is the shape of the leaves that have the widest part of the leaf is a zone in the middle 1/3 of the long axis where the opposite margins are roughly parallel. As we see in the Fig. 15 a and b, the oblong shape is determine in point 3, 4 and 5, 6 which have more point in the middle of the leaf and roughly parallel for threshold 1.5.

In point 1, 2 respectively be formed as the apex of the leaf shape and 7, 8 be formed as the base of the leaf shape which have less broader than the middle 1/3 of the long axis.

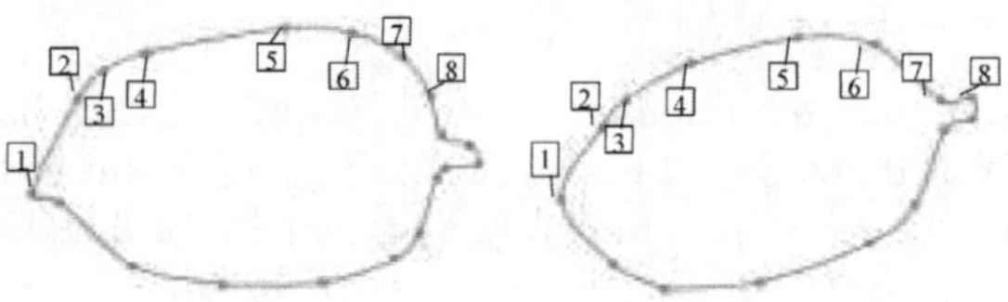

Figure 15. Oblong Shape Analysis

4. Conclusions

This study presents shape representation of the leaf based on landmark using polygonal approximation method. Based on the experimental results, polygonal approximation method can represent landmarks of four leaf shapes such as eliptic, obovate, ovate, and oblong. To get the best landmark, necessary to determine the value of static threshold manually, it is not effective so that required dynamic threshold.

5. References

[1] Dilip K. Prasad, Chai Quek, Maylor K.H. Leung, and Siu-Yeung Cho. 2011. "A parameter independent line fitting method," 1st Asian Conference on Pattern Recognition (ACPR 2011), Beijing, China, 28-30 Nov,

[2] Dilip K. Prasad and Maylor K. H. Leung. 2012. "Polygonal Representation of Digital Curves," Digital Image Processing, Stefan G. Stanciu (Ed.), InTech.

[3] Douglas, D. H., & Peucker, T. K. 1973. Algorithms for the reduction of the number of points required to represent a digitized line or its caricature. *Cartographica: The International Journal for Geographic Information and Geovisualization, 10*(2), 112-122.

[4] Grigore O *et al.* 2003. On the Implementation of Polygonal Approximation Algorithms. Department of Information and Computing Sciences, Utrecht University. technical report UU-CS-2003-005.
[5] Gwo., *et al.* 2013, Plant Identification Through Images: Using Feature Extraction of Key Points on Leaf Contours, Botanical Society of America
[6] Harlow WM, Harrar ES. 1969. *Textbook of Dendrologi*. Mc.Graw-Hill: United States of America
[7] Joao Camargo Neto, George E. Meyer, David D Jones, Ashok K. Samal. 2006. Plant species identification using Elliptic Fourier leaf shape analysis.
[8] Laga H, Kurtek S, Srivastava A, Miklavcic S J. 2014. Landmark-Free Statistical Analysis of The Shape of Plant Leaves. Journal of Theoretical Biology. Volume 363, 21 December 2014, Pages 41–52.
[9] Lowe, D. G. 1987. Three-dimensional object recognition from single two-dimensional images. *Artificial Intelligence, 31*(3), 355-395.
[10] Lukasiewicz , T. U. Straccia . 2009 . Description logic programs under probabilistic uncertainty and fuzzy vagueness. *International Journal of Approximate Reasoning* 50 : 837 – 853 .
[11] Ramer, U. 1972. An iterative procedure for the polygonal approximation of plane curves. *Computer Graphics and Image Processing, 1*(3), 244-256.
[12] Takemura *et al*. 2002. Shape Analysis and Classification using Landmarks: Polygonal Wavelet Transform
[13] Wing *et al.* 1999 Manual of Leaf Architecture - morphological description and categorization of dicotyledonous and net-veined monocotyledonous angiosperms by Leaf Architecture Working Group. 65p.

Dynamics of Agent-Based Growing Network

Budhi Suryo Jatmiko[1], Hendradi Hardhienata[2], Husin Alatas[2]

[1]Theoretical Physics Division, Department of Physics, Jalan Meranti Kampus IPB Darmaga, Bogor 16680, Indonesia
[2]Research Cluster for Dynamic and Modelling of Complex System, Faculty of Mathematics and Natural Sciences, Jalan Meranti Kampus IPB Darmaga, Bogor 16680, Indonesia

E-mail: budhiargomiko@gmail.com,

Abstract. In this work we describe the dynamics of growing network using an agent-based model. We restrict our analysis for the case of a closed spatial boundary and apply certain rules that must be followed by each agent. We show the emergence of an outbreak in the number of agents and connectivity degree before reaching a saturated state. Using the concept of Gibbs-Network entropy we show that the saturated connectivity entropy always converge to the same constant independent of the initial agent number and boundary dimension.

1. Introduction

The dynamics of growing network has been an attracting topic for investigation due to its importance for explaining many phenomena from natural to social complex systems. It is a well known fact that any object/entity interaction and relation phenomena can be described as graph or network. In reality, some network tends to change time, either in number of vertices or number of vertices edge respectively. Their changes tend to continually grow, such as the World Wide Web and social network. This kind of network called growing network. Some research has been conducted on growing network, such as its topology and properties, as growing network more likely evolve to complex network [1]. Many complex networks have a degree distribution $P(k)$ that follows a power law, whereas some other network exhibit exponential distribution [2]. Moreover, in order to connect two indirect connected nodes, typically a few nodes needs to be passed (small-world) and their shortest path length is small [3].

The dynamics of growing network has been an attracting topic for investigation due to its importance for explaining many phenomena from natural to social complex systems [4]. In this paper, we created growing network simulation based on agent interaction which obeys certain rules. Later, we investigated our simulated network by Gibbs-Network-Entropy which adopted from the concept of Gibbs's Entropy in statistical physics [5], as follows:

$$S = -\sum_{k} p_k \log p_k$$

which p_k denotes the probability of an agent to be in a degree of connectivity k.

Our simulation rules govern the interaction between agents. Initially some agents are placed randomly on an element in the matrix. Later, each agents evolve simultaneously. If there are two agents that are not connected to each other, both agents will be connected each other and new agents will be spawned between them and connected to both agents respectively. Also there is restriction which limits each agent only interacts with one agent each iteration.

2. Algorithm

In our model, we used three main matrices, which are Identity (abbreviated as IDEN), Display (DISP), and Edge (EDG). IDEN is $n \times n$ sized square matrix which any elements in the matrix contains integer i, which $i > 0$ defined that the element is occupied by an agent i and it should be unique, whereas $i = 0$ defined that the element is unoccupied. DISP is representation of degree of each agent, which its element position references degree of agent with same position. Moreover, EDG is $n^2 \times n^2$ sized adjacency matrix which defined network of agent interaction.

Simulation runs with integer t as iteration counter (as simulation time). Initially some agent will be placed randomly in IDEN elements. Within one iteration, to simulate agents' random motion, elements value of IDEN will be swapped with its adjacent elements randomly or stay not swapped. Agent interaction scanning performed after all agents' motion processed, which illustrated in Figure 1. First in Figure 1(a), scanning focused around IDEN(2,2). There are 2 unconnected agents in shaded area. Based on interaction rule, a randomly placed new agent will appear in these shaded area, named Agent 4 (Figure 1(b)). We continue the scan to the next cell (Figure 1(c)). In Figure 1(d) and Figure 1(e), any interacted agent (Agent 4) cannot interact with her un-interacted agent (Agent 3) since within one iteration, Agent 2 and Agent 4 has interacted with other agent (Agent 1). Later, in Figure 1(f) each agent's degree of connectivity will be described as DISP matrix, which each element in DISP with same coordinate as element in IDEN represent its degree of connectivity.

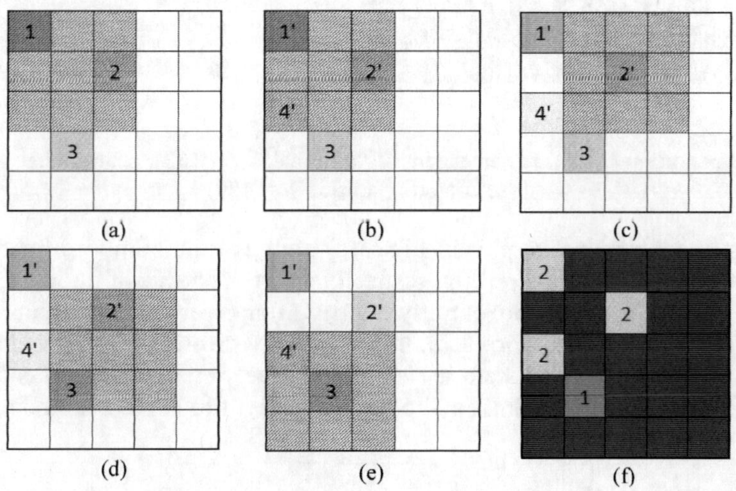

Figure 1. Agent interaction scheme in IDEN matrix as depicted in Figure 1(a) to Figure 1(e), and DISP matrix in Figure 1(d).

3. Gibbs-Network-Entropy of The Simulated Network

We consider an initial condition with 15 agents at random location inside cells as shown in Fig. 2(a). Each agent walks randomly until two of them, according to above mentioned generate a newborn agent. It is observed that at certain critical condition occur emergent outbreak explosion by rapidly increasing amount of agent as depicted in Fig. 2(b) and Fig. 3(c). The growth is becoming saturated at longer time as demonstrated in Fig. 2(d).

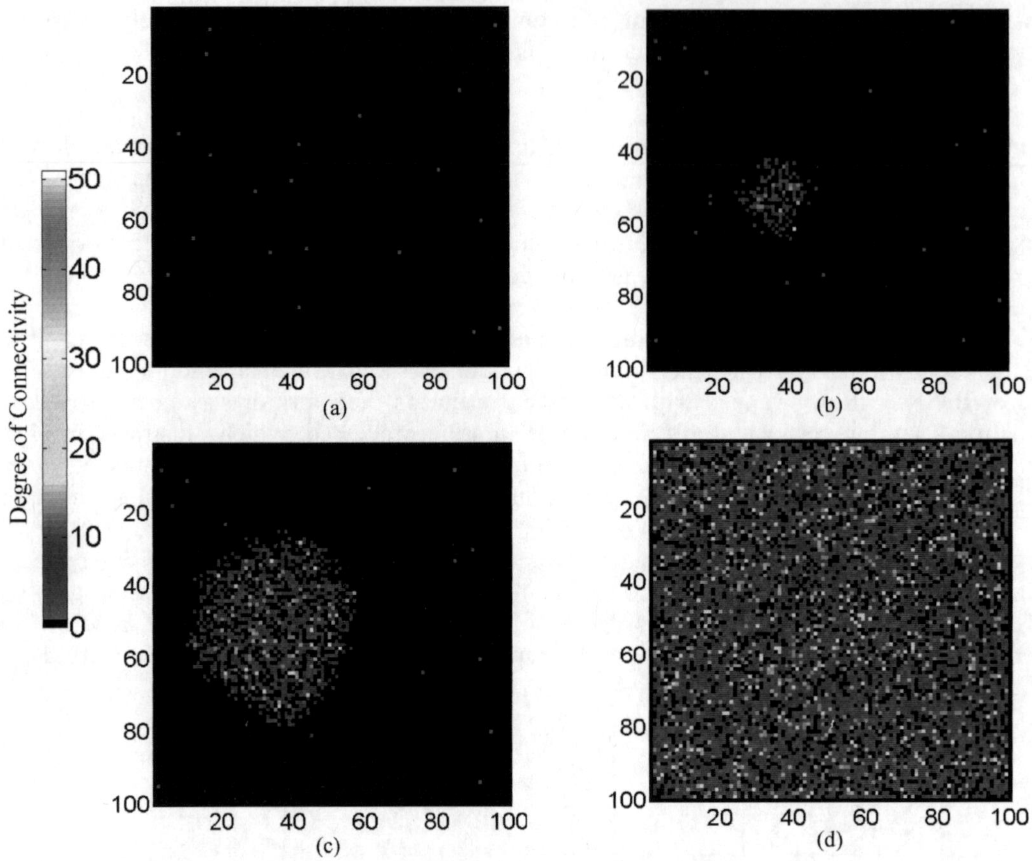

Figure 2. (a) Initial condition (t = 0), (b) emergent of "outbreak" (t = 168), (c) outbreak of agents (t = 198) (d) saturated condition (t = 420)

To explain this phenomenon, we plot in Fig. 3(a) and 3(c) the number of growing agents (N) and the associated Gibbs-Network-Entropy, respectively. It is shown in the Fig. 3(a), as expected that the amount of agents grows exponentially shortly after emergence of "outbreak" and become saturated at longer time. The corresponding Gibbs-Network-Entropy exhibits fluctuating characteristic whilst at the saturated state it become relatively steady. This phenomenon indicates that the network tends to be self-organized although visually it looks like a random pattern.

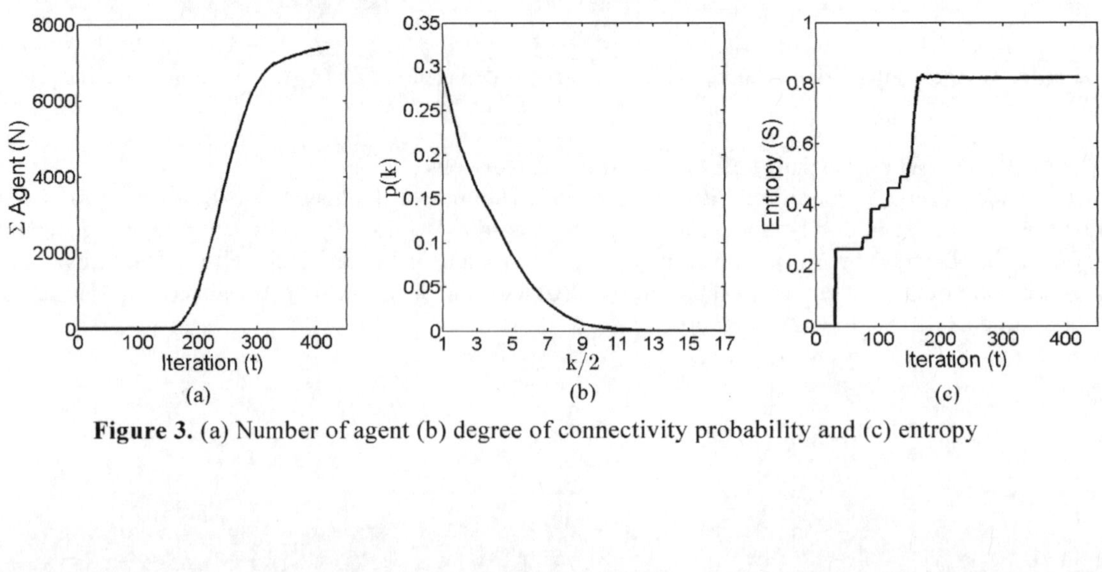

Figure 3. (a) Number of agent (b) degree of connectivity probability and (c) entropy

IOP Conf. Series: Earth and Environmental Science **31** (2016) 012019 doi:10.1088/1755-1315/31/1/012019

An interesting feature that occurs from the simulation can be seen in Fig. 3c which is the emergence of a stable total entropy at around 0.8. This number always occur from the same sum of individual entropies related to each degree of connectivity regardless of the choice of initial agent position and amount of initial agent as long as the computational area is large and given enough time to reach equilibrium. Although it is not fully clear, we believe at present that this feature might arise from the interaction rule and topology of the mesh. Further work is required (e.g. choosing different mesh such as triangular or hexagonal) to investigate this emergence.

References

[1] Albert, Réka, and Albert-László Barabási. "Topology of evolving networks: local events and universality." Physical review letters 85.24 : 5234 (2000)

[2] Newman, Mark EJ. "The structure and function of complex networks." SIAM review 45.2 167-256. (2003)

[3] Klemm, Konstantin, and Victor M. Eguiluz. "Growing scale-free networks with small-world behavior." Physical Review E 65.5: 057102 (2002)

[4] M. C. González and A. L. Barabási, Nat. Phys. **3** 224-225 (2007)

[5] K. Stowe, An Introduction to Thermodynamics and Statistical Mechanics, Cambridge Univ. Press, Cambridge (2007)

IOP Conf. Series: Earth and Environmental Science **31** (2016) 012020 doi:10.1088/1755-1315/31/1/012020

The role of Kleinman Symmetry in the Simplified Bond Hyperpolarizability Model

Adalberto Alejo-Molina[1*] **and Hendradi Hardhienata**[2]

[1] CONACYT Research Fellow - Center for Research in Engineering and Applied Science (CIICAp), Institute for Research in Pure and Applied Science (IICBA), UAEM Cuernavaca, Mor. 62209, Mexico
[2] Theoretical Physics Division, Department of Physics, Bogor Agricultural University, Jl. Meranti, Gedung Wing S, Kampus IPB Darmaga, Bogor 16680, Jawa Barat, Indonesia

*E-mail: adalberto.alejo@uaem.mx

Abstract. We have shown previously in [J. Opt. Soc. Am. **31**, 526-533 (2014)] that Kleinman symmetry plays an important role when comparing the third rank tensor associated with Second Harmonic Generation (SHG) obtained from the Simplified Bond Hyperpolarizability Model (SBHM) and from group theory (GT) for diamond and zincblende lattice. In this work we show that the third rank tensor derived from the SBHM has inherently the assumption of Kleinman symmetry..

1. Introduction

In 2002 Powell *et. al.* [1] developed the phenomenological model first proposed by Sipe *et. al.* [2] to calculate the second harmonic susceptibilities of silicon surface layers. Their new approach, called the Simplified Bond Hyperpolarizability Model (SBHM) is based on the core assumption that the SHG susceptibility tensor can be obtained by direct product of the bond vector rather than using crystal group theory (GT). It has been shown by us in Ref. [3] that there is a close connection between the third rank tensor obtained from SBHM and GT especially if Kleinman symmetry is assumed to hold and that this model can also be expanded for the case of third harmonic generation and electric field induced second harmonic generation [4]. Very recently, SBHM was introduced in calculating the bulk dipolar SHG contribution in zincblende lattice yielding very good agreement with experiment [5] and also successfully reproduced Rotational Anisotropy Spectroscopy (RAS) SHG intensity data using bulk quadrupular and interface dipolar from a flat Si(111) facet [6]. In this work we show how this symmetry is automatically included in SBHM.

In undergraduate textbooks [7,8] it is assumed that the material has a susceptibility derived via an oscilator model and takes the form of a scalar because the focus is in obtaining its estimate strength of and its dependence to the frequency and damping. Thus the often oversimplified assumption is taken that the polarization vector and the driving field is the same which is however just a specific case. In reality, the tensorial nature of the susceptibility cannot be neglected because the polarization and the field can in general be pointing in a different direction depending on the symmetry of the crystal. The tensor indices used throughout this work is explained here for the case of one monochromatic incomming field $E(\omega)$. The tensor for a general field input is not discussed here.

In general when we consider nonlinear effects, $P(\omega)$ can be modeled in terms of power series of the field $\mathbf{E}(\omega)$:

$$\mathbf{P} = \chi^{(1)}\mathbf{E}(\omega) + \chi^{(2)}\mathbf{E}(\omega)\mathbf{E}(\omega) + \chi^{(3)}\mathbf{E}(\omega)\mathbf{E}(\omega)\mathbf{E}(\omega) + \cdots \quad (2)$$

where $\chi^{(2)}$ and $\chi^{(3)}$ are respectively the second- and third-order susceptibilities. Eq. (2) can be rewritten in terms of the components:

$$\mathbf{P} = \chi_{ij}{}^{(1)}\mathbf{E}(\omega) + \chi_{ijk}{}^{(2)}\mathbf{E}(\omega)\mathbf{E}(\omega) + \chi_{ijkl}{}^{(3)}\mathbf{E}(\omega)\mathbf{E}(\omega)\mathbf{E}(\omega) + \cdots \quad (3)$$

Eq. (3) contains the second-order nonlinear susceptibility and is a third rank tensor denoted by three subscripts "*i*", "*j*" and "*k*".

2. Group Theory, Neumann principle, and Kleinman Symmetry

Group theory can be regarded as a general theory that can be applied to obtain the nonlinear susceptibility of a crystal. Three necessary properties must be fulfilled by a set of matrices to form a group under a particular operation, in this case the usual product of matrices. Neumann's principle relates group theory and physics [9,10]:

> *"The symmetry elements of any physical property of a crystal must include the symmetry elements of the point group of the crystal."*

In other words, after applying a symmetry operation belonging to a certain crystal, a particular property of this tensor should remain. These symmetry operations must preserve the crystal properties, which includes the optical ones.

Symmetry operations such as rotations and reflections can be represented mathematically by a matrix and for a particular crystal only rotation for specific angles are allowed for some specific existing mirror planes. Depending on the crystal symmetry various symmetry operation (rotation, reflection, etc.) belonging to the point group of the crystal to the susceptibility tensor can be applied. The tensorial elements of the material susceptibility which is indeed a physical property must fulfill this principle. After these transformations, Neumannn's principle implies that the two tensors must remain the same.

In addition, in nonlinear optics there is often a symmetry condition that is fulfilled for the case when anomalous dispersion can be ignored, then the susceptibility remains unchanged when the frequencies of the two input signals and the resulting beam are permuted. This has as a consequence that the indices in the susceptibility tensor must be invariant to all permutations [2]. Mathematically this can be stated as

$$d_{ijk} = d_{jki} = d_{kij} = d_{ikj} = d_{jik} = d_{kji}. \quad (4)$$

This symmetry is known as Kleinman symmetry. As an example, for Second Harmonic Generation (SHG) only one input frequency is used, therefore Kleinman symmetry requires:

$$\begin{aligned} &d_{jkl}^{(2)}(2\omega = \omega + \omega) = d_{jlk}^{(2)}(2\omega = \omega + \omega) = d_{kjl}^{(2)}(\omega = 2\omega - \omega) = \\ &d_{klj}^{(2)}(\omega = -\omega + 2\omega) = d_{ljk}^{(2)}(\omega = 2\omega - \omega) = d_{lkj}^{(2)}(\omega = -\omega + 2\omega) \end{aligned} \quad (5)$$

This kind of symmetry generally holds if the frequency of the laser is far below the resonance frequency ω_0. Fortunately, many optical processes in material characterization applies an input frequency bandwith that is far below the material resonant frequency. A more detailled discussion about the group theoretical properties of certain diamond lattice surfaces can be seen in Refs. [2,11].

3. Simplified Bond Hyperpolarizability Model (SBHM)

The main assumption of SBHM is to assume that the nonlinear polarization source produces an anharmonic dipole oscillation along the atomic bonds. This model is phenomenological in the sense that the hyperpolarizabilities must be deriverd from experiment fit. In addition it assigns different values for the polarizabilities and hyperpolarizabilities according to their direction and is therefore affected by crystal symmetry. In its original formulation [1], SBHM assumes that the electron moves only along the bond which must be reconsidered if transversal motion is possible such as in third harmonic generation. Another additional view of SBHM is that the bond orientations in the surface have no molecular reconstruction, however such effects can also be incorporated in the extended model.

In SBHM the polarization up to the second order is calculated by [1,12]:

$$\mathbf{P} = \frac{1}{V}\sum_j \left[\alpha_{1j}\left(\mathbf{R}^{(z)}(\phi)\cdot\hat{\mathbf{b}}_j\right)\otimes\left(\mathbf{R}^{(z)}(\phi)\cdot\hat{\mathbf{b}}_j\right)\right]\cdot\mathbf{E} +$$
$$\frac{1}{V}\sum_j \left[\alpha_{2j}\left(\mathbf{R}^{(z)}(\phi)\cdot\hat{\mathbf{b}}_j\right)\otimes\left(\mathbf{R}^{(z)}(\phi)\cdot\hat{\mathbf{b}}_j\right)\otimes\left(\mathbf{R}^{(z)}(\phi)\cdot\hat{\mathbf{b}}_j\right)\right]\cdot\mathbf{E}\otimes\mathbf{E} \quad (6)$$

therefore, the susceptibility third rank tensor, is calculated by the formula:

$$\chi^{(2)} = \frac{1}{V}\sum_j \left[\alpha_{2j}\left(\mathbf{R}^{(z)}(\phi)\cdot\hat{\mathbf{b}}_j\right)\otimes\left(\mathbf{R}^{(z)}(\phi)\cdot\hat{\mathbf{b}}_j\right)\otimes\left(\mathbf{R}^{(z)}(\phi)\cdot\hat{\mathbf{b}}_j\right)\right] \quad (7)$$

where V is the volume, α_{1j} are the linear polarizabilities, α_{2j} are the hyperpolarizabilities and $\hat{\mathbf{b}}_j$ are the unit vectors in the direction of the atomic bonds. Note that the summation is going over all the bonds in the conventional cell. In the following we will show that Eq. (7) always obeys Kleinman symmetry if a monochromatic incoming field is applied. However there are also cases especially near the resonant frequency where this symmetry breaks down [13]

4. Results and Discussions

To prove that Eq (7) obey Kleinman symmetry for certain cases. Indeed it can be shown e.g. using brute force by applying the most general bond vector components and construct the tensor from SBHM that the final tensor always symmetric. This is due to the following fact: the tensor is generated by direct product of the same bonds $\hat{b}_j$ and the hyperpolarizabilities in SBHM are just constants. Here we give a formal proof that this is indeed the case. Let us think that we have a number of vectors

$$\mathbf{a} = (a_1, a_2, a_3)$$
$$\mathbf{b} = (b_1, b_2, b_3)$$
$$\mathbf{c} = (c_1, c_2, c_3)$$
$$\mathbf{d} = (d_1, d_2, d_3) \quad (8)$$

and we want to calculate a tensor using direct product in the following way. We can generate a tensor of rank n doing the direct product

$$\prod_{i=1}^{n} \mathbf{x}_i \quad (9)$$

where $\mathbf{x}_i$ represents any fixed vector, multiplied by itself n times, under the understanding that it is a direct product. In particular for the fourth rank tensor:

$$\mathbf{b}\otimes\mathbf{b}\otimes\mathbf{b}\otimes\mathbf{b} \quad (10)$$

or in terms of the components

$$B_{ijkl} = b_i b_j b_k b_l. \quad (11)$$

If we permutated any two factors in Eq. (11), we get

$$B_{ijlk} = b_i b_j b_l b_k \quad (12)$$

but the subindices are free suffix and can be renamed

$$B_{mnop} = b_m b_n b_o b_p. \tag{13}$$

Moreover, there is nothing that forbid to repeat the same procedure, thus

$$B_{ijkl} = b_i b_j b_k b_l, \tag{14}$$

and as can be seen, the last expression Eq. (14) is identical to Eq. (11), so the permutation of whatever two factors generated exactly the same tensor:

$$B_{ijkl} = B_{ijlk}. \tag{15}$$

Therefore, the result is that any permutation of the subindices generates the same tensor and this is the equivalent to Kleinman symmetry. The result is that any permutation of the subindices generates the same tensor and this is the equivalent to Kleinman symmetry. In fact, for the summation in SBHM over the different bonds, also generates a tensor with Kleinman symmetry. We are going to show it only for two bonds but can be proved in general for any number of bonds and for a tensor of rank *n*.

In a pedestrian way, we have two vectors and after generating the tensor doing the direct product four times, we can sum each other

$$\Omega_{ijkl} = A_{ijkl} + B_{ijkl} \tag{16}$$

the free suffix now are the same (dummy suffix) in both vectors because we are summing component to component in each tensor, or

$$\Omega_{ijkl} = a_i a_j a_k a_l + b_i b_j b_k b_l \tag{17}$$

and of course, it does not matter how we named the result of this sum, as tensors A_{ijkl} and B_{ijkl} have Kleinman symmetry, the sum also will have this symmetry, because the permutation of any two subindices will result on

$$\Omega_{ikjl} = a_i a_k a_j a_l + b_i b_k b_j b_l \tag{18}$$

but for the reason explained above each term in the sum results in itself under the permutation of its indices. Then

$$\Omega_{ikjl} = a_i a_j a_k a_l + b_i b_j b_k b_l = \Omega_{ijkl}. \tag{19}$$

The same argumentation can be used to add more tensors generated by the direct product of any number of the remaining bonds:

$$\Sigma_{ikjl} = a_i a_j a_k a_l + b_i b_j b_k b_l + c_i c_j c_k c_l + d_i d_j d_k d_l + \ldots \tag{20}$$

References

[1] G. D. Powell, J.-F. Wang, and D. E. Aspnes, Simplified bond-hyperpolarizability model of second harmonic generation, Phys. Rev. **B** 65, 205320 (2002).

[2] V. Mizrahi and J. E. Sipe, "Phenomenological treatment of surface second-hamonic generation," J. Opt. Soc. Am. **B**, 5, 233402 (1988).

[3] A.Alejo-Molina, H. Hardhienata, and K. Hingerl, " Simplified bond-hyperpolarizability model of second harmonic generation, group theory, and neumann's principle," J. Opt. Soc. Am. **B**, Vol. 31, 526-533 (2014)

[4] Alejo-Molina, K. Hingerl, and H. Hardhienata,"Model of third harmonic generation and electric field induced optical second harmonic using the simplified bond-hyperpolarizability model," J. Opt. Soc. Am. **B**., Vol. 32, 562 - 570 (2015).

[5] H. Hardhienata, A. Alejo-Molina, C. Reitböck, A. Prylepa, D. Stifter, and K. Hingerl, "Bulk dipolar contribution to second-harmonic generation in zincblende," J. Opt. Soc. Am. **B**., Vol. 33, 195 - 201 (2016).

[6] C. Reitböck, D. Stifter, A. Alejo-Molina, K. Hingerl, and H. Hardhienata,"Bulk quadrupole and interface dipole contribution for second harmonic generation in Si(111)," J. Opt., **18**, 035501 (2016).

[7] G. New. Introduction to Nonlinear Optics. (Cambridge University Press, 2011).
[8] R. W. Boyd, Nonlinear Optics, 2nd ed. (Academic Press, San Diego, 2003).
[9] J. F. Nye, Physical Properties of Crystals, Their Representations by Tensors and Matrices (Clarendon, Oxford, 1957).
[10] R. C. Powell, Symmetry, Group Theory, and the Physical Properties of Crystals (Lecture Notes in Physics) (Springer, New York, 2010).
[11] E.S. Jatirian-Foltides, J.J. Escobedo-Alatorre, P.A. Marquez-Aguilar, H. Hardhienata, K. Hingerl, and A. Alejo-Molina, "About the calculation of the second-order $\chi(2)$ susceptibility tensorial elements for crystals using group theory " Rev. Mex. Fis. **E** 62,5-13 (2016).
[12] H. Hardhienata, A. Prylepa, D. Stifter, and K. Hingerl,"Simplified bond-hyperpolarizability model of second- harmonic-generation in Si(111): theory and experiment," J. Phys. 423, 012046 (2013).
[13] V. Mizrahi and D. P. Shelton, "Deviations from Kleinman symmetry measured for several simple atoms and molecules," Phys. Rev. **A** 31, 3145–3154 (1985).

IOP Conf. Series: Earth and Environmental Science **31** (2016) 012021 doi:10.1088/1755-1315/31/1/012021

Estimating Origin-Destination Matrix of Bogor City Using Gravity Model

I Ekowicaksono[1], F Bukhari[1], and A Aman[1]

[1]Department of Mathematics, Bogor Agriculture University,
Jl. Meranti, Kampus IPB Dramaga Bogor 16680, Indonesia

Abstract. Origin-Destination (O-D) Matrix describes people movement in a certain area. An O-D matrix is necessary for planning a good public transportation system. However, the exact values of O-D matrix are difficult to measure. There are several ways to estimate O-D matrix such as gravity model, gravity opportunity model, etc. In this study, gravity model was used to estimate the O-D matrix in Bogor city. The following assumptions were used to estimate the O-D matrix: (i) forces between two different zones are related to some existing parameters such as population, social-economic condition, etc. (ii) the people movements are influenced by accessibility from origin to destination, and the accessibility affected by distance, time, and/or cost.

1. Introduction

Nowadays, the world is growing fast, including Bogor city. In 2013, Bogor city has more than 1 million citizen who lived, but they still encounter the same transportation problem. The main transportation problem is about making a good transportation system in Bogor. The transportation system including the comfort infrastructure for people, the effective and efficient route for mass rapid transit, the number of bus or another transportation infrastructure and its frequency in every route. The most important thing is the transportation system should have to fulfill the demand of transportation.

In order to fulfill the demand of transportation, the demand of transportation should be known[11]. The demand of transportation can be presented by origin-destination matrix[7]. The origin-destination matrix is a matrix which is each cell represent the number of trip from origin (row) to the destination (column). There are several ways to estimate the origin-destination matrix. In conventional way, we can count the people trip from their origin to the destination by conducting a survey. But, conducting a survey would spend a lot of resources[8].

The scientist have conducted research about estimating the origin-destination matrix[3,9]. They used synthetic model to estimate the origin-destination matrix. This study used synthetic model, the gravity model for transportation, to estimate the origin-destination matrix.

2. Gravity Model

This study used gravity model to estimate the O-D matrix. The idea of gravity method originally came from Newton's gravitational law. Casey was the first who suggested such an approach to synthesise shopping trips and catchment areas between towns in region[9]. In its simplest formulation to the model has the following functional form:

$$T_{ij} = \frac{\alpha P_i P_j}{d_{ij}^2} \tag{1}$$

where P_i and P_j were the following populations of the origin and destination, d_{ij}^2 was the distance between origin and destination, and α was a proportional factor. These gravity model was not ready to use for transportation purpose. So, there were some improvements to be applied including the use of total trip ends instead of population and several parameter to calibrate. After the improvement have been applied to equation (1), the gravity model for transportation can be used for transportation purpose[6].

The gravity model for transportation uses some existing parameter such as population and social-economic condition. The formulation of gravity model for transportation is shown below.

$$T_{ij} = O_i D_j A_i B_j f(c_{ij}) \tag{2}$$

Here, T_{ij} is a people movement from zone i to zone j, O_i the number of people from zone i, D_j the number of fieldwork in zone j, $f(c_{ij})$ the deterrence function, A_i and B_j the balancing factor which are defined as follow:

$$A_i = \left(\sum_j \left(B_j D_j f\left(c_{ij}\right)\right)\right)^{-1} \tag{3}$$

$$Bj = \left(\sum_i \left(A_i O_i f\left(c_{ij}\right)\right)\right)^{-1} \tag{4}$$

In equation (1), we use negative exponential function for the deterrence function[2],

$$f(c_{ij}) = e^{-\beta c_{ij}} \tag{5}$$

The deterrence function in equation (1) is affected by the value of β. β means the average cost trip in the research area. If the value of β is increasing, the average cost trip is decreasing. There are some calibration techniques to get the value of β, such as Hyman method. This study used Hyman method to calibrate the value of β.

3. Hyman Method

The Hyman method is one of the techniques to calibrate the parameter in deterrence function[4]. Williams was compared several method to calibrate the parameter, and the Hyman method was robust and effective than other methods[10]. We would describe the Hyman method by the following paragraph.

Define $T(\beta)$ the O-D matrix of the current estimate of β. The matrix also defined by the equation:

$$\sum_{i,j} T_{ij}(\beta) = T(\beta) \tag{6}$$

The method based on the requirement of $c(\beta) = c^*$, where:

$$c(\beta) = (T(\beta))^{-1} \sum_{ij} T_{ij}(\beta) c_{ij} \tag{7}$$

$$c^* = \left(\sum_{ij} N_{ij}\right)^{-1} \sum_{ij} N_{ij} c_{ij} \tag{8}$$

c^* is the mean cost from observation and N_{ij} is the number of observed trip form each origin destination pair. The Hyman method can be described as follows:

1. Start the first iteration by making $m = 0$ and an initial estimate of $\beta_0 = (c^*)^{-1}$.
2. Using the value of β_0 calculate a trip matrix using the standard gravity model. Obtain the mean modelled trip cost c_0 and estimate a better value for β as follows:
$$\beta_m = \beta_0 c_0 (c^*)^{-1}$$
3. Make $m = m+1$. Using the latest value for β (i.e. β_{m-1}) calculate a trip matrix using standard gravity model and obtain the new mean modelled trip cost c_{m-1} and com[are it with c^*. if they are sufficiently close, stop and accept β_{m-1} as the best estimate for this parameter; otherwise go to step 4.
4. Obtain the better estimate of β as:
$$\beta_{m+1} = \frac{(c^* - c_{m-1})\beta_m - (c^* - c_m)\beta_{m-1}}{c_m - c_{m-1}}$$
5. Repeat steps 3 and 4 as necessary, i.e. until the last mean modelled cost c_{m-1} is sufficiently close to the observed value of c^*.

4. Numerical Result

This study was applied to Bogor city. In 2013, Bogor city had more than 1 million population including 403,628 workers An assumption was applied to this study which was the worker trip used to estimate the origin-destination matrix of Bogor city. Bogor city assumed to be a closed system, so that the worker trip had the origin and destination within Bogor city.

From the observation, we got the trip matrix of the worker's trip as shown in table 1 below. The distance between subdistrict was shown in table 2. There were assumed no distance between the same subdistrict. We use distance for deterrence function.

Table 1. Observed worker's trip matrix

	Central	West	South	East	North	Tn. Sareal
Central	24680	6250	1068	8380	511	2759
West	50950	12903	2204	17299	1054	5697
South	42105	10663	1822	14296	871	4708
East	22624	5730	979	7681	468	2530
North	41674	10554	1803	14150	862	4660
Tn. Sareal	46193	11699	1998	15684	956	5165

Table 2. Distances between subdistrict (km)

	Central	West	South	East	North	Tn. Sareal
Central	0	1.047	2.852	3.442	1.833	2.486
West	1.047	0	2.988	4.005	2.914	3.138
South	2.852	2.988	0	1.645	3.818	5.194
East	3.442	4.005	1.645	0	3.54	5.274
North	1.833	2.914	3.818	3.54	0	1.913
Tn. Sareal	2.486	3.138	5.194	5.274	1.913	0

IOP Conf. Series: Earth and Environmental Science **31** (2016) 012021 doi:10.1088/1755-1315/31/1/012021

Numerical simulation using Fortran[5] yield the value of $\beta = 1.1679\times10^{-7}$ and the calculated origin-destination matrix of Bogor city shown in table 3. In figure 1 below, we served the calculated origin-destination matrix in a map of Bogor city to facilitate the government for determine the best policy to solve any problem that occurred in transportation in Bogor city.

Table 3. Calculated origin-destination matrix in Bogor city

	Central	West	South	East	North	Tn. Sareal
Central	24680	6250	1068	8379	511	2759
West	50950	12904	2204	17299	1054	5697
South	42105	10663	1822	14296	871	4708
East	22623	5729	979	7682	468	2530
North	41674	10554	1803	14150	862	4660
Tn. Sareal	46193	11698	1998	15683	956	5165

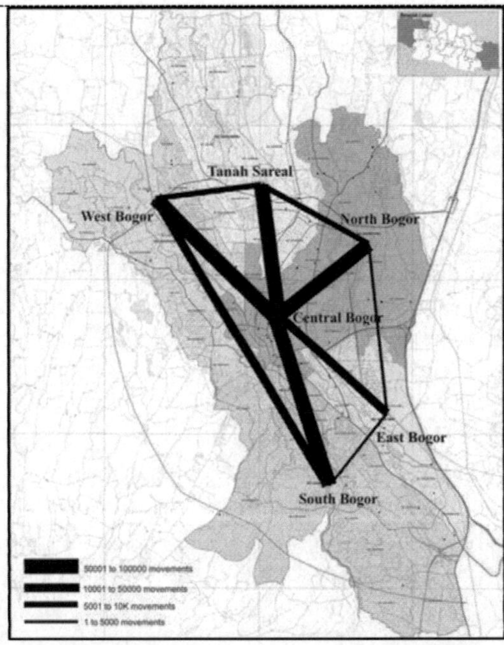

Figure 1. Worker's trip pattern in Bogor city.

5. Conclusion

This study estimated the worker's trip in the morning. The worker's trip would presented as origin-destination matrix. The origin-destination matrix cam be estimated by gravity model for transportation. In gravity model for transportation, there was a parameter which was need to calibrate. The parameter was in the deterrence function. There were some calibration techniques that have been used in different study, one of them was Hyman method. This study used Hyman method as the calibration method.

In the numerical simulation, the parameter which was calibrated by Hyman method, has been evaluated. This study got the value of $\beta = 1.1679\times10^{-7}$, and the origin-destination matrix of Bogor city as shown in table 3 above. For the further study, the value of β can be used for estimating the origin-destination matrix of Bogor city for coming years.

IOP Conf. Series: Earth and Environmental Science **31** (2016) 012021 doi:10.1088/1755-1315/31/1/012021

References

[1] Casey H J 1955 *Traffic Quarterly* **9** 23-35
[2] Evans A W 1971 *Transpn. Res.* **5** 15-38.
[3] Fathoni M 2005 *Estimasi matriks asal dan tujuan perjalanan penumpang angkutan umum trans jawa – sumatera melalui lintasan merak – bakauheni* (Yogyakarta: Gadjah Mada University).
[4] Hyman G M 1969 *The calibration of trip distribution models* (London: Environment and Planning A) pp 105-112.
[5] Lemmon D R and Schafer J L 2005 *Developing Statistical Software in Fortran 95* (New York: Springer).
[6] Ortúzar J de D and Willumsen L G 2011 *Modelling Transport 4th Edition* (New Delhi: John Wiley & Sons).
[7] Peterson A 2007 *The Origin-Destination Matrix Estimation Problem – Analysis and Computation* (Norrköping: University of Linköping).
[8] Roziqin C 2012 *J. Tek. Sipil*. **12** 28-34.
[9] Tamin O Z 2008 *Perencanaan dan Pemodelan Transportasi* (Bandung: Penerbit ITB).
[10] Williams I 1976 *Transpn. Res.* **10** 91-104.
[11] Willumsen L G 1078 *Estimation of an O-D Matrix From Traffic Counts – A Review* (Leeds: Institute of Transport Studies).

IOP Conf. Series: Earth and Environmental Science **31** (2016) 012022 doi:10.1088/1755-1315/31/1/012022

Optimal intervention strategies for cholera outbreak by education and chlorination

Toni Bakhtiar

Division of Operation Research, Department of Mathematics, Bogor Agricultural University, Kampus IPB Dramaga, Bogor 16680, INDONESIA

E-mail: tbakhtiar@ipb.ac.id

Abstract. This paper discusses the control of infectious diseases in the framework of optimal control approach. A case study on cholera control was studied by considering two control strategies, namely education and chlorination. We distinct the former control into one regarding person-to-person behaviour and another one concerning person-to-environment conduct. Model are divided into two interacted populations: human population which follows an SIR model and pathogen population. Pontryagin maximum principle was applied in deriving a set of differential equations which consists of dynamical and adjoin systems as optimality conditions. Then, the fourth order Runge-Kutta method was exploited to numerically solve the equation system. An illustrative example was provided to assess the effectiveness of the control strategies toward a set of control scenarios.

1. Introduction

Cholera is still a major public health problem in underdeveloped and developing countries causing morbidity and mortality. According to WHO [12], a more than 190,000 cholera cases with more than 2,200 deaths were reported by 42 countries in 2014, resulting in an overall case fatality rate (CFR) of 1.17 per cent, a 47 per cent increase compared to 2013. Most of cholera outbreaks were caused by serogroup O1 of *Vibrio cholerae*. While serogroup O139 which first identified in Bangladesh in 1992, is confined to South-East Asia. Non-O1 and non-O139 *V. cholerae* can cause mild diarrhea but do not generate epidemics. Cholera outbreak in Haiti, appeared ten months after the devastating earthquake of 2010, claimed over 4,500 lives and sickened almost 300,000 people, was the first case in nearly a century; revealing that cholera is an ancient-fatal disease that continues to cause epidemic and pandemic infection. Another fact that cases were reported from all regions with 55 per cent of all reported cases originated from Africa, 30 per cent from Asia and 15 per cent from Hispaniola [12] indicates that cholera needs global awareness.

As an acute bacterial infection of the intestine caused by bacterium *V. cholerae*, cholera transmitted through sanitary channel, i.e., contaminated food and drinking-water, as well as by person-to-person contact through the fecal-oral route. Thus, the dynamics of cholera involve multiple interactions between the human host, the pathogen and the environment [6]. A number of mathematical models have been developed for gaining better knowledge of the complex dynamics of cholera, where a brief introduction to the basics of ordinary differential equation models and its modification was given by [2]. Several deterministic cholera epidemic models were introduced by [9] to analyze the global stability of the models. A general compartmental model is proposed by [8] by incorporating both direct and indirect transmission, non-linear incidence, multiple infectious states of the pathogen, and multiple infection stages of infectious individuals. A study by [4] attempted to estimate the reproductive numbers for the 2008-2009 cholera outbreaks in Zimbabwe. In the framework of intervention model, a mathematical model that captures some essential dynamics of cholera transmission to study the impact of public health educational campaigns, vaccination and treatment as control strategies in curtailing the disease was formulated by [5]. In [11], a cholera epidemiological model with control measures including the effects of vaccination, therapeutic treatment, and water sanitation was investigated.

In this paper we discuss the control of cholera transmission in the framework of optimal control approach, where we take into account two intervention strategies, namely education and chlorination. The former is intended to educate the community about the importance of improving sanitation and hygienic practices in order to reduce direct and indirect contacts. The latter is referred to the use of chlorine to degrade *V. cholerae* concentration in the water source.

2. Mathematical model

2.1. Model

To assess the effectiveness of the control strategies we adopt a model introduced by [1], where a different approach of analysis was considered. Firstly, instead of analyzing the effects of education and chlorination as constant parameters of the model, we represent both efforts as dynamic variables. In the framework of optimal control, we consider two parameters as control variables should be determined optimally with respect to a certain performance index. Secondly, beyond the control by chlorination, we introduce two educational control variables to respectively quantify the effect of education on both direct human-to-human and indirect environment-to-human transmission pathways.

We consider a model with two interacted populations: human population which follows an SIR model and pathogen population (Figure 1). We denote by $S(t)$, $I(t)$, and $R(t)$, the number of susceptible, infected and recovered individuals at time t, respectively, and by $P(t)$ the concentration of pathogen *V. cholerae* in the aquatic environment, i.e., the contaminated water source, at time t. For a relatively short period of time and for low rate of mortality we assume that the total population N is constant, i.e., $N = S + I + R$. Susceptible individuals enter the model with constant recruitment rate Λ and are exposed to a force of infection λ, where $\lambda = \beta_1 P/(K_1 + P) + \beta_2 I(K_2 + I)$. Here, K_1 and K_2 respectively denote the pathogen concentration level at which half of all contacts with contaminated water produce infection with per capita contact rate β_1 and the contact between susceptible and infected individuals with rate β_2. Parameters μ_h, μ_d and μ_p denote the natural mortality rate of human, cholera-induced mortality rate of infected human and mortality rate of pathogen, respectively. The average contribution of each infected individual to the pathogen population is given by θ. Infected individuals may recover at rate δ and become susceptible at a rate α. We assume a logistic growth for pathogen with per capita growth rate b and carrying capacity K.

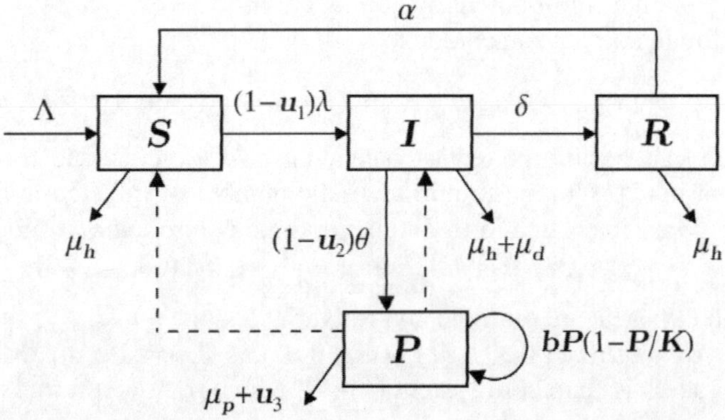

Figure 1. Compartmental model of cholera dynamics [1].

The compartmental model of cholera dynamics described by Figure 1 is then represented by the following ordinary differential equation system.

$$\frac{dS}{dt} = \Lambda + \alpha R - (1 - u_1)\left(\frac{\beta_1 P}{K_1 + P} + \frac{\beta_2 I}{K_2 + I}\right)S - \mu_h S, \tag{1}$$

$$\frac{dI}{dt} = (1 - u_1)\left(\frac{\beta_1 P}{K_1 + P} + \frac{\beta_2 I}{K_2 + I}\right)S - (\mu_h + \delta + \mu_d)I, \tag{2}$$

$$\frac{dR}{dt} = \delta I - (\alpha + \mu_h)R, \tag{3}$$

$$\frac{dP}{dt} = bP\left(1 - \frac{P}{K}\right) + \theta(1 - u_2)I - (\mu_p + u_3)P, \tag{4}$$

where the following initial values are applied: $S(0) = S_0 > 0$, $I(0) = I_0 > 0$, $R(0) = R_0 > 0$ and $P(0) = P_0 > 0$. To intervene the model during the control period of $[0, T]$ we introduce three control variables $u_1(t)$, $u_2(t)$ and $u_3(t)$. If $u_1 = u_2 = u_3 = 0$, then we refer to a model without control, where its stability analysis has been undertaken by [1].

2.2. Control problem

A multidisciplinary approach is key for reducing cholera outbreaks, controlling cholera in endemic areas and reducing deaths. A series of prevention and control actions include water and sanitation interventions; treatment using oral rehydration salt, intravenous fluid, and antibiotic; surveillance as part of an integrated disease surveillance system that includes feedback at the local level and information-sharing at the global level; social mobilization through health education campaigns; and application of cholera vaccines.

In model (1)-(4), the first control u_1 refers to the direct cholera-related education aimed to behaviorally change the interaction among susceptible and infected individuals. This control strategy seems sensible since a cholera outbreak in a Singapore psychiatric hospital indicated that the direct human-to-human transmission was a driving force [3]. It is also discovered that the cholera outbreak in Papua, Indonesia, in 2008 primarily caused by misconduct during funeral of cholera victim, i.e., family members are hugging and kissing the dead-body [7]. The second control u_2 relates to the indirect education whose objective is to improve the person-to-environment conduct. This action comprises interventions at the household level such as the importance of water filtration, water chemical or solar disinfection, and safe water storage containers as well as advising community with the so-called Five Basic Cholera Prevention Messages (drink and use safe water, wash hands, use latrines or bury feces, cook food well and clean up safely). The third control u_3 represents the disinfection of the source of water by adding chlorine to kill harmful microorganisms such as bacteria, viruses, protozoa and helminths. In our setting, u_1 and u_2 can also be interpreted as the portion of human population which can be influenced by control actions. While, u_3 is the portion of *V. cholerae* which can be removed from the environment by adding chlorine.

We consider the following performance index:

$$J(u_1, u_2, u_3) := \int_0^T (A_0 I(t) + A_1 u_1^2(t) + A_2 u_2^2(t) + A_3 u_3^2(t))\, dt, \tag{5}$$

where A_k $(k = 0,1,2,3)$ can be interpreted as balancing cost weights due to the size and relative importance of each term in (5). We aim to minimize the number of infected individual as well as the control efforts. Hence, we are interested in finding an optimal control pair (u_1^*, u_2^*, u_3^*) such that

$$J(u_1^*, u_2^*, u_3^*) = \min_{(u_1,u_2,u_3)\in\mathcal{U}} J(u_1, u_2, u_3), \tag{6}$$

where $\mathcal{U}$ is the set of all Lebesgue measurable control variables given by

$$\mathcal{U} := \{u_i(t) \mid 0 \le u_i(t) \le \bar{u}_i,\ 0 \le t \le T, i = 1,2,3\}. \tag{7}$$

The minimization process is subject to system (1)-(4), which re now referred to as the state system. Correspondingly, the unknowns S, I, R and P are now called the state variables, in contrast to the control variables u_1, u_2, u_3. In (7), bounds $\bar{u}_i$ $(i = 1,2,3)$ reflect practical limitations on the maximum rates of controls in period $[0, T]$.

3. Optimality conditions

In this section we provide the necessary conditions for the existence of optimal control variables and corresponding state variables. Pontryagin maximum principle [10] was exploited by introducing the adjoin functions and represents an optimal control in terms of the state and adjoin functions. First we assign the problem of minimization (5) into one of minimization the following Hamiltonian:

$$\begin{aligned}\mathcal{H} = A_0 I + A_1 u_1^2 + A_2 u_2^2 + A_3 u_3^2 + p_1 \left[\Lambda + \alpha R - (1 - u_1)\left(\frac{\beta_1 P}{K_1 + P} + \frac{\beta_2 I}{K_2 + I}\right) S - \mu_h S\right] \\ + p_2 \left[(1 - u_1)\left(\frac{\beta_1 P}{K_1 + P} + \frac{\beta_2 I}{K_2 + I}\right) S - (\mu_h + \delta + \mu_d) I\right] \\ + p_3 [I - (\alpha + \mu_h) R] + p_4 \left[bP\left(1 - \frac{P}{K}\right) + \theta(1 - u_2) I - (\mu_p + u_3) P\right],\end{aligned} \tag{8}$$

where p_i $(i = 1,2,3,4)$ are adjoin functions corresponding to S, I, R and P, respectively. According to Pontryagin maximum principle, optimality conditions for the control problem are given by the following set of requirements:

$$\frac{\partial \mathcal{H}}{\partial u_i} = 0, \quad i = 1,2,3, \tag{9}$$

$$\frac{\partial x_i}{\partial t} = \frac{\partial \mathcal{H}}{\partial p_i}, \quad i = 1,2,3, \;\; x_i \in \{S, I, R, P\}, \tag{10}$$

$$\frac{\partial p_i}{\partial t} = -\frac{\partial \mathcal{H}}{\partial x_i}, \quad i = 1,2,3, \;\; x_i \in \{S, I, R, P\}. \tag{11}$$

By considering that we have a bounded optimal control problem, i.e., $0 \le u_i(t) \le \bar{u}_i$, condition (9) provides the optimal control pair (u_1^*, u_2^*, u_3^*) as follows

$$u_1^* = \min\left\{\bar{u}_1, \max\left\{0, \frac{1}{2A_1}(p_2 - p_1)\left(\frac{\beta_1 P}{K_1 + P} + \frac{\beta_2 I}{K_2 + I}\right) S\right\}\right\}, \tag{12}$$

$$u_2^* = \min\left\{\bar{u}_2, \max\left\{0, \frac{\theta}{2A_2} p_4 I\right\}\right\}, \tag{13}$$

$$u_3^* = \min\left\{\bar{u}_3, \max\left\{0, \frac{1}{2A_3} p_4 P\right\}\right\}. \tag{14}$$

Indeed, condition (10) provides the state system (1)-(4) alongside with its initial values. Further, condition (11) gives the following adjoin system:

$$\frac{dp_1}{dt} = (1 - u_1)(p_1 - p_2)\left(\frac{\beta_1 P}{K_1 + P} + \frac{\beta_2 I}{K_2 + I}\right) + \mu_h p_1, \tag{15}$$

$$\frac{dp_2}{dt} = (1 - u_1)(p_1 - p_2)\frac{\beta_2 K_2 S}{(K_2 + I)^2} + \delta p_3 - \theta(1 - u_1) p_4 + (\mu_h + \delta + \mu_d + A_0), \tag{16}$$

$$\frac{dp_3}{dt} = -\alpha p_1 + (\alpha + \mu_h) p_3, \tag{17}$$

$$\frac{dp_4}{dt} = (1 - u_1)(p_1 - p_2)\frac{\beta_1 K_1 S}{(K_1 + P)^2} - b\left(1 - \frac{2}{K} P\right) p_4 + (\mu_p + u_3) p_4. \tag{18}$$

Since in this control problem we assume a free terminal time, then the adjoin system (15)-(18) should satisfy the so-called transversality condition

$$p_1(T) = p_2(T) = p_3(T) = p_4(T) = 0. \tag{19}$$

Therefore, the optimality condition for the existence of optimal solution are represented by optimal controls (12)-(14), the state system (1)-(4) and adjoin system (15)-(18), which all should be solved simultaneously. Since the admissible control set $\mathcal{U}$ is closed and convex, and the integrand of performance index (5) is also convex in I and u_i, then the optimal controls (12)-(14) minimize (5), i.e., Pontryagin maximum principle constitutes both necessary and sufficient conditions for optimality.

4. Numerical solution

4.1. Iterative method

For a model without control, i.e., $u_1 = u_2 = u_3 = 0$ and thus the adjoin system is not exist, we applied a forward-in-time iterative method over state system (1)-(4) under initial conditions $S(0) = S_0$, $I(0) = I_0$, $R(0) = R_0$, dan $P(0) = P_0$. However, for a model with control whose optimality conditions include a set of differential equations with initial conditions and another set with terminal conditions, we implemented the forward-backward sweep method based on the fourth-order Runge-Kutta algorithm as did in [11]:

a. Set an initial guess for the control variables u_i^0 $(i = 1,2,3)$.
b. Solve forward-in-time the initial value problem of state system (1)-(4).
c. Solve backward-in-time the terminal value problem of adjoin system (15)-(18).
d. Calculate the new controls (12)-(14) with the new values of the state and adjoin solutions and then update the controls. The update of the controls can be the average between old and new controls.
e. Iterate the process until the solutions converge with a sufficiently small level of tolerance.

4.2. Control scenarios

A set of control strategies were examined to quantify the effectiveness of the control actions. The first three scenarios relate to the application of a single control action, i.e., we set either u_1, u_2 or u_3 to be active, while the last scenario makes all controls active. For parameter values we followed those given in [1]. In addition, we also use following values: $A_0 = A_1 = A_2 = A_3 = 0.25$, $\bar{u}_1 = \bar{u}_2 = \bar{u}_3 = 1$, and $T = 60$ days.

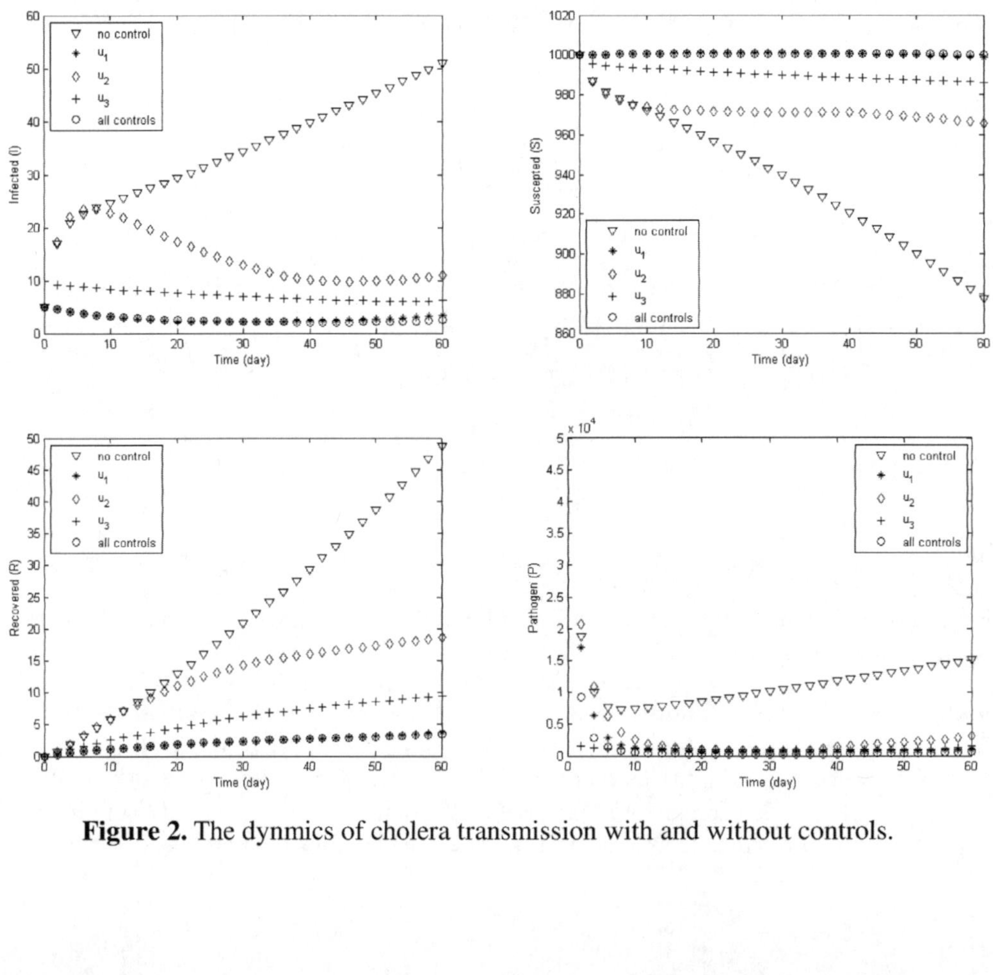

Figure 2. The dynmics of cholera transmission with and without controls.

Figure 2 provides the dynamics of cholera transmission in each compartment. It can be seen that without intervention a cholera outbreak occurs, where the number of infected individuals increase from 5 people in the beginning of period to more than 50 people in two months. An intervention in the form of human-to-human related education (u_1) prevents the occurrence of outbreak by maintaining the number of infected individuals no more than 5 people. Other types of intervention (u_2 and u_3) contributed the similar effect in avoiding the community from the outbreak, even though with lower level of effectivity. Surprisingly, when all control actions were simultaneously applied, the effect in infection reduction is not different with that of only single control u_1 was applied. This, however, indicates that the human-to-human related education is quite effective in directly controlling not only the cholera infection but also indirectly the biomass of pathogen in aquatic environment. In fact, the application of chlorine in reducing the concentration of *V. cholerae* is extremely effective as the number of pathogen reduced up to 96 per cent in few days.

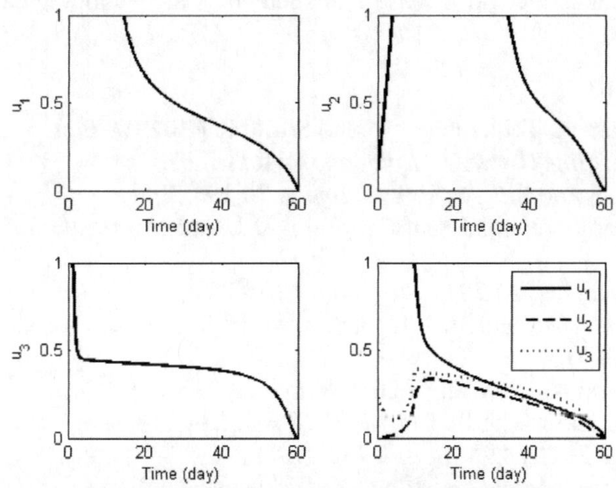

Figure 3. The optimal interventions.

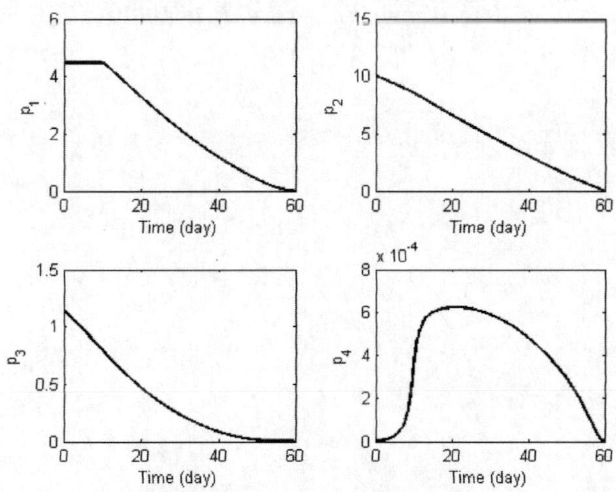

Figure 4. The adjoin functions.

Figure 3 shows that an intervention in the form of human-to-human related education should be implemented at maximum rate during the first 15 days in order to provide the optimal contribution. While that in the form human-to-environment related education should be maximally applied during a longer time, i.e., the 4-34th days. Chlorination needs to be implemented at maximum rate during the first two days. When all interventions were applied concurrently then the first control were active at maximum rate during the first 10 days as there were additional contributions from others actions of control. Figure 4 illustrates the adjoin functions. It is confirmed that transversality conditions (19) are satisfied.

5. Conclusion

We have presented in this present work a cholera transmission model representing an interaction between human and pathogen populations. Three types of intervention were attached to the model, namely human-to-human related education, human-to-environment related education and chlorination. It is revealed that the first control performs the most effective contribution. In comparison with other controls, the effect of the first control dominates the action in reducing the number of infected individuals as well as the control effort during the period of two months. This result, however, exhibits similarities with that drawn by [1], where direct education is a critical factor in cholera control that has a greater and longer-lasting effect on disease management than technological interventions such as chlorination.

References

[1] Al-Arydah M, Mwasa A, Tchuenche J M and Smith R J 2013 *J. Biol. Syst.* **21(4)** 1340007-1-20.
[2] Fung I C 2014 *Emerging Themes in Epidemiol.* **11(1)** 1-11.
[3] Goh K, Teo S, Lam S and Ling M 1990 *J. Infect.* **20** 193-205.
[4] Mukandavire Z, Liao S, Wang J, Gaff H, Smith D L and Morris JG 2011 *PNAS* **108(21)** 8767-8772.
[5] Mwasa A and Tchuenche J M 2011 *BioSystems* **105** 190-200.
[6] Nelson E J, Harris J B, Morris J G, Calderwood S B and Camilli A 2009 *Nature Rev.: Microbiol.* 7 693-702.
[7] Puspandari N, Sariadji K and Wati M 2010 *Widyariset* **13(2)** 69-74.
[8] Shuai Z and van den Driessche P 2011 *Math. Biosci.* **234** 118-126.
[9] Tian J P and Wang J 2011 *Math. Biosci.* **232** 31-41.
[10] Vinter R 2010 *Optimal control* (Boston, NY: Birkhauser).
[11] Wang J and Modnak C 2011 *Canad. App. Math. Quart.* **19(3)** 255-273.
[12] World Health Organization 2015 *Weekly Epidemiol. Rec.* **40(90)** 517-544.

Measuring happiness in large population

Annabelle Wenas[1], Smita Sjahputri[1,2], Bagus Takwin[2], Alfindra Primaldhi[2], and Roby Muhamad[1,2]

[1]ProveticLab, Jl. Kerinci I No. 2, Jakarta, 12120 Indonesia
[2]Faculty of Psychology, University of Indonesia, Depok, 16424 Indonesia

Corresponding Author's E-mail: annabellewenas@provetic.com

Abstract. The ability to know emotional states for large number of people is important, for example, to ensure the effectiveness of public policies. In this study, we propose a measure of happiness that can be used in large scale population that is based on the analysis of Indonesian language lexicons. Here, we incorporate human assessment of Indonesian words, then quantify happiness on large-scale of texts gathered from twitter conversations. We used two psychological constructs to measure happiness: valence and arousal. We found that Indonesian words have tendency towards positive emotions. We also identified several happiness patterns during days of the week, hours of the day, and selected conversation topics.

1. Introduction

Governing complex modern societies requires some basic measurements in the societal level. These measurements will ensure that public policies are effective and meet the ever changing demand. However, currently, the most common aggregate measures of societies are economic measures such as economic growth. Yet, as modern societies grow more complex, there is a need to develop other measurements beyond economic measures especially for psychological measures that can capture subjective well-being [7]. It is reasonable to think that a combination of economic and psychological measures can provide more comprehensive view of a society which, in turn, will be useful for formulating better public policies and their evaluations. In this paper we propose an approach to measure psychological characteristics for large population that based on text data.

Our focus is on the measurement of emotional states and we follow Russell [9] who asserted that emotion, mood, and other emotionally charged events are states that are simply combination between feeling good or bad and energized or worn out. Russell addressed these emotional states as core affect, and mapped its structure into circumplex model [4-6, 8-10]. Horizontal axis of circumplex model is valence, which is a measure of emotion ranging from negative to positive emotions. Whereas its vertical axis is arousal, a measure of emotional intensity. Thus, for example, anger is a negative emotion with high intensity and lethargic is a negative emotion with low intensity. On the other end of the spectrum, excited and calm are positive emotions with high and low intensity respectively. Note that happiness is a positive emotion with moderate intensity.

Core affect is the fundamental of happiness measurement in present study. Moreover we propose a measure for happiness that can be used in large scale population that is based on the analysis of

Indonesian language lexicons. Similar works for the measurement of happiness for English and other languages have been done extensively [1-3], but their measure of happiness based solely on the valence dimension.

Here, in addition to measuring word valence, we also included the measurement of arousal dimension. The reason to include arousal is because positive valence is necessary but not sufficient component of happiness – Simply because there are other states that have positive valence like excitement and calm. Thus, level of arousal is the key to differentiate excitement, happiness, and calm; Three of them indeed have positive valence, yet their arousal level are variable from high, moderate, to low respectively. Hence, we propose a more comprehensive way to measure happiness, using combination of valence and arousal measurement.

2. Measures and Methods

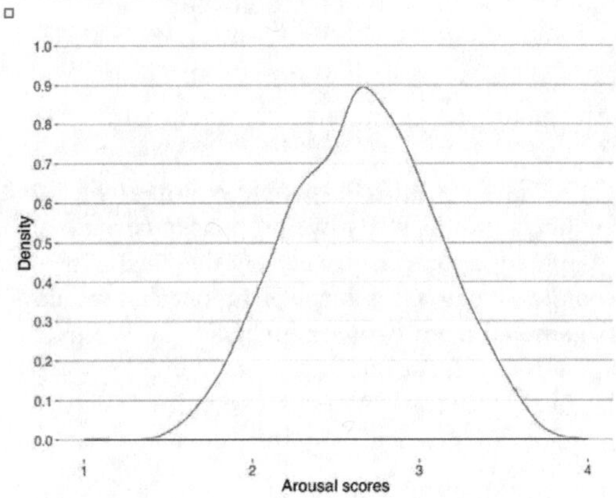

Figure 2. Density of arousal scores

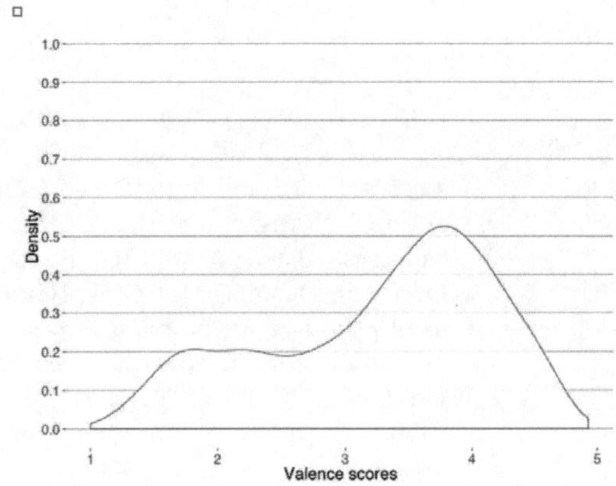

Figure 3. Density of valence scores

In order to explore arousal and valence contents in Indonesian language, we constructed a corpus consisting 3,000 Indonesian words. The source of corpus is the most frequently used words in twitter conversation [1–3] from 2010-2012. Then we measured how those words perceived by individuals in terms of arousal and valence. Therefore 83 students were recruited to rate each word – Rating process was administered using online panel. Arousal rating scale ranging from 1 (calm) to 4 (excited), whereas valence ranging from 1 (negative) to 5 (positive), both with 1 incremental point. Average rating scores then calculated for each word, see figure 2 and 3 for the results.

In terms of arousal, rating average is near symmetrically distributed. Whereas for valence, rating average is negatively skewed. This valence result is consistent with study which reveals that human language has positivity bias [3]. Further, we found that there is no significant correlation between valence and arousal ($R^2 = 0.028$, $F(1, 2981) = 86.89$, $p < .001$). This result is consistent with Russell's core affect theory [7-8], in which valence and arousal dimensions are orthogonal to each other.

IOP Conf. Series: Earth and Environmental Science **31** (2016) 012023 doi:10.1088/1755-1315/31/1/012023

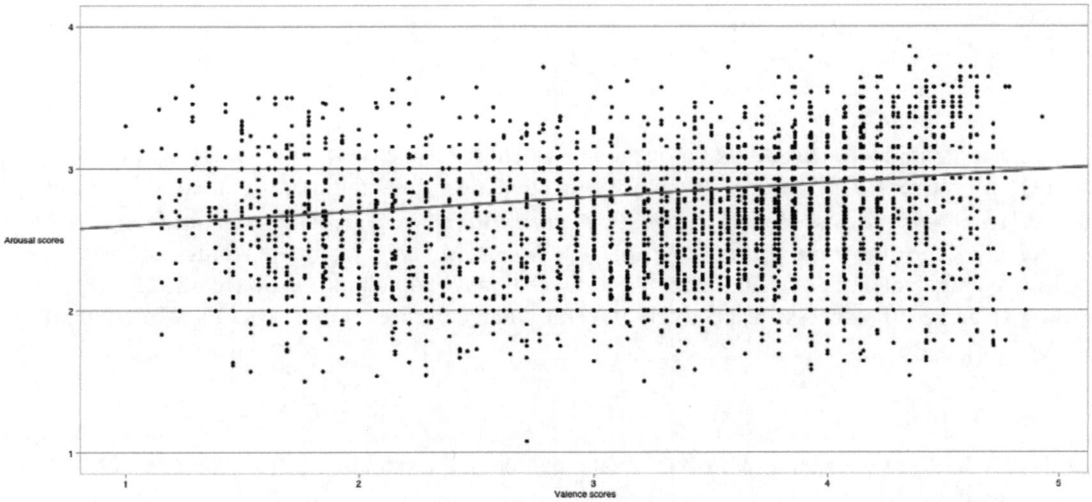

Figure 4. Valence and arousal scores distribution

From figure 5, notice that word with highest arousal is *"spektakuler"* (spectacular), and with lowest arousal is *"sunyi"* (silence). Whereas word with most positive valence is *"optimistis"* (optimistic), and with most negative valence is *"membunuh"* (murder). Thus in a sense, structural consistency with circumplex model is evident in this result.

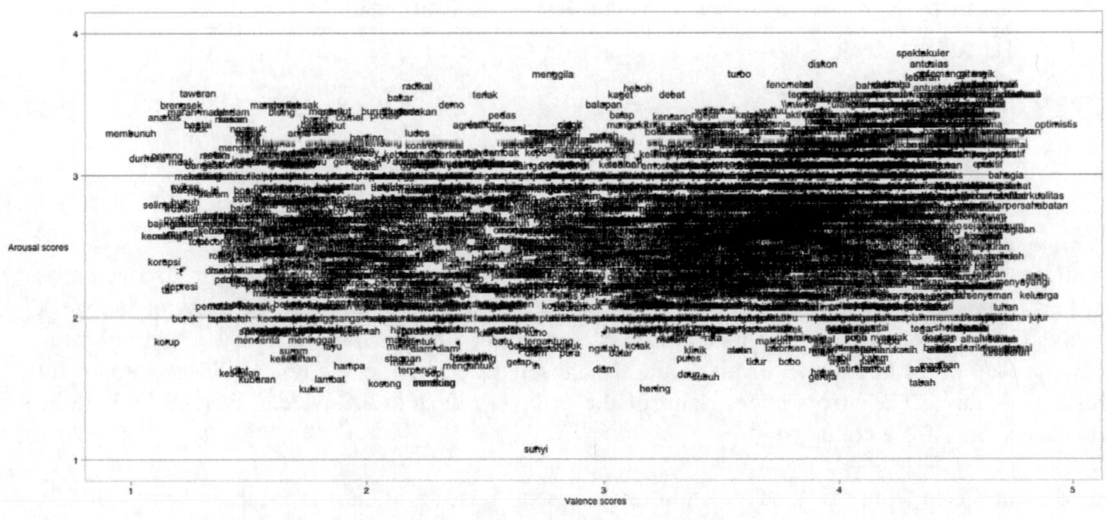

Figure 5. Distribution of words based on valence and arousal scores

Valence text score estimation [1] used to estimate both arousal (a_{text}) and valence (v_{text}) score for a text, which in this case tweets. We identify frequency of word(s) from corpus that mentioned in the text, and compute a weighted average of the arousal and valence of words as

$$a_{text} = \frac{\sum_{i=1}^{n} a_i f_i}{\sum_{i=1}^{n} f_i} \quad (1)$$

$$v_{text} = \frac{\sum_{i=1}^{n} v_i f_i}{\sum_{i=1}^{n} f_i} \quad (2)$$

Note that a_i and v_i are recorded average arousal and valence, respectively, from every word of our corpus (i). Moreover f_i is lexicon i frequency in text. As illustration let's take this tweet "*@MerryRiana: Mungkinkah di balik tembok keputusasaan yang sedang menghadang Anda terletak sebuah jalan besar menuju kesuksesan? #MR*" The seven underlined words recorded in corpus with average valences of 2.50 (under), 2.07 (wall), 2.07 (something), 2.58 (road), 2.93 (big), 2.66 (toward), and 3.30 (success) respectively. Overall valence score for the tweet therefore would be

$$v_{text} = \frac{1}{7}(1\times2.50 + 1\times2.07 + 1\times2.07 + 1\times2.58 + 1\times2.93 + 1\times2.66 + 1\times3.30) \simeq \mathbf{2.6}$$

From calculation above we found that tweet's valence text score falls in 2.6 from the scale of 1 to 5, making it moderately positive.

3. Results

Using this method, we calculated arousal and valence scores for various topics in public debates and also how it vary within temporal and topical dimensions. We randomly collected tweets and analyze it within three contexts, which are: day of the week, hour of the day, and topic (see Table 1).

Table 1. Explanation of data contexts

Context	Number of Variables	Number of Tweets
Day of the week	7	1,513,085
Hour of the day	24	1,564,505
Topic	8	738,967

3.1. Day of The Week

We collected data from Indonesia's local twitter conversations in 2014. In order to overcome limitations of the server, hence randomly chosen 1,200 timestamps points for every day of the week (Monday, Tuesday, Wednesday, Thursday, Friday, Saturday, and Sunday). Suppose all Mondays in 2014 had a total of 74,880 hours, of all randomly chosen 1,200 hours; We repeat this procedure for the rest of the days. Then from every point of these days, selected 200 tweets. In total, 1,513,085 tweets have successfully collected.

As we can see in figure 6, conversation on Thursday has the highest arousal, and immediately followed with lowest arousal on Friday. Whereas for valence, Monday has the highest valence compared to the rest of days. This is contrary to popular belief "Monday blues", which says back to work day is identical to sadness. Although one thing to keep in mind is Eid al-Fitr was on Monday (July 28th, 2014); The day when social media crowded with positive words of greetings, like *"selamat"* or congratulation ($v_{selamat} = 4.4$). This fact might explain why Monday has the very high positive valence compared to the other days' measurement. Monday's moderate arousal and high valence indicate happiness is evident.

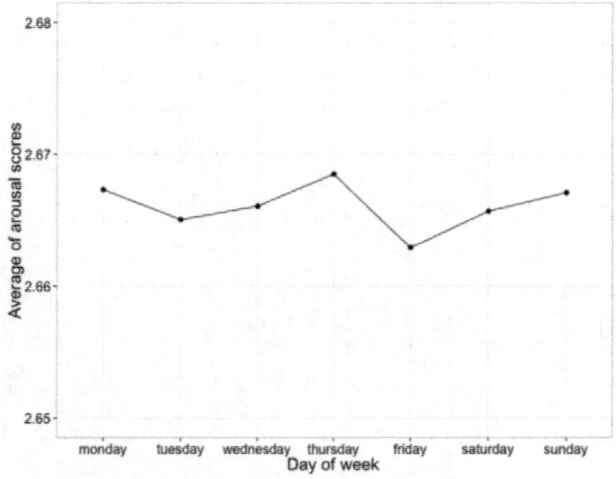

Figure 6. Arousal scores distribution per day of the week

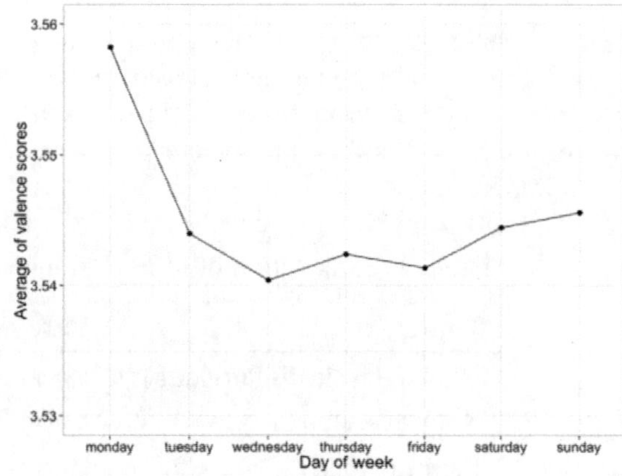

Figure 7. Valence scores distribution per day of the week

3.2. Hour of The Day

For the hour of the day, we use the same method as for the days, yet this time we chose 365 time stamps points for each hour. Therefore total points from all hours are 8,760. Further, from each point of these hours selected 200 tweets. Total tweets that we have collected are 1,564,505.

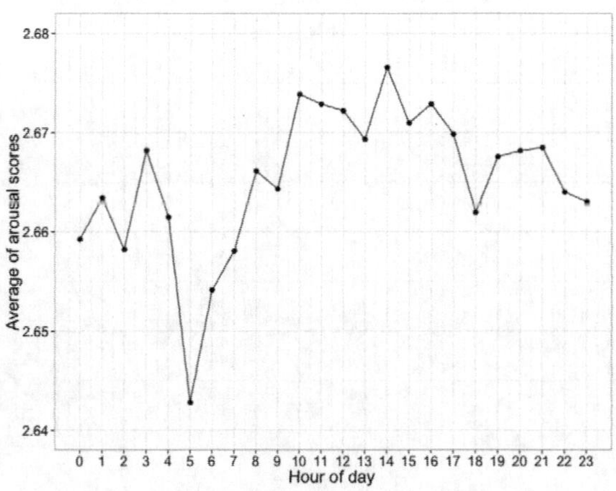

Figure 8. Arousal scores distribution per hour the of day

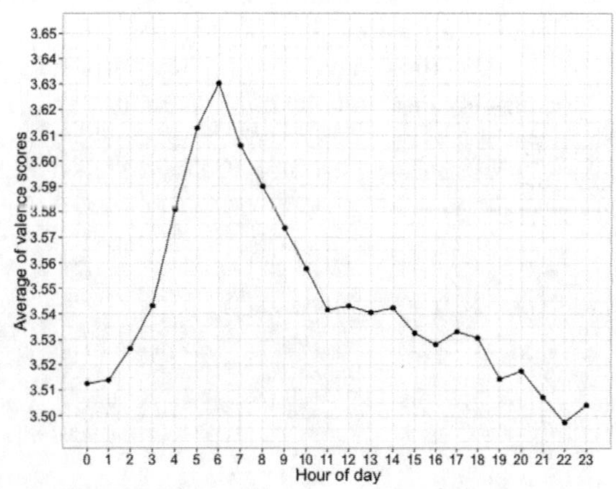

Figure 9. Valence scores distribution per hour of the day

Variations are more pronounced for both arousal and valence in hour of the day compared to previous analysis for day of the week. Lowest arousal hour observed on 5 am, whereas highest peak of arousal hour on 14 pm. Moreover 6 am is the hour with highest valence, the peak is constantly decreased until reaches its lowest on 10 pm. Apparently on early hours of the day, 6 am to 8 am, moderate arousal and positive words are frequently used in social media. This finding indicates that upon those hours happiness is most strongly present. Further, there is a unique reversed pattern of data points of arousal and valence. Arousal significantly decreased from 4 am and elevated back on 6 am, whereas valence significantly increased from 2 am and shifts to decline on 7 am.

3.3. Topic

Next, we selected various twitter conversations and analyze their valence and arousal scores. We chose eight of which to compare arousal and valence of each topic. Selected topics along with further details explained in Table 2 below. From each topic we randomly chose 2,000 timestamps points, and from each of these points selected 50 tweets. Total tweets that have been successfully collected are 738,967.

Table 2. Explanation of topical dimensions

Topic	Category	Number of Tweets
Ahok (Basuki T. Purnama)	Local Leader	93,409
Ridwan Kamil	Local Leader	87,536
Banjir (*Flood*)	City Issue	92,032
Macet (*Traffic jam*)	City Issue	95,811
Golkar	Political Party	87,394
PDIP	Political Party	90,625
McDonald's	Fast Food Brand	96,524
KFC	Fast Food Brand	95,636

The order of topic in figure 10 and figure 11 below is based on the average valence score from the highest to the lowest. Thus, we see that Ridwan Kamil (Mayor of Bandung) has the highest valence, whilst both flood and traffic jam are identically have the lowest valence. Moreover both political parties have the highest arousal, with Golkar has the highest arousal level, and presumably moderate level of valence.

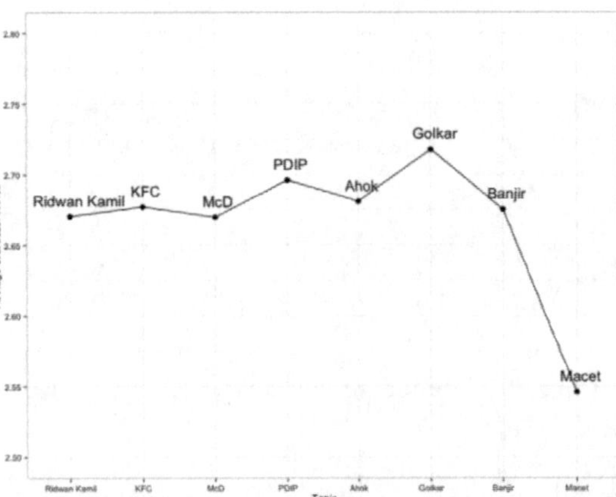

Figure 10. Arousal scores distribution per topic

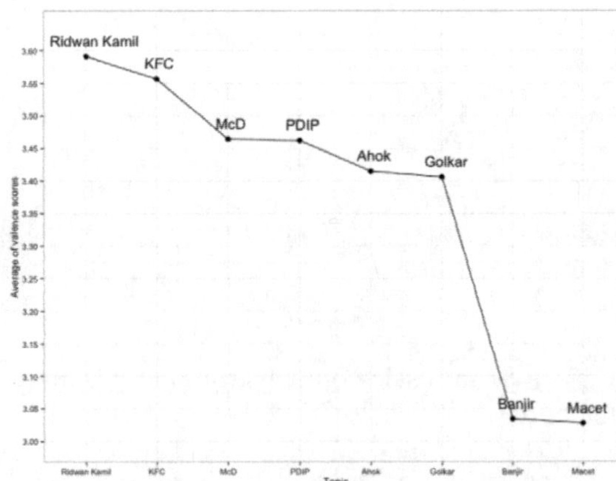

Figure 11. Valence scores distribution per topic

Same lowest valence observed for both city issues, although there is large gap between their arousal average scores. Flood has the third highest arousal, whereas traffic jam is the eight or lowest among all. KFC and McDonald's are the two most popular fast food brands in Indonesia, have almost similar level of arousal, yet KFC has higher valence compared to McDonald's. Ahok (Governor of Jakarta) has slightly higher arousal compared to Ridwan Kamil (Mayor of Bandung). However, Ridwan Kamil clearly dominates valence chart, not only upon Ahok but also among any other topic. This finding indicates that Ridwan Kamil and KFC are the happiest topics compared to others.

4. Discussion

Core affect mapped in circumplex model has opened a new direction to the study of emotions [4-6, 8-10]. For instance to research about happiness as presented in this paper. Whereas most study measure happiness only through valence component [1-3] here we also measure arousal; Because positive valence isn't always equal to happiness, and vice versa on negative valence. Arousal measurement serves as indicator of emotional intensity, providing more comprehensive insight. As explained in introduction, positive valence with high arousal is excitement, moderate arousal is happiness, and low arousal is calm.

Inspired by the study initiated by Dodds [1-3], we constructed corpus comprising 3,000 Indonesian words, each has rating scores of arousal (on the scale of 1-4), and valence (on the scale of 1-5). We found that there is no significant correlation between valence and arousal (R^2 = 0.028, $F(1, 2981) = 86.89, p < .001$). As seen in figure 5, emotional states are present across the combination of arousal and valence level [8-9]. Our particular finding might be useful as new method in the future to label unknown emotional states. This proposed method isn't limited to social media's corpora, but any text source like speech and song lyric. More integrated corpora, which combine various text sources would be ideal.

Moreover, valence scores are negatively skewed toward positive range. Whereas, arousal is symmetrically distributed with dominance on moderate range of the score 2 and 3. Therefore in accordance to finding from study by Dodds et al. and Pollyanna hypothesis [3], tendency toward positivity bias is also present in corpus used in this study. Note that, our result shows that words used in social media conversation (in this case twitter) are indeed biased toward positive emotion and contradict the assumption that conversations in social media are dominated by negative expression.

We have calculated arousal and valence scores for various public conversations in Indonesian language and from this nation's spatial location, within temporal and topical dimensions. We randomly collected tweets and analyze it within three contexts, which are: day of the week, hour of the day, and conversation topics. Using combination of arousal and valence average scores we indicate that several data points as the happiest among others. First, from day of the week context, Monday is presumably has the highest level of happiness. Second, from hour of the week context, 6 am to 8 am are indicated of having highest level of happiness. Lastly, Ridwan Kamil (Mayor of Bandung) and fast food brand KFC, are the first and second most topics with highest level of happiness. Notice that we reduce score range up close to enhance differences between points.

We conclude by noting that our approach has the potential to be used as a measure of emotions for large population in multitude domains. Further development of this approach will include tests for sensitivity and robustness, and also the inclusion of other psychological measures such as moral judgments, values, and personality.

Keywords: arousal, computational psychology, emotion, happiness, hedonometer, psychology, valence

Acknowledgment
We thank Judotens Budiarto, Muhammad Robee, and Heri Purnomo who have developed the technology for data collection and analysis. We are also grateful to Jehan Amanda, Steffi Hartanto, Prasakti Tenri, and Karina Suhenda who have helped collecting the data and performing initial data analysis.

References
[1] Dodds, P. S., & Danforth, C. M. (2009). Measuring the happiness of large-scale written expression: Songs, blogs, and presidents. *Journal of Happiness Studies*, *11*(4), 441-456.
[2] Dodds, P. S., Harris, K. D., Kloumann, I. M., Bliss, C. A., & Danforth, C. M. (2011). Temporal patterns of happiness and information in a global social network: Hedonometrics and Twitter. *PloS one*, *6*(12), e26752.
[3] Dodds, P. S., Clark, E. M., Desu, S., Frank, M. R., Reagan, A. J., Williams, J. R., Mitchell, L., Harris, K.D., Kloumann, I. M., Bagrow, J. P., Megerdoomian, K., McMahon, M. T., Tivnan, B. F., & Danforth, C. M. (2015). Human language reveals a universal positivity bias. *Proceedings of the National Academy of Sciences*, *112*(8), 2389-2394.
[4] Feldman, L. A. (1995a). Valence focus and arousal focus: Individual differences in the structure of affective experience. *Journal of personality and social psychology*, *69*(1), 153.
[5] Feldman, L. A. (1995b). Variations in the circumplex structure of mood. *Personality and Social Psychology Bulletin*, *21*(8), 806-817.
[6] Gregg, V. H., & Shepherd, A. J. (2008). Factor structure of scores on the state version of the four dimension mood scale. *Educational and Psychological Measurement*.
[7] Helliwell, J. F., Layard, R., & Sachs, J. (2015). World happiness report 2015. New York: *Sustainable Development Solutions Network*.
[8] Russell, J. A. (1980). A circumplex model of affect. *Journal of personality and social psychology*, *39*(6), 1161.
[9] Russell, J. A. (2003). Core affect and the psychological construction of emotion. *Psychological review*, *110*(1), 145.
[10] Widhiarso, W., & Prawitasari, J. E. (2010). Struktur Semantik Kata Emosi dalam Bahasa Indonesia. *Jurnal Psikologi*, *37*(2), 153-164.

Numerical solution for option pricing with stochastic volatility model

Andi Mariani, Endar H Nugrahani, Donny C Lesmana

Department of Mathematics, Bogor Agricultural University, Jalan Meranti, Kampus IPB Dramaga, Bogor 16680, Indonesia

E-mail: andi.marianidatu@gmail.com

Abstract. The option pricing equations derived from stochatic volatility models in finance are often cast in the form of nonlinear partial differential equations. To solve the equations, we used the upwind finite difference scheme for the spatial discretisation and a fully implicit time-stepping scheme. The result of this scheme is a matrix system in the form of an M-Matrix and we proof that the approximate solution converges to the viscosity solution to the equation by showing that the scheme is monotone, consistent and stable. Numerical experiments are implemented to show that the behavior and the order of convergence of upwind finite difference method.
Keywords: stochastic volatility model, option pricing, upwind finite difference method, convergence, nonlinear partial differencial equation

1. Introduction

Stock option is a contract that give its holder the right but not the obligation to buy (*call option*) or sell (*put option*) stocks at a specified price [1]. The time when the contract ends is known as the expiration date or the maturity date of the option. There are two types of options, the European option that can be exercised at maturity date and the American options that can be exercised on or before the maturity date. Fisher Black and Myron Scholes [2] showed that the price of the option is the solution of partial differential equations (PDE) called the standar Black-Scholes model [2].

In the Black-Scholes model assumes that the volatility is constant. This assumption is incompatible with the market price movement, where the volatility has a tendency to go down and will go up again at some point [3]. The volatility in the Black-Scholes can depend on the time, the price of the underlying stock, and/or derivative of the option price. In this case, the standard Black-Scholes equation becomes the following Nonlinear Black-Scholes equation:

$$V_\tau = \frac{\sigma^2(t,S,V_S,V_{SS})}{2} S^2 V_{SS} + rSV_S - rV \quad (1)$$

where σ^2 is a modified volatility as a function of t, S, V_S, and V_{SS}

Some models assume volatility surface across underlying asset prices and time [4, 5], the model assumes that the volatility follows the random process [6], and stochastic volatility models developed by Avellaneda, Levy and Paras [7]. [8], formulated partial differential model for option pricing with stochastic volatility is given by

$$V_\tau = \frac{\sigma^2(\Gamma)}{2} S^2 V_{SS} + rSV_S - rV \tag{2}$$

where $\Gamma = V_{SS}$ is the second partial derivative of V to S, $\sigma^2(\Gamma)$ is the volatility as a function of Γ. Volatility is assumed to be in the interval

$$\sigma_{min} \leq \sigma(\Gamma) \leq \sigma_{max}.$$

With such a range of possible volatility value, the equation (2) is nonlinear and does not have a unique solution. This values are obtained by either maximizing or minimizing equation (2) by selecting $\sigma^2(\Gamma)$ according to the value of $\Gamma = V_{SS}$. Specifically, if we consider the worst case for an investors with a long position in the option, then the value of $\sigma^2(\Gamma)$ is:

$$\sigma^2(\Gamma) = \begin{cases} \sigma_{max}^2 & \text{if } \Gamma \leq 0 \\ \sigma_{min}^2 & \text{if } \Gamma > 0. \end{cases} \tag{3}$$

On the other hand, the best case for an investor with a long position is determined by:

$$\sigma^2(\Gamma) = \begin{cases} \sigma_{max}^2 & \text{if } \Gamma > 0 \\ \sigma_{min}^2 & \text{if } \Gamma \leq 0. \end{cases} \tag{4}$$

Prices for the investors with short positions are given by the negative of the solutions when applying equations (3) and (4). With a transformation $\tau = T - t$, equation (2) can be written as

$$U_\tau = \frac{\sigma^2(\Gamma)}{2} S^2 U_{SS} + rSU_S - rU \tag{5}$$

for $S > 0$ and $0 < \tau \leq T$, where $\sigma^2(\Gamma)$ followed equations (3) and (4).

This model involves nonlinear partial differential equations that required a numerical method to determine the option price solution. Several numerical approach can be carried out to solve it, including finite difference method, upwind finite difference method and finite volume methods. Upwind finite difference method and finite volume method proved to be a consistent, stable and monotonous [9, 10]. Finite difference method with fully implicit discretization method is monotone and converges to the viscosity solution, while the Crank-Nicolson is only monotone conditionally [8]. In this work, we develop the upwind finite difference method for pricing the stochastic volatility model. We show theoretically that the upwind finite difference scheme in space and the implicit scheme in time is consistent, stable, and monotone, so it converges to the viscocity solution. We also derive conditions which ensure that the discrete method is monotone.

2. Initial and Boundary Condition

The nonlinear Black-Scholes model has a domain $S \in (0, \infty)$, but for computational purpose, it is necessary to truncate it into $S \in (0, S_{max})$, where S_{max} denotes a sufficiently large positive number to ensure the accuracy of the solution. Then the initial and boundary conditions for nonlinear Black-Scholes equations as follows:

$$U(S,0) = u_0(S), \quad S \in (0, S_{max}), \tag{6}$$
$$U(0,\tau) = g_1(\tau), \quad \tau \in (0, T], \tag{7}$$
$$U(S_{max}, \tau) = g_2(\tau), \quad \tau \in (0, T], \tag{8}$$

with u_0, g_1, and g_2 are given functions $u_0(0) = g_1(0)$ and $u_0(S_{max}) = g_2(0)$. The choices of u_0, g_1, and g_2 depend on type of option. Popular European options are vanilla, butterfly, dan cash or nothing (CoN) with the initial and boundary conditions are given by:

$$u_0 = \begin{cases} \max(S - K, 0) & \text{Vanilla call} \\ \max(K - S, 0) & \text{Vanilla put} \\ \max(S - K_1, 0) - 2\max(S - K_2, 0) + \max(S - K_3, 0)) & \text{butterfly spread} \\ B \times \mathcal{H}(S - K) & \text{CoN} \end{cases}$$

$$g_1 = \begin{cases} 0 & \text{Vanilla call} \\ Ke^{-r\tau} & \text{Vanilla put} \\ 0 & \text{butterfly spread} \\ 0 & \text{CoN} \end{cases}$$

$$g_2 = \begin{cases} S_{max} - Ke^{-r\tau} & \text{Vanilla call} \\ 0 & \text{Vanilla put} \\ 0 & \text{butterfly spread} \\ Be^{-r\tau} & \text{CoN} \end{cases}$$

with $\mathcal{H}$ is a Heaviside functions, B is constanta, K, K_1, K_2, and $K_3$is a strike price.

3. Discretization

Price discretization, let $I = (0, S_{max})$ divided to sub-interval M, where $I_i = (S_i, S_{i+1}), i = 0,1, \dots, M-1$ with $0 = S_0 < S_1 < \cdots < S_M = S_{max}$, and for each $i = 0,1, \dots, M-1$ let $h_i = h = S_{i+1} - S_i$. For time discretization, let $\tau = (0, T)$ divided to sub-interval N, where $\tau_n = (\tau_n, \tau_{n+1})$, $n = 0,1, \dots, N-1$ with $0 = \tau_0 < \tau_1 < \cdots < \tau_n = T$ and for each $n = 0,1, \dots, N-1$ being let $\Delta\tau_n = \tau_{n+1} - \tau_n$. For any $W^n = (W_0^n, W_1^n, \dots, W_M^n)^{\mathrm{T}}$ and $W_i = (W_i^0, W_i^1, \dots, W_i^N)^{\mathrm{T}}$ with $i = 0,1, \dots, M$ and $n = 0,1, \dots, N$, the first and second derivatives followed finite difference operator:

$$(\delta_\tau W_i)(n) = \frac{W_i^{n+1} - W_i^n}{\Delta\tau_n},\ (\delta_S^+ W^n)(i) = \frac{W_{i+1}^n - W_i^n}{h},\ (\delta_S^- W^n)(i) = \frac{W_i^n - W_{i-1}^n}{h} \tag{9}$$

$$(\delta_{SS} W^n)(i) = \frac{W_{i-1}^n - 2W_i^n + W_{i+1}^n}{h^2} \tag{10}$$

By using operators (9-10), nonlinear Black-Scholes model (5) was approximated to the upwind finite difference method:

$$\delta_\tau U_i(n) - \frac{1}{2}\sigma^2((\delta_{SS}U^{n+1})(i)S_i^2(\delta_{SS}U^{n+1})(i) - rS_i(\delta_S^+ U^{n+1})(i) + rU_i^{n+1} = 0, \tag{11}$$

We get:

$$U_{i-1}^{n+1}\left(-\frac{1}{2h^2}\sigma^2(\delta_{SS}U^{n+1})(i)S_i^2\right) + U_i^{n+1}\left(\frac{1}{\Delta\tau_n} + \frac{1}{h^2}\sigma^2(\delta_{SS}U^{n+1})(i)S_i^2 + \frac{rS_i}{h} + r\right) + U_{i+1}^{n+1}\left(-\frac{1}{2h^2}\sigma^2(\delta_{SS}U^{n+1})(i)S_i^2 - \frac{rS_i}{h}\right) = \frac{U_i^n}{\Delta\tau_n}. \tag{12}$$

For simplicity, equation (12) can be written into the following form:

$$\alpha_i^{n+1}(U^{n+1})U_{i-1}^{n+1} + \beta_i^{n+1}(U^{n+1})U_i^{n+1} + \gamma_i^{n+1}(U^{n+1})U_{i+1}^{n+1} = \frac{1}{\Delta\tau_n}U_i^n, \tag{13}$$

for $i = 1, \dots, M-1$ and $n = 1, \dots, N-1$, where:

$$\alpha_i^{n+1}(U^{n+1}) = -\frac{1}{2h^2}\sigma^2(\delta_{SS}U^{n+1})(i)S_i^2, \tag{14}$$

$$\beta_i^{n+1}(U^{n+1}) = \frac{1}{\Delta\tau_n} + \frac{1}{h^2}\sigma^2(\delta_{SS}U^{n+1})(i)S_i^2 + \frac{rS_i}{h} + r, \tag{15}$$

$$\gamma_i^{n+1}(U^{n+1}) = -\frac{1}{2h^2}\sigma^2(\delta_{SS}U^{n+1})(i)S_i^2 - \frac{rS_i}{h}. \tag{16}$$

Using (6-8), the initial and boundary conditions for equation (13) is defined as follows

$$U_i^0 = g_1(S_i) \qquad U_0^n = g_2(\tau_n) \qquad U_M^n = g_3(\tau_n) \tag{17}$$

for $i = 1, 2, \ldots, M-1$ and $n = 1, \ldots, N$, the equation of (13) can be written into the following matrix form

$$A^{n+1}(U^{n+1})\,\widehat{U}^{n+1} = \frac{1}{\Delta\tau_n}\widehat{U}^n + B^{n+1}, \tag{18}$$

for $n = 1, \ldots, N-1$, where

$$A^{n+1}(U^{n+1}) = \begin{bmatrix} \beta_1^{n+1} & \gamma_1^{n+1} & 0 & \cdots & 0 & 0 & 0 \\ \alpha_2^{n+1} & \beta_2^{n+1} & \gamma_2^{n+1} & \cdots & 0 & 0 & 0 \\ 0 & \alpha_3^{n+1} & \beta_3^{n+1} & \cdots & 0 & 0 & 0 \\ \vdots & \vdots & \vdots & \ddots & \vdots & \vdots & \vdots \\ 0 & 0 & 0 & \cdots & \beta_{M-3}^{n+1} & \gamma_{M-3}^{n+1} & 0 \\ 0 & 0 & 0 & \cdots & \alpha_{M-2}^{n+1} & \beta_{M-2}^{n+1} & \gamma_{M-2}^{n+1} \\ 0 & 0 & 0 & \cdots & 0 & \alpha_{M-1}^{n+1} & \beta_{M-1}^{n+1} \end{bmatrix}$$

$\widehat{U}^k = (U_1^k, U_2^k, \ldots, U_{M-1}^k)'$ for $k = n, n+1$
$B^{n+1} = (-\alpha_1^{n+1}U_0^{n+1}, 0, \ldots, 0, -\gamma_{M-1}^{n+1}U_N^{n+1})'$
$A^{n+1}(U^{n+1})$ is a M matrix for any given U^{n+1}.

3.1. Theorem 1. M-Matrix

For any $n = 0, 1, \ldots, N$, $A^n = (A_{ij}^n)$ is a M matrix for any U^n that being given.

Proof:
To prove Theorem 1, it must be shown that for $i = 1, 2, \ldots, M-1$:

$$\alpha_i^n < 0, \quad \beta_i^n > 0, \quad \gamma_i^n < 0 \tag{19}$$
$$\beta_i^n \ge |\alpha_i^n| + |\gamma_i^n| \tag{20}$$

For Matrix A^{n+1}, from models (14-16) it can be seen that the condition (19) is obviously true. Further conditions (20), because $r \ge 0$ and $\frac{1}{\Delta\tau_n} \ge 0$ so:

$$\begin{aligned} \beta_i^{n+1} &\ge |\alpha_i^{n+1}| + |\gamma_i^{n+1}| + r + \frac{1}{\Delta\tau_n} \\ &\ge |\alpha_i^{n+1}| + |\gamma_i^{n+1}| \end{aligned} \tag{21}$$

From definition of $A^n = (A_{ij}^n)$ and depend on (21), we got:

$$A_{ij}^n \le 0, \quad i \ne j, \quad A_{ii}^n > 0, \quad A_{ii}^n > \sum_{j=1}^{M-1} |A_{ij}^n|$$

It is obvious that A^n is irreducible.Thus, A^n is a M-matrix for any given U^n.

4. Convergence of The Numerical Scheme

[11] showed that a discretisation scheme is convergence to the viscosity solution if it is a stable, consistent, and monotone discretization. For $1 \le i \le M-1$ and $0 \le n \le N-1$, a functional F_i^{n+1} defined by:

$$F_i^{n+1} = \left(-\frac{rS_i}{h}\right) U_{i+1}^{n+1} + \left(\frac{1}{\Delta\tau_n} + \frac{rS_i}{h} + r\right) U_i^{n+1} - \frac{U_i^n}{\Delta\tau_n} - \frac{1}{h^2}\sigma^2((\Gamma^{n+1})(i))(\Gamma^{n+1})(i). \tag{22}$$

Then, equation (22) becomes

$$F_i^{n+1}\left(U_i^{n+1}, U_{i+1}^{n+1}, U_{i-1}^{n+1}, U_i^n\right) = 0.$$

4.1. *Lemma 1. Monotonicity*

Discretization scheme (13) is monotone, i.e., for any $\varepsilon > 0$ and $1 \le i \le M-1$,

$$F_i^{n+1}\left(U_i^{n+1}, U_{i+1}^{n+1} + \varepsilon, U_{i-1}^{n+1} + \varepsilon, U_i^n + \varepsilon\right) \le F_i^{n+1}\left(U_i^{n+1}, U_{i+1}^{n+1}, U_{i-1}^{n+1}, U_i^n\right) \tag{23}$$

And

$$F_i^{n+1}\left(U_i^{n+1}, U_{i+1}^{n+1} + \varepsilon, U_{i-1}^{n+1} + \varepsilon, U_i^n + \varepsilon\right) \le F_i^{n+1}\left(U_i^{n+1}, U_{i+1}^{n+1}, U_{i-1}^{n+1}, U_i^n\right). \tag{24}$$

Proof

Since $-\frac{rS_i}{h} \le 0, \frac{1}{\Delta\tau_n} > 0$ and $\frac{1}{\Delta\tau_n} + \frac{rS_i}{h} + r > 0$, the first three (linear) terms on the right hand side of equation (22) are respectively non-increation in U_{i+1}^{n+1}, increasing in U_i^{n+1} and decreasing in U_i^n. Let $E_k = \left(0, 0, \ldots, \underbrace{1}_{k}, 0, \ldots, 0\right)'$ be the $(M-1) \times 1$coloum vector. From the definition δ_{SS} (11), we have

$$\begin{aligned}\delta_{SS}(U^{n+1} + \varepsilon E_{i-1} + \varepsilon E_{i+1})(i) &= \frac{(U_{i-1}^{n+1} + \varepsilon) - (2U_i^{n+1}) + (U_{i+1}^{n+1} + \varepsilon)}{h^2} \\ &= (\Gamma^{n+1})(i) + \frac{\varepsilon}{h^2}\end{aligned}$$

and

$$\begin{aligned}\delta_{SS}(U^{n+1} + \varepsilon E_i)(i) &= \frac{(U_{i-1}^{n+1}) - 2(U_i^{n+1} + \varepsilon) + (U_{i+1}^{n+1})}{h^2} \\ &= (\Gamma^{n+1})(i) - \frac{2\varepsilon}{h^2}\end{aligned}$$

For the nonlinear term on the right hand side $-\frac{1}{h^2}\sigma^2((\Gamma^{n+1})(i))(\Gamma^{n+1})(i)$, we consider case 1 and case 2 are the worst case for long position in option, case 3 and case 4 are the best case for long position.

Case 1

For $\Gamma_i^{n+1} \leq 0$, implies that $\sigma^2(\Gamma_i^{n+1}) = \sigma_{max}^2 > 0$

Then we have;

$$-\frac{\sigma^2}{h^2}\left(\Gamma^{n+1}(i)\right)\left(\Gamma^{n+1}(i)\right) = -\frac{\sigma_{max}^2}{h^2}\left(\Gamma^{n+1}(i)\right)$$

$$\sigma^2(S,\Gamma).\Gamma = \sigma^2(S,Z).Z$$

For any S, Z_1 and Z_2, which Z_1 and $Z_2 \leq 0$,

$$-\frac{1}{h^2}\left((\sigma_{max}^2)Z_1 - (\sigma_{max}^2)Z_2\right) = \begin{cases} -\dfrac{1}{h^2}\sigma_{max}^2(Z_1 - Z_2) \geq 0; & if\ Z_1 < Z_2 \quad Z_1, Z_2 \leq 0 \\ -\dfrac{1}{h^2}\sigma_{max}^2(Z_1 - Z_2) \leq 0; & if\ Z_1 > Z_2 \quad Z_1, Z_2 \leq 0 \end{cases}$$

Similarity, we can prove that it also holds for:

Case 2
For $\Gamma_i^{n+1} > 0$, implies that $\sigma^2(\Gamma_i^{n+1}) = \sigma_{min}^2 > 0$
Case 3
For $\Gamma_i^{n+1} > 0$, implies that $\sigma^2(\Gamma_i^{n+1}) = \sigma_{max}^2 > 0$
Case 4
For $\Gamma_i^{n+1} \leq 0$, implies that $\sigma^2(\Gamma_i^{n+1}) = \sigma_{min}^2 > 0$

Thus, for any $\varepsilon > 0$ and $1 \leq i \leq M-1$ we have

$$\begin{aligned} F_i^{n+1}&\left(U_i^{n+1}, U_{i+1}^{n+1} + \varepsilon, U_{i-1}^{n+1} + \varepsilon, U_i^n + \varepsilon\right) \\ &= \left(-\frac{rS_i}{h}\right)\left(U_{i+1}^{n+1} + \varepsilon\right) + \left(\frac{1}{\Delta\tau_n} + \frac{rS_i}{h} + r\right)U_i^{n+1} - \frac{1}{\Delta\tau_n}(U_i^n + \varepsilon) \\ &- \frac{1}{h^2}S_i^2\left[\sigma^2(u_i^{n+1} + \varepsilon)\right]\left(u_i^{n+1} + \varepsilon\right) \\ &\leq F_i^{n+1}\left(U_i^{n+1}, U_{i+1}^{n+1}, U_{i-1}^{n+1}, U_i^n\right) \end{aligned}$$

This is (23), Similarly, it is easy to show that (24) also hold true. So, the discretization scheme is monotone.

4.2. Lemma 2. Stability

For each $0 \leq n \leq N-1$ let $U^{n+1} = \left(U_0^{n+1}, (\widehat{U}^{n+1})^T, U_M^{n+1}\right)^T$ let $\widehat{U}^{n+1}$ is a solution (13). Then, U^{n+1} satisfies

$$\| U^{n+1} \|_\infty \leq max\{\| u_0 \|_\infty, \| g_1 \|_\infty, \| g_2 \|_\infty\}$$

where u_0, g_1, g_2 is initial and boundary conditions (6–8), and $\|.\|_\infty$ is norm l_∞.

Proof

For any $0 \leq n \leq N-1$, (13) can be written as:

$$\beta_i^{n+1} U_i^{n+1} = -\alpha_i^{n+1} U_{i-1}^{n+1} - \gamma_i^{n+1} U_{i+1}^{n+1} + \frac{1}{\Delta\tau_n} U_i^n$$

For each $1 \leq i \leq M-1$. Remember that $\alpha_i^{n+1} < 0, \gamma_i^{n+1} < 0$ and $\beta_i^{n+1} > 0$.

If $\| U^{n+1} \|_\infty = |U_k^{n+1}|$ for $k \in \{1,2,\ldots,M-1\}$, then the equation

$$\beta_i^{n+1} \|U_i^{n+1}\|_\infty \leq -\alpha_i^{n+1} \|U_{i-1}^{n+1}\|_\infty - \gamma_i^{n+1} \|U_{i+1}^{n+1}\|_\infty + \frac{1}{\Delta\tau_n} \|U_i^n\|_\infty$$

with $i = k$ becomes $(\alpha_i^{n+1} + \beta_i^{n+1} + \gamma_i^{n+1}) \|U^{n+1}\|_\infty \leq \frac{1}{\Delta\tau_n} \|U^n\|_\infty$

Thus, since $\alpha_i^{n+1} < 0$ and $\gamma_i^{n+1} < 0$ then we obtain

$$\begin{aligned} \|U^{n+1}\|_\infty &\leq \frac{1/\Delta\tau_n}{(\alpha_i^{n+1} + \beta_i^{n+1} + \gamma_i^{n+1})} \|U^n\|_\infty \leq \|U^n\|_\infty \\ &\leq \|U^{n-1}\|_\infty \leq \cdots \leq \|U^0\|_\infty \leq \|g1\|_\infty. \end{aligned} \quad (25)$$

If $\|U^{n+1}\|_\infty = |U_0^{n+1}|$ or $\|U^{n+1}\|_\infty = |U_M^{n+1}|$, from equations (), (7) and (8) we can see that

$$\| U^{n+1} \|_\infty \leq max\{|U_0^{n+1}|, |U_M^{n+1}|\} \leq max\{\| g_1 \|_\infty, \| g_2 \|_\infty\} \quad (26)$$

Combining equations (25) and (26), gives

$$\begin{aligned} \| U^{n+1} \|_\infty &\leq max\{\| U^n \|_\infty, |U_0^{n+1}|, |U_M^{n+1}|\} \\ &\leq max\{\| u_0 \|_\infty, \| g_1 \|_\infty, \| g_2 \|_\infty\} \end{aligned}$$

Hence, the discretization scheme (13) is stable.

4.3. Lemma 3. Consistency

Discretization scheme (13) is consistent.

Proof:
Lax equivalence theorem [12] states that the finite difference method is consistent for a given initial value problem.

4.4. Theorema 2. Convergence

Discretization scheme (13) is convergen to solutions (5) with boundary conditions (6-8) as $(h, \Delta\tau) \to 0$, where $h = max_{0 \leq i \leq M-1} h_i$ and $\Delta\tau = max_{0 \leq n \leq N-1} \Delta\tau_n$.

Proof:
[11], proved that if a discretization of the nonlinear 2nd order PDE is consistent, stable and monotone, then the solution of the fully discretized system converge to the vicosity solutionto the PDE. Since discretization (13) proved to be a consistent, stable and monotone, then discretizing (13) converges. Theorem 4 is just a consequence of Lemma 1, 2 and 3.

IOP Conf. Series: Earth and Environmental Science **31** (2016) 012024 doi:10.1088/1755-1315/31/1/012024

5. Numerical Solution

The calculation of the option value by using the parameter $r = 0.1$, $T = 1, K = 40, S_{max} = 80, \sigma_{max}^2 = 0.0625, S_{min}^2 = 0.0225$, with a uniform mesh $h = 2$, $M = 40$ and $\Delta\tau = 0.05$ N = 20. Comparison of the option price on the best case and worst case for a position as an option buyer (long position) can be seen in Figure 1. Since the exact solution is unknown, we use the numerical solution on the uniform mesh with h = 0.03125, M = 2560 and $\Delta\tau$ = 0.00078125, N = 1280 as the "exact" solution, V_{exact}.

(a) (b)

Figure 1 The price of the European call option for a position as a buyer for option with, (a) the best case and (b) the worst

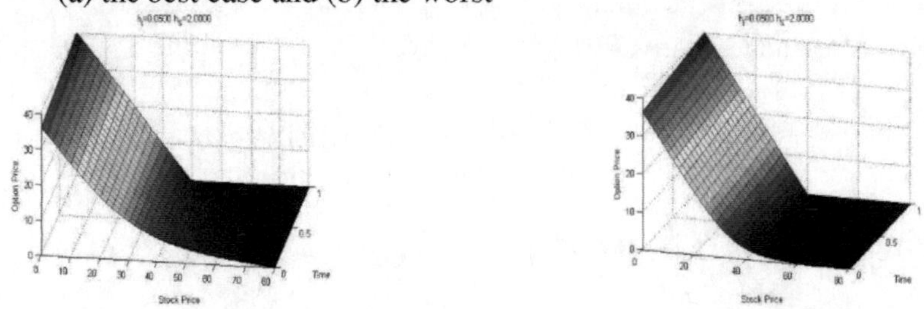

(a) (b)

Figure 2 The price of the European put option for a position as an option to, (a) the best case and (b) the worst case

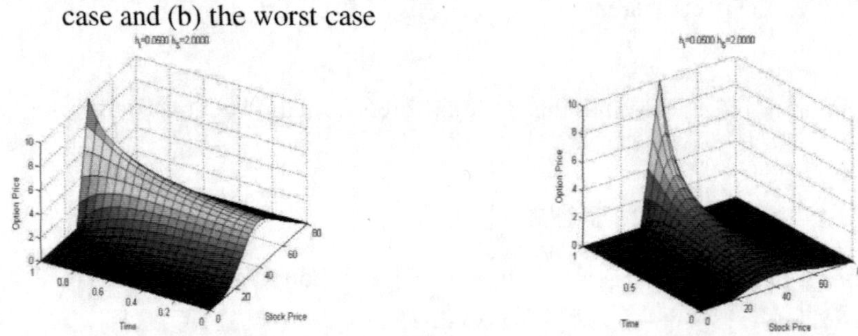

(a) (b)

Figure 3 The price of the European butterfly option for a position as an option to, (a) the best case and (b) the worst case

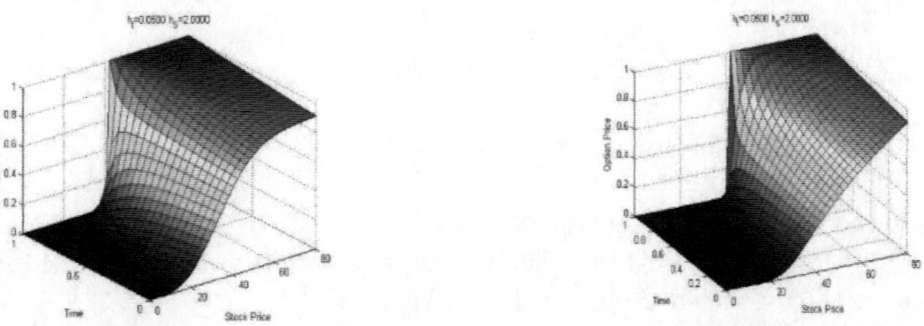

(a) (b)

Figure 4 The price of the European cash or nothing option for a position as a buyer option to, (a) the best case and (b) the worst case

Table 1 Computed rated of convegence for the *call option*

M	N	The Best Case		The Worst Case	
		$\|\|\cdot\|\|_{h,\infty}$	Ratio	$\|\|\cdot\|\|_{h,\infty}$	Ratio
41	21	2.1583e-01		1.7887e-01	
81	41	1.2705e-01	1.70	9.9884e-02	1.79
161	81	7.7695e-02	1.64	5.6592e-02	1.76
321	161	4.7794e-02	1.63	3.2573e-02	1.74
641	321	2.7615e-02	1.73	1.8008e-02	1.81
1281	641	1.2437e-02	2.22	7.9266e-03	2.27

Using the "exact" solution, we can calculate the ratio of the numerical solution from the mesh with

$$\text{Ratio} = \frac{\left\|V_h^{\Delta\tau} - V_{exact}\right\|_{h,\infty}}{\left\|V_{h/2}^{\Delta\tau/2} - V_{exact}\right\|_{h,\infty}}$$

Where $V_h^{\Delta\tau}$ denotes a solution on the mesh with h is mesh size of stock dan $\Delta\tau$ mesh size of time, and

$$\left\|V_h^{\Delta\tau} - V_{exact}\right\|_{h,\infty} := \max_{1\le i\le M;1\le n\le N} |V_i^n - V_{exact}(S_i,\tau_n)|.$$

Ratio calculation results in Table 1 shows the convergence order of upwind method of call option with the best case and worst case are about 1.6 and 1.7 respectively.

Futhermore, calculating rates of convergence from butterfly option, put option, and cash-or-nothing (CoN) option are the same as calculating the rates of convergence of call option. Therefore, we get that the rates of convergence of put option with the best case and the worst case are about 1.6 and 17 respectively, butterfly option with the best case and the worst case are about 1.6 and 17 respectively, and cash or nothing (CoN) option with the best case and the worst case is about 1.3 and 1.3 respectively.

6. Conclusions

This study used upwind finite difference method for the discretization of space and implicit method for time discretization of nonlinear PDE in option pricing with stochastic volatility model, where this discretization scheme proved monotone, consistent and stable. Based on the results of numerical simulations, it showed that the order of convergence for upwind finite difference method with stochastic volatility models is about 1.6 to 1.7 for the worst case and best case, with a position as a buyer of the option (long position).

IOP Conf. Series: Earth and Environmental Science **31** (2016) 012024 doi:10.1088/1755-1315/31/1/012024

References

[1] A J. Hull, *Option, Futures and Other Derivatives*, Sixth Edition, New Jersey: Prentice-Hall, 2009.

[2] F. Black and M. Scholes, The Pricing of options and corporate liabilities, *J. Polit. Econ.*, **81** (1973), 637 - 659.

[3] B. Dupire, Pricing with smile. *Risk.* **7** (1994), 18 - 20.

[4] L. B. G. Anderson and R. Brotherton-Ratcliffe. The equity option volatility smile: An implicit finite-difference approach. *J Comput. Finance,* **1** (1998), 5 - 37.

[5] T. F. Coleman, Y. Li, and A. Verma, Reconstructing the unknown local volatility function, *Journal of Computational Finance,* **2** (1999), 77 – 102.

[6] S. L.Heston, A closed-form solution for option with stochastic volatility with applications to bond and currency option. *Rev. of Financial Stud.* **6** (1993), 327 - 343.

[7] M. Avellaneda, A. Levy, and A. Paras, Pricing and hedging derivative securities in markets with uncertain volatilities. *Appl. Math. Finance,* **2** (1995), 73 - **88**.

[8] D. M. Pooley, P. A. Forsyth, K.R. Vetzal, Numerical convergence properties of option pricing PDEs with uncertain volatility. *IMA J. Numer. Anal*, **23** (2001), 241 - 267.

[9] K. Zhang and S. Wang, A computational scheme for uncertain votality model in option pricing. *Appl. Numer. Math*, **59** (2009), 1754 – 1767.

[10] D. C. Lesmana and S. Wang, An upwind finite difference method for a nonlinear Black-Scholes equation governing European option valuation under transaction cost. *Appl. Math. Comput.* **219** (2013), 8811 - 8828.

[11] G. Barles, *Convergence of numerical schemes for degenerate parabolic equations arising in finance*, in: L.C.G. Rogers, D. Talay (Eds.), Numerical Methods in Finance, Cambridge University Press, Cambridge, 1997.

[12] J. C. Strikwerda, *Finite Difference Schemes and Partial Differential Equations* (1st ed.), Chapman & Hall, 1989.

Inhibition of α-glucosidase activity by ethanolic extract of *Melia azedarach* L. leaves

Sulistiyani[1, 2]**, Mega Safithri**[1]**, Yoana Puspita Sari**[1]

[1]Departement of Biochemistry, Bogor Agricultural University, Jl. Agathis Dramaga, Bogor, Indonesia
[2]Biopharmaca Research Center, Bogor Agricultural University, Jl. Taman Kencana No.3, Bogor Indonesia

E-mail: sulistyanisoemardi@gmail.com

Abstract. Development of α-glucosidase inhibitor derived from natural products is an opportunity for a more economic management of diabetes prevention. The objective of this study was to test the activity of α-glucosidase with or without potential inhibitor compounds. By *in vitro* method, α-glucosidase hydrolizes *p*-nitrophenyl-α-D-glucopiranoside to glucose and the yellow of *p*-nitrophenol which can be determined with spectrophotometry at 400 nm. The ability of ethanolic leaf extract of *Melia azedarach* L. as α-glucosidase inhibitor was compared with that of commercial acarbose (Glucobay®). Acarbose showed strong inhibitory activity against α-glucosidase with IC_{50} values of 2.154 µg/mL. The crude ethanolic leaf extract of *M. azedarach,* however, showed less inhibitory activity with IC_{50} value of 3,444.114 µg/mL. Total phenolics of *M. azedarach* leaves EtOH extract showed 17.94 µg GAE/mg extract and flavonoids total compound of 9.55 µg QE/mg extract. Based on the published wide range of IC_{50} values of extracts reported as α-glucosidase inhibitor which were between 10,000 ppm-0.66 ppm, our result suggests that extract of *M.azedarach* leaves is potential candidate for development of anti-hyperglycemic formulation.

1. Introduction

α-Glucosidase is an intestinal enzyme that catalyzes the break of α-1.4-glycosidic bond in oligosaccharides into α-glucose molecules which can be absorbed by the intestine[1]. Development of α-glucosidase inhibitor derived from natural products is an opportunity for a more economic management of diabetes mellitus prevention. Diabetes mellitus is a disease characterized by hyperglycemia which increased levels of sugar in the blood exceeds normal levels with fasting glucose levels ≥ 126 mg / dL and 2 hours after eating ≥ 200 mg / dL. The disease is caused by lack of pancreas β cells to produce insulin or the cell resistance of insulin. One therapeutic approach to decreasing postprandial hyperglycemia is to retard the absorption of glucose via inhibition of carbohydrate-hydrolyzing enzymes, such as glucosidase, in the intestine[2]. The glucosidase enzymes are located in the brush border of the small intestine and are required for the breakdown of carbohydrates before monosaccharide absorption. The α-glucosidase inhibitors delay the absorption of ingested carbohydrates, reducing the postprandial glycemia and insulin peaks[3].

The objective of this research was to test the activity of α-glucosidase with or without potential inhibitor compounds. By in vitro method, α-glucosidase hydrolizes p-nitrophenyl-α-D-glucopiranoside to glucose and the yellow of p-nitrophenol which can be determined with spectrophotometry at 400 nm. The ability of ethanolic leaf extract of *Melia azedarach* L. as α-glucosidase inhibitor was compared with that of commercial acarbose (Glucobay®). *M. azedarach* L., known as mindi, is a forest plant that has been used traditionally as medicinal plant for diabetes mellitus. Jo et al.[4] reported that flavonoid which is one of its phytochemical compounds can inhibit α-glucosidase by hydroxylating C-3 of the flavonol ring carbon. Acarbose is structurally similar to oligosaccharides, but it has a higher affinity around 10^4-10^5 to bind α-glucosidase, thus acarbose is a competitive inhibitor, which resulted in the decreased formation of monosaccharides from oligosaccharides[5]. Therefore, the results will provide scientific informations on the potency of *M.azedarach* L. leaf extract in the development of antidiabetic natural products formulation with antihyperglicemic mechanism of action.

2. Materials and Methods

2.1 Total phenolic compound determination[6]

Plant extract stock solution with concentration of 1000 μg/mL was made with methanol. Five mL of extract solution was added with 2.5 mL of 10% Folin-Ciocalteu reagent and 2.5 mL of 7.5% Na_2CO3. The mixture in triplicate were then incubated in the waterbath at 45°C for 45 minutes. Absorbancy was determined with spectrophotometer at 765 nm wavelength. Similar protocol was done to prepare the galic acid standard curve with the following concentrations of 10, 20, 30, 40, and 50 μg/mL. The extract total phenolic content was expressed as mg galic acid equivalent (GAE)/g of extract.

2.2 Total flavonoid determinatiorn[7]

Plant extract solution with concentration of 1000 μg/mL was prepared in methanol. Five mL of extract solution in triplicate were added with 0.3 mL of 5% $NaNO_2$ and 0.3 mLof 10% AlCl3 and was kept at room temperature for 5 minutes. These mixtures were then added with 2 mL of 1M NaOH and the volume was made up to 10 ml with distilled water. Absorbancy was determined with spectrophotometer at 510 nm wavelength. Similar protocol was used to prepare quercetin standard curve in varying concentrations. The extract total flavonoid content was expressed as mg quercetin equivalent (QE)/g of extract.

2.3 *In vitro assay of α-glucosidase activity (modified from Lelono & Tachibana* [8])

Enzyme solution was prepared by dissolving 1 mg of α-glucosidase in 100 mL of phosphate buffer (pH 7) which contained 200 mg of bovine serum albumin. Prior to use, 1 mL of enzyme solution was diluted 25 times with phosphate buffer (pH 7). The reaction mixture was prepared in the microplate wells which consisted of 25 μl of 10 mM *p*-nitrophenyl-D-glucopyranose as substrate and 50 μl of 100mM phosphate buffer (pH 7). Briefly, plant extract was dissolved in DMSO and aliquots of extract samples (10 μL) was added to the reaction mixture to final concentrations of: 50, 100, 200, 500, 1000, 5000, 7500, and 10,000 μg/mL. Solution of 1% acarbose (Glucobay®) was prepared with phosphate buffer pH 7. Then it was mixed with 2N HCl of equal volume (1:1) and was centrifuged. Aliquots of supernatant (10 μL) was taken and added into the reaction mixture at final concentration of 0.1, 0.5, 1, 5, and 10 μg/mL. Blanks, controls and each concentration of samples were done in triplicate. Following incubation at 37°C for 5 minutes, 25 μl of enzyme solution was added into the reaction mixture and incubated further for 15 minutes. Enzyme reaction was stopped by adding 100 μl of 200mM Na_2CO_3.. Blanks, controls, and samples absorbance of the *p*-nitrophenol product was measured by microplate reader spectrophotometer at 400 nm wavelength.

Percent of inhibition of the enzyme activity was calculated using the following formula:

$$\% \text{ of inhibition} = \frac{\text{absorbance of control-absorbance of sample}}{\text{Absorbance of control}} \text{ x } 100$$

All data analyses were done using Microsoft Excel and expressed as the average of triplicate.

3. Results and Discussion

Acarbose, a commercially known α-glucosidase inhibitor[9], showed strong inhibitory activity against α-glucosidase with IC_{50} values of 2.154 μg/mL (Figure 1). This data is consistent with study by Permasku [10] who reported smaller acarbose's IC_{50} value of 1.46 μg/mL. Using similar in vitro system, Septiawati[11] reported that acarbose at concentration of 10,000 μg/mL inhibited αglucosidase activity by 99.34%. Acarbose is an oligosaccharide derived from the *Actinoplanes* strain of fungi[9]. Due to its similarity to the structure of oligosaccharide, acarbose acted as competitive inhibitor of the enzyme in this study. According to Arungarinthan *et al.*[9] the mechanism of action is predominantly through competitive, reversible inhibition of intestinal brush border α-glucosidase, with a weaker effect on pancreatic α-amylase. Acarbose, the first α-glucosidase inhibitor to be identified, is currently used for the treatment of type 2 diabetes [9].

On the other hand, crude ethanolic leaf extract of *M. azedarach* showed less inhibitory activity with IC_{50} value of 3444.114 μg/mL (Figure 2). The inhibitory activity of the ethanolic extract of *M.azedarach,* however, was higher than that of ethanolic extract of *Graptophyllum pictum Griff* (Daun Wungu)[12] and *Orthosiphon stamineus* Benth (Kumis Kucing)[13] which inhibited 50% of the α-glucosidase at concentration of 10,000 ppm. The IC_{50} of ethanolic bark extract of *Toona sinensis* Merr. (suren) was 0.66 ppm [14] and that of ethanolic seed extract of *Swietenia mahagony* Jacq was 100 ppm [15]. Based on these IC_{50} values, *M.azedarach* ethanolic extract was better and more potential enzyme inhibitor than *Graptophyllum pictum Griff* and *Orthosiphon stamineus* Benth. The inhibition of this enzyme can be achieved by various natural compounds such as the phenolic group, the flavonoids, luteolin, miricetin, and quercetin[16]. Therefore, inhibitory activity of mindi leaf extracts was likely due to phytochemical compounds contained in the extract.

Quantitative analysis of total flavonoids content showed that there was 9.55 μg QE/mg extract. Previous study by Purnama[17] showed that the total flavonoid of ethanolic extract was 5.99 mg *QE/g* extract. Marek et al.[18] reported higher flavonoids content in 70% ethanolic extract (15.91 mg QE/g extract). This is consistent with the fact that our extraction was using 96% ethanol which is less polar compare to the 70% ethanol, thus resulted in less amount of flavonoids that can be extracted. This amount of flavonoids is relatively small compared to other medicinal plant such as the rhizome of bawang dayak that had been reported by Febrinda *et al.*[19]. which contained as much as 65.35 mg QE/g extract. Flavonoids are naturally occurring phenolic compounds that are widely distributed in plants and some of them have been described as glucosidase inhibitors[20,21].

Total phenolics of *M. azedarach* leaves ethanolic extract was as much as 17.94 μg GAE/mg extract. This data is consistent with that of Purnama [17] which reported total phenolic compound of 17.77 mg GAE/g extract. Nahak & Rajani[22] reported higher content of total phenolics in ethanolic extract of *M.azedarach* as much as 360 μg CE/mg extract, and also that of Ahmed *et al.*[23] was as much as 492 μg CE/mg ekstrak. As comparison, the rhizome of bawang dayak contained as much as 217.71 mg GAE/g extract[19].

Our result suggests that extract of M.azedarach leaves is potential candidate for development of anti-hyperglycemic formulation.

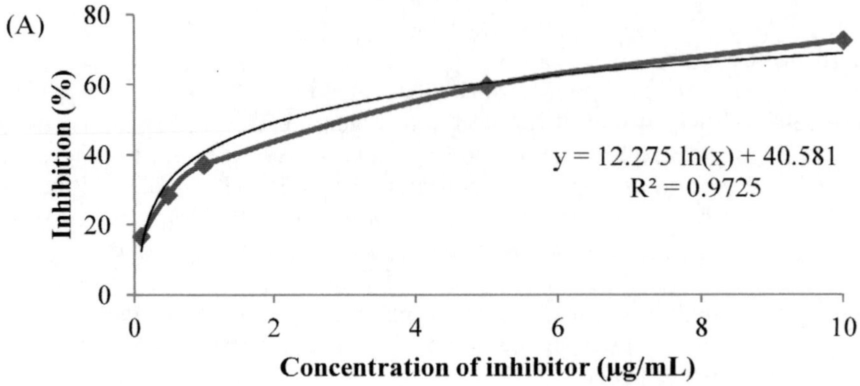

Figure 1. Inhibition of α-glucosidase by acarbose

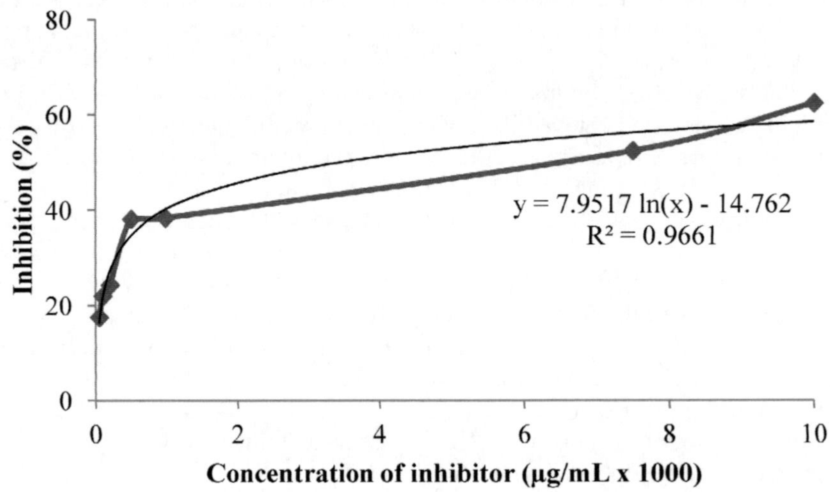

Figure 2. Inhibition of α-glucosidase by *M. azedarach* leaves EtOH extract

References

[1] Gao H, Huang Y, Gao B, Kawabata J. 2008. Chebulagic acid is a potent a-glucosidase inhibitor. *Biosci Biotechnol Biochem* **72** 601-603.

[2] Holman RR, Cull CA, Turner RC. A randomized double-blind trial of acarbose in type 2 diabetes shows improved glycemic control over 3 years. Diabetes Care 1999;22:960–4.

[3] Stuart AR, Gulve EA, Wang M. Chemistry and biochemistry of type 2 diabetes. *Chem Rev* 2004;104:1255–82.

[4] Jo SH, Ka EH, Lee HS, Apostolidis E, Jang HD, Kwon YI. 2009. Comparison of antioxidant potential and rat intestinal alpha glucosidases inhibitory activities of quercetin, rutin, and isoquercetin. *International Journal of Applied Research in Natural Products* **4** 52-60.

[5] Rosak C, Gabrielle M. 2012. Critical evaluation of the role of acarbose in the treatment of diabetes: patient consideration. *Diabetes, Metabolic Syndrome and Obesity:Targets and Therapy* **5** 357-367.

[6] Pourmorad F, Hosseinimehr SJ, Shahabimajd N. 2006. Antioxidant activity, phenol and flavonoid contents of some selected Iranian medicinal plants. *African J. Biotech.* **5** 1142-1146.
[7] Sahu R, Jyoti S. 2013. Screening of total phenolic and flavonoid content in conventional an non-conventional species of curcuma. *J. Pharmacog and Phytochem.* **2** 176-179.
[8] Lelono RAA, Tachibana S. 2013. Preliminary studies of Indonesian *Eugenia polyantha* leaf extracts as inhibitory of key enzymes for type 2 diabetes. *Journal of Medical Sciences* **13** 103-110.
[9] Arungarinathan G, McKay GA, Fisher M. 2011. Drugs for diabetes: part 4 acarbose. *Br. J Cardiol.* **18** 78-81
[10] Permasku G. 2014. Aktivitas inhibisi enzim α-glukosidase dan sitotoksisitas ekstrak kurkuminoid rimpang temulawak dari berbagai aksesi *(in vitro)* [skripsi]. Bogor (ID): Institut Pertanian Bogor.
[11] Septiawati T. 2008. Daya hambat ekstrak etanol buah mahkota dewa terhadap α-glukosidase secara *in vitro* [skripsi]. Bogor (ID): Institut Pertanian Bogor.
[12] Irwan F. 2011. Aktivitas antidiabetes dan analisis fitokimia ekstrak air dan etanol daun wungu (*Graptophyllum pictum* Griff) [skripsi]. Bogor (ID): Institut Pertanian Bogor.
[13] Kurniawati A, Sulistiyani, Mega S. 2013. Identification of morphological characteristic, bioactive content, and alfa-glucosidase inhibitory activity of *Orthosiphon stamineus* Benth accesion. *The proceding of PPM IPB Research Seminar* **2** 493-509.
[14] Ichsan SA. 2011. In vitro activity of suren (*Toona sinensis* Merr.) bark extract as antioxidant and antidiabetic. [skripsi]. Bogor (ID): Bogor Agricultural University.
[15] Febriyany V. 2014. Assay of alfa-glucosidase inhibitor and hypoglycemic potency of ekstrak mahoni seed (*Swietenia mahagony* Jacq.) extract as antidiabetic agent [skripsi]. Bogor (ID): Bogor Agricultural University
[16] Tadera K, Yuji M, Kouta T, Tomoko M. 2006. Inhibition of α-glucosidase and α-amylase by flavonoids. *J. Nutr Sci Vitaminol.* **52** 149-153.
[17] Purnama RL. 2015. Aktivitas antioksidan, kandungan total fenol, dan flavonoid lima tanaman hutan yang berpotensi sebagai obat alami [skripsi]. Bogor (ID): Institut Pertanian Bogor.
[18] Marek CB, Ana MI, Tereza CM, Rubiani AP, Natalie RT, Alana MR. 2012. Influence of leaf extracts from *Melia azedarach* L. on butyrylcholinesterase activity in rat liver. *J. Med Plants Res.* **6** 3931-3938.
[19] Febrinda AE, Made A, Tutik W, Nancy DY. 2013. Kapasitas antioksidan dan inhibitor alfa glukosidase ekstrak umbi bawang dayak. *J. Teknol dan Industri Pangan* **24** 161-167.
[20] Cazarolli LH, Zanatta L, Alberton EH, Figueiredo MSRB, Folador P, Damazio RG, et al. 2008. Flavonoids: prospective drug candidates. Mini Rev Med Chem **8** 1429–40.
[21] Cazarolli LH, Zanatta L, Alberton EH, Figueiredo MSRB, Folador P, Damazio RG, et al. 2008. Flavonoids: cellular and molecular mechanism of action in glucose homeostasis. *Mini Rev Med Chem* **8** 1032–8.
[22] Nahak G, Rajani KS. 2010. *In vitro* antioxidative activity of *Azadirachta indica* and *Melia azedarach* leaves by DPPH scavenging assay. *Nat and Sci.* **8** 22-28
[23] Ahmed MF, Rao AS, Shaik RA, Mohammed I. 2012. Phytochemical studies and antioxidant activity of *Melia azedarach* Linn leaves by DPPH scavenging assay. *International Journal of Pharmaceutical Applications* **3** 271-276.

Screening of Endophytic Fungi from Chlorophyta and Phaeophyta for Antibacterial Activity

A.C Rahaweman[1], J Pamungkas[2], H Madduppa[3], C Thoms[3], K Tarman[1]*
Department of Aquatic Products Technology, Faculty of Fisheries and Marine Sciences Bogor Agricultural University, Bogor 16680 West Java, Indonesia

E-mail: kustya@gmail.com

Abstract. Chlorophyta and Phaeophyta macroalgae are important sources of secondary metabolites with pharmaceutically relevant antibacterial, antifungal and antiviral bioactivities. Oftentimes, these algae-derived compounds are, in fact, produced by endophytic fungi living inside the macroalgal tissue. Numerous studies have shown that endophytic fungi can produce a broad range of active metabolites such as terpenes, alkaloids, and quinones. The aim of the present study was to screen fungal strains isolated from a variety of *Caulerpa* spp., *Halimeda* spp., and *Sargassum* spp. for antibacterial activity against *Staphylococcus aureus* and *Escherichia coli*. Thirteen morphologically different isolates were tested. Two of them showed pronounced activity against *S. aureus* in agar diffusion assays.
Keywords: Antibacterial activity, Antibiotics, Chlorophyta, Endophytic fungi, Macroalgae, Phaeophyta

1. Introduction
Marine macroalgae are highly biodiverse with more than 10,000 extant species [1]. They are a prolific source of secondary metabolites with antibacterial, antiviral, and antifungal activities [2]. For green algae (Chlorophyta) such as *Caulerpa* spp. phenols with antimicrobial properties against bacteria, fungi, and the DENV-2 virus have been described [3,4]. *Halimeda* spp. have been reported to yield compounds with antimicrobial activity against *Vibrio* sp. and a variety of fungal pathogens as well with antioxidant activity [5,6]. Antibacterial and antiviral activity has also been reported for polyphenolic compounds derived from brown algae (Phaeophyta) such as *Sargassum wightii* and *Turbinaria ornata* [7], and fucosterol from *Turbinaria* sp. [8,9]. It has been shown that antibacterial activity in macroalgae can be caused by the presence of sulphate polysaccharides (SPs) produced by the seaweeds themselves [10]. But there is also evidence that in some cases the ability to inhibit bacterial growth is due to the metabolic activities of fungal endophytes in the macroalgae. Oftentimes, endophytic fungi produce bioactive compounds similar in their chemical structure and their functional properties to the bioactive compounds produced by their hosts [11]. A number of studies have been conducted on bioactive compounds with antibacterial activity isolated from Chlorophyta such as

1 Department of Aquatic Products Technology, Faculty of Fisheries and Marine Sciences Bogor Agricultural University, Bogor 16680 West Java, Indonesia
2 Primate Research Center and Faculty of Veterinary Medicine at Bogor Agricultural University
3 Department of Marine Science and Technology, Faculty of Fisheries and Marine Sciences Bogor Agricultural University, Bogor 16680 West Java, Indonesia

Caulerpa spp. and *Halimeda* spp. and Phaeophyta such as *Sargassum* spp. However, the exploration of endophytic fungi from Chlorophyta and Phaeophyta has been explored in a lesser extent. The aim of the research was to explore the potential of endophytic fungi isolated from macroalgae to produce metabolites with antibacterial activity against *Staphylococcus aureus* and *Escherichia coli.*

2. Material and Method

Endophytic fungi were obtained by direct isolation (direct inoculation) of seaweed *Caulerpa* spp., *Halimeda* spp., and *Sargassum* spp. that were collected off Karya Island, Kepulauan Seribu. This procedure involved taking seaweed sample and isolation of endophytic fungi using the surface sterilization method. The isolate was then inoculated on Potato Dextrose Agar (PDA) medium in fresh water and sea water (using NaCl 3% of the volume of media) and incubated at room temperature (27-29°C) for 5-7 days. The cultures were checked once per day until the fungal growth was visible. Then purification was performed to obtain single isolate. Endophytic isolates were screened for their antibacterial activity against *S. aureus* and *E. coli*. Antibacterial assays were conducted by culturing the bacteria on semisolid NA medium (NA 1.2 grams/100 ml of water) to which then the fungal mycelia were added on a disc with ± 1 cm diameter on the surface of the culture medium. The inoculated cultures were incubated for 24 hours at 37°C. Antibacterial activity was detected by the presence of inhibition zone (clear zone) around the mycelia. This inhibition zone indicates that the fungus produces compounds which are able to inhibit the bacterial growth. The inhibition zones were measured using a caliper. The extent of bacterial growth inhibition was classified in four categories based on the diameter of the inhibition zones, namely (1) weak if the inhibition zones were smaller than 5 mm, (2) intermediate for a range of 5-10 mm, (3) strong for 10-20 mm, and (4) very strong if the inhibition zones were more than 20 mm in diameter [12]. Selected fungi were then characterized macroscopically and microscopically.

3. Results and Discussions

3.1. Endophytic fungi of macroalgae Chlorophyta and Phaeophyta

Thirteen morphologically distinct endophytic fungal strains were successfully isolated from algal tissues. Of these, 6 isolates were obtained with freshwater media (sample codes KHC0003, KHC0008 A, KHC0009, KHC0016, KHC0026 A, and KHCOO26 B) and 7 isolates with seawater media (KHC0008 B, C KHC0008, KHC0012, KHC0015 A, B KHC0015, KHC0016, and KHC0026 C). The endophytic fungi isolated from *Caulerpa* sp. were written wuth the sample codes KHC0008, KHC0016, KHC0026; one strain isolated from *Halimeda* sp. written as KHC0003, and the strains derived from *Sargassum* sp. labelled as KHC0009, KHC0012, and KHC0015. The morphology of the endophytic fungi isolated from *Caulerpa* sp., *Halimeda* sp. and *Sargassum* sp. is depicted in Table 1.
Based on their morphological characteristics, the isolated endophytic fungi were tentatively identified as *Aspergillus* sp. (KHC0016, KHC0026 A, and KHC0026 C), *Fusarium* sp.(KHC0026), *Penicillium* sp. (KHC0003, KHC0009, KHC0015 A, B; KHC0015, KHC0008 A, B; KHC0008, KHC0026 P). Another isolates could not been identified yet. This identification based on the fungal characteristics such as the shape of the conidiophores, the colony color (green or yellow), and their feathery texture with straight conidiophores [13].

IOP Conf. Series: Earth and Environmental Science **31** (2016) 012026 doi:10.1088/1755-1315/31/1/012026

Table1 Endophytic fungi isolated from *Caulerpa* sp., *Halimeda* sp. and *Sargassum* sp.

Isolate	
Freshwater medium	**Seawater medium**
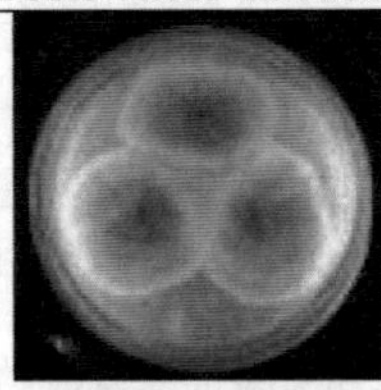 KHC0003	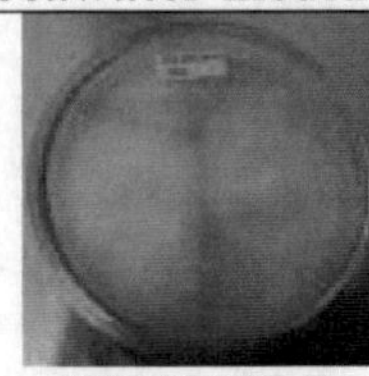 KHC0008 B
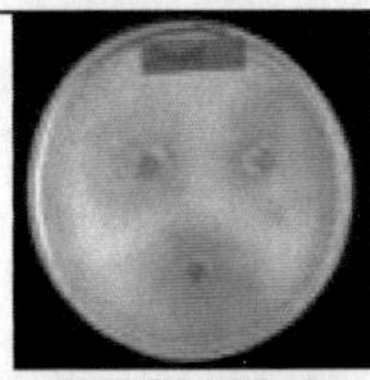 KHC0008 A	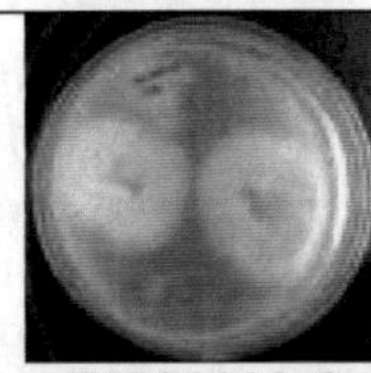 KHC0008 C
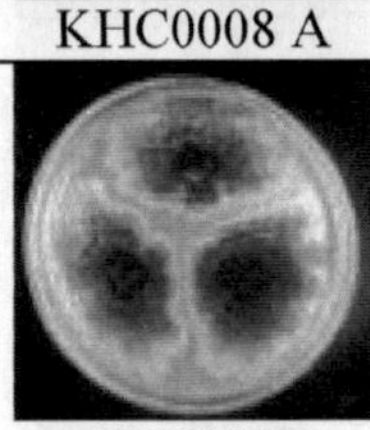 KHC0009	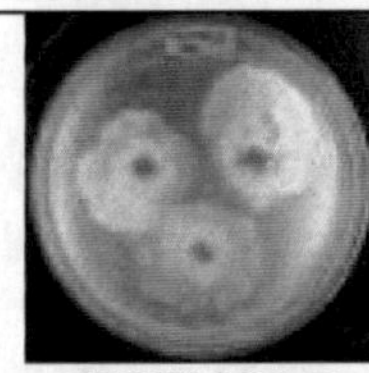 KHC0026 B
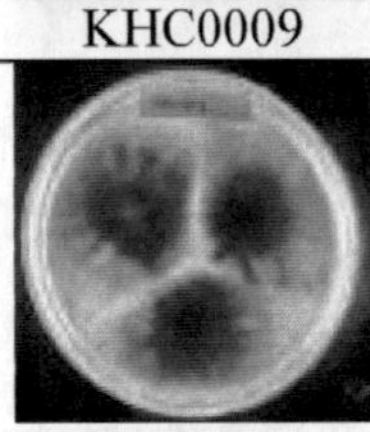 KHC0016	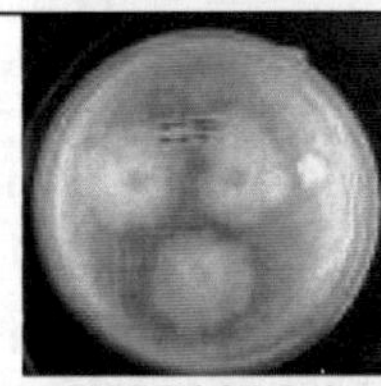 KHC0015 A
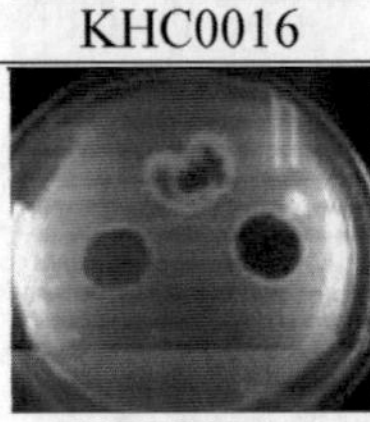 KHC0026 A	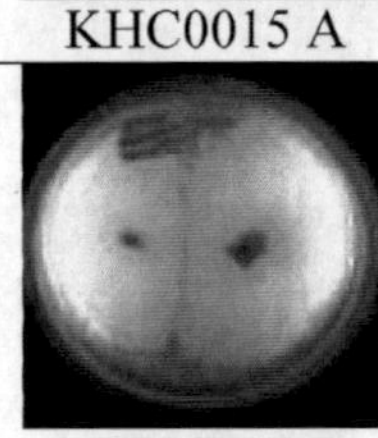 KHC0015 B
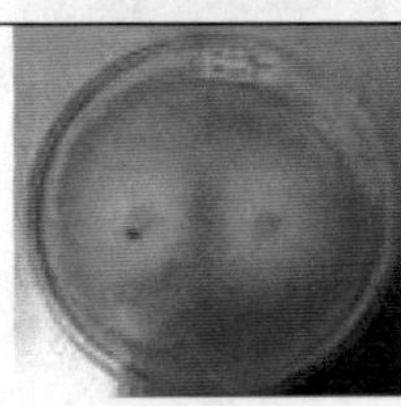 KHC0012	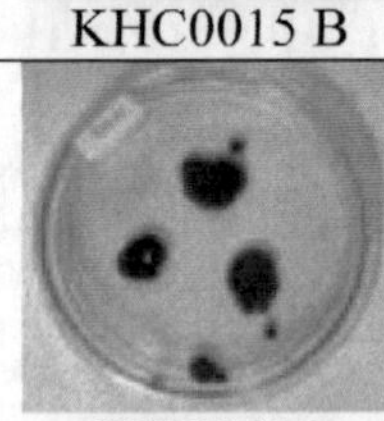 KHC0016
	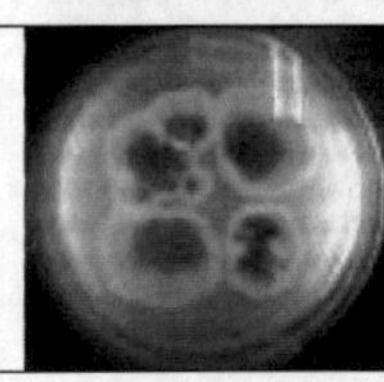

IOP Conf. Series: Earth and Environmental Science **31** (2016) 012026 doi:10.1088/1755-1315/31/1/012026

Isolate	
Freshwater medium	**Seawater medium**
	KHC0026 C

3.2. Antibacterial Activities of Endophytic Fungi

The screening of the endophytic fungi for antibacterial activity of their metabolites was performed by an antagonism test [14]. The screening results revealed antagonistic properties for endophytic fungi isolated from all three algal genera against the tested bacteria. The results are shown in Figure 3.

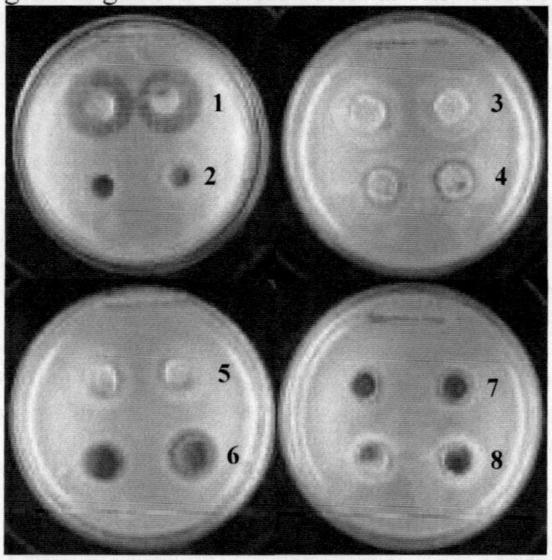

Figure 1 Antibacterial screening of endophytic KHC0003 (1), KHC0016 (2), KHC0008 A (3), KHC0012 (4), KHC0009 (5), KHC0015 A (6), KHC0026 B (7), KHC0015 B (8) against *Staphylococcus aureus*

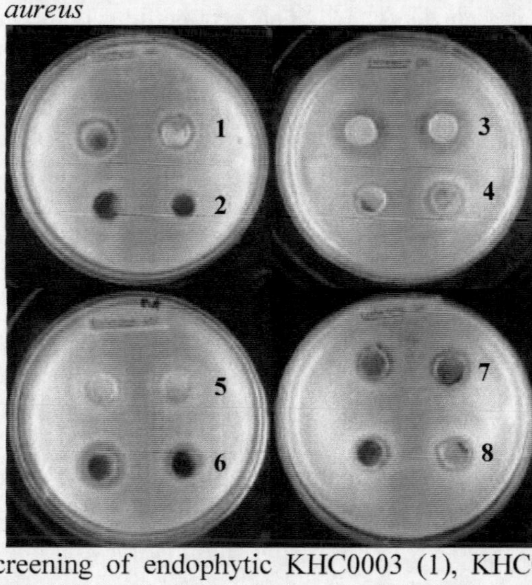

Figure 2 Antibacterial screening of endophytic KHC0003 (1), KHC0016 (2), KHC0008 A (3), KHC0012 (4), KHC0009 (5), KHC0015 A (6), KHC0026 B (7), KHC0015 B (8) against *Escherichia coli*

The results showed that fungal isolates KHC0003 and KHC0026 B had antibacterial activity against *S. aureus* and *E. coli.* Isolate KHC003 produced the diameter of inhibition zone of 11.0 ± 0.0 mm and

4.0 ± 0.0 mm against *S. aureus* and *E. coli*, respectively. For strain KHC0026 the diameter of the inhibition zone was 6.5 ± 0.7 mm and 4 ± 0.0 mm, respectively (Figure 3). It shows that both isolates inhibited the growth of *S. aureus* strongly and relatively weak in inhibiting *E. coli.*

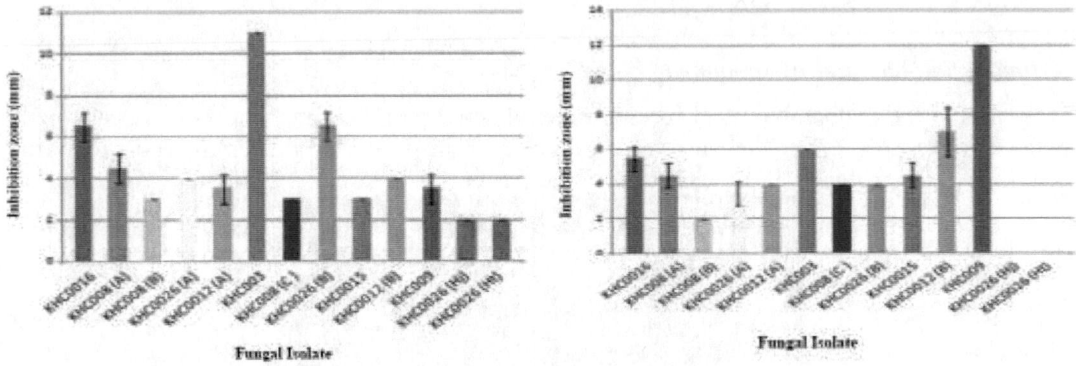

(A) (B)

Figure 3 Diameters of inhibition zones resulting from endophytic fungal isolates tested against *Staphylococcus aureus* (A) and *Escherichia coli* (B)

Extracellular metabolites produced by KHC0003 and KHC0026 B were suspected to be responsible for the ability of both fungi to inhibit bacteria. Observation of the inhibition zones after 20 hours incubation revealed that the fungal compounds killed the bacterial cells (bactericidal effect). The ability of the endophytic fungi to produce bioactive compounds is oftentimes beneficial to their hosts (i.e., the plants) by improving their resistance to diseases caused by microbial pathogens such as bacteria [15]. Antagonism mechanisms of fungi typically are: 1) mycoparasitism; i.e., the hyphae of antagonistic fungi twist or stick to the target microorganisms, and then penetrate the cell wall, entering the cell to take up the nutrients causing the microorganisms to die, 2) production of antibiotics that destroy microbial cells through the destruction of the target cell membrane structure, and 3) competition for living space and nutrients [16]. The extracellular compounds released by the fungal isolates KHC0003 and KHC0026 showed antibacterial activity without being extracted with inorganic solvents. This indicates that the usages of bioactive compounds may be used directly or simply extracted using water solvent. The use of water as a solvent keeps the bioactive compounds from the damage [17].

3.3. Characterization of the Selected Endophytic Fungi

The characterization of the selected endophytic fungal isolates was performed by macroscopic and microscopic observation. The results of fungal characterization are presented in Figures 4 and 5.

3.3.1. Characteristics of fungal isolate KHC0003

Fungal isolate KHC0003 cultivated on PDA medium produced pinkmycelia and orange for the reverse. The colonies had a velvety surface and whitish edges (Figure 4A). When the isolate was examined under a microscope at a magnification of 400x it showed long and septate hyphae, and rounded conidia (Figure 4B). Because of these features, the isolate was assumed to belong to the genus *Fusarium* [13].

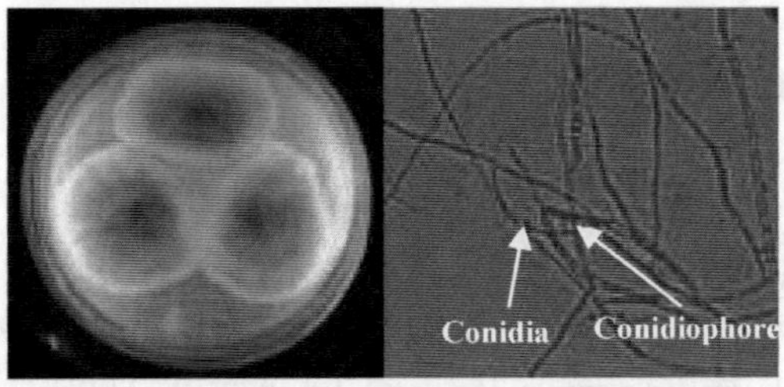

A B

Figure 4. Fungal isolate KHC0003; macroscopic (A) and microscopic (B) image

3.3.2. Fungal isolate KHC0026

Fungal isolate KHC0026 cultivated on PDA medium had white mycelia and the reverse, a thin mycelia surface, flat ledges and spreading hyphae (Figure 5A). Under the microscope at a magnification of 400x it showed long conidiophores and no septa and rounded conidia at the tips (Figure 5B). This isolate has not been identified yet.

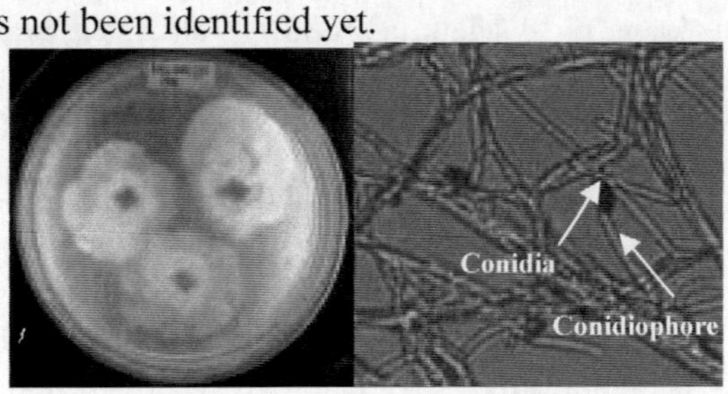

A B

Figure 5. Fungal isolate KHC0026; macroscopic (A) and microscopic (B) image

4. Conclusion

The endophytic fungi isolated from *Caulerpa* sp., *Halimeda* sp., and *Sargassum* sp. consisted of 13 morphologically different isolates. Isolates KHC0003 and KHC0026 were selected based on their antibacterial activity against *Staphylococcus aureus*. Based on their morphological features the isolates KHC0003 and KHC0026 were tentatively identified as *Fusarium* spp.

5. Acknowledgment

This study was financially supported by the Ministry of Research, Technology, and Higher Education through a grant within the funding scheme "International Research Collaboration and Scientific Publication" awarded to Dr. Kustiariyah Tarman.

6. References

[1] Phillips J. 2001. Marine macroalgal biodiversity hotspots: why is there high species richness and endemism in southern Australian marine benthic flora? *Biodivers Conserv*. **10**(9):1555–1577.

[2] Smith AJ. 2004. Medicinal and Pharmaceutical Uses of Seaweed Natural Products: A review. Journal of Applied Phycology. 16:245-262

[3] Dimara L, Yenusi TNB. 2011. Uji aktivitas antibakteri dan antioksidan ekstrak pigmen klorofil rumput laut *Caulerpa racemosa* (Forsskal). J.Agardh. 3(2): 53-58

[4] Pujol CA, Rayb S, Rayb B, Damote EB. 2012. Antiviral activity against dengue virus of diverse classes of algal sulfated polysaccharides. International Journal of Biological Macromolecules. 51:412– 416

[5] Purnama R, Melki, Wike AEP, Rozirwan. 2011. Potensi ekstrak rumput laut *Halimeda rencii* dan *Euchema cotonii* sebagai antibakteri *Vibrio* sp. Maspari Journal. 02: 82-88.

[6] Widiastuti HN. 2003. Uji aktivitas metabolit sekunder dari rumput laut *Halimeda macroloba* sebagai senyawa bioaktif antijamur *Candida albicans* dan *Torula histolutica*. Ilmu Kelautan. 8(2): 114-118.
[7] Vijayabakar P, Shiyamala V. 2011. Antibacterial Activities of Brown Marine Algae (*Sargassum wightii* and *Turbinaria ornata*) from The Gulf of Mannar Biosphere Reserve. Advances in Biological Research. 5(2): 99-102.
[8] Kumar SS, Kumar Y, Khan MSY, Anbu J, and De Clercq E. 2009. Antihistaminic, anticholinergic and antiviral activities of fucosterol from *Turbinaria conoides*. J. Agardh Kutzing. Pharmacologyonline. 1: 1104-1112.
[9] Sridharan MC, Dhamotharan R. 2012. Antibacterial Activity of marine brown alga *Turbinaria conoides*. Journal of Chemical and Pharmaceutical Research. 4(4): 2292-2294.
[10] Castro R, Zarrab I, Lamas J. 2004, Water-soluble Seaweed Extracts Modulate the *Pantoea agglomerans* lipopolysaccharide (LPS). Fish Shellfish Immunol. 10: 555–558.
[11] Tan RX, Zou WX. 2001. Endophytes: a rich source of functional metabolites. Natural Product Reports.18: 448-459.
[12] Harborne JB. 1987. Physiochemical Methods. 2nd ed. Padmawinata K, Soediro I (translator). Bandung: ITB.
[13] Gandjar I, Samson RA, Vermeulen, Oetari A, Santoso I. 1999. Pengenalan Kapang Tropik Umum. Jakarta (ID): Yayasan Obor Indonesia.
[14] Pereira E, Santos A, Reis F, Tavares RM, Baptista P, Lino-Neto T, Almeida-Aguiar C. 2013. A New Effective Assay to Detect Antimicrobial Activity of Filamentous Fungi. Microbiological Research. 168(1): 1-5.
[15] Lumyong S, Lumyong P, Hyde KD. 2004. Endophytes. In Jones, E. B. G., M. Tantichareon and K. D. Hyde (Ed.), Thai Fungal Diversity. BIOTEC Thailand and Biodiversity Research and Training Program (BRTI/TRF. Biotec). 197–212.
[16] Schubert M, Siegfried F, dan Francis WMRS. 2008. *In Vitro* Screening of an Antagonistic *Trichoderma* Strains Against Wood Decay Fungi. Arboricultural Journal. 31:227–248.
[17] Garmus TT, Paviani LC, Queiroga CL, Magalhães PM, Cabral FA. Extraction of phenolic compounds from pitanga (*Eugenia uniflora* L.) leaves by sequential extraction in fixed bed extractor using supercritical CO_2, ethanol and water as solvents. 2014. Journal of Supercritical Fluids, 86:4-14.

IOP Conf. Series: Earth and Environmental Science **31** (2016) 012027 doi:10.1088/1755-1315/31/1/012027

Changes in water quality and trophic status associated with cage aquaculture in Lake Maninjau, Indonesia

C Henny[1] and S Nomosatryo
Research Center for Limnology, Indonesian Institute of Sciences
Cibinong Science Center, Jl. Raya Bogor km 46, Cibinong 16911, Indonesia

E-mail: cynthia@limnologi.lipi.go.id

Abstract. The cage aquaculture unquestionably has been degrading lake water quality by increasing nutrients and organic carbon in lake water and sediments. The question is to what extend this condition affects other key indictors such as the temporal changes in trophic status and the thickness of anoxic hypolimnion layer where the anoxic water column is moving upward pushing up the oxic epilimnion layer. The condition in Lake Maninjau could be worse since the lake is steadily producing sulfide which can cause not only oxygen depletion in the water column but also the phosphate release from the sediments. The study is based on the long term monitoring data from on going research for about 8 years observation. The results indeed show the anoxic water column is moving upward increasing the thickness of anoxic hypolimnion layer and decreasing epilimnion layer from 30 m to 10 m depth. The trophic status of the lake also has changed from mesotrophic to eutrophic decreasing the water transparency to even a critical level < 1m. The months of July to September with prolonged hot season could be the critical time for trophic condition for the lake. The results suggest that determination of these conditions further could help identify and predict the critical time for possibility of fish kill.

1. Introduction

Cage aquaculture has been a major pressure to Lake Maninjau contributing major nutrients especially phosphorus (P) and organic carbon loading [1,2]. The main source of nutrient inputs in Lake Maninjau has known from the cage aquaculture counted around 400 ton/year. Ninety five percents of nutrient waste load as nitrogen and phosphorus that enter the lake were from cage aquaculture [3]. Increase P and organic carbon inputs has degraded water quality and reduced oxygen level in the water column [2,4].

The eutrophication problem and indication of toxic cyanobacteria (*Microcystis*) blooming also has beeen reported by previous study [2,5,6,7]. The fact that the lake is a tectono-volcanic containing fair amount of sulfur, the organic carbon inputs has increased hydrogen sulfide production in the hypolimnion subsequently could trigger the phosphorus release from sediment [4,8]. These findings indicate that the source of phosphorus not only from the external but also from the internal suggesting that P is not subject to limiting factor for the algal growth. Excessive algal biomass also contributes to more nutrient and organic carbon inputs to the lake. Furthermore combined aerobic and anaerobic microbial degradation and sulfate reduction bacteria (SRB) activity could cause the oxygen depletion in lake water column.

[1] To whom any correspondence should be addressed.

The eutrophication and degrading water quality related to the oxygen depletion has also adversely affected the fish. Frequent fish kill incident due to degrading lake water quality has been reported each year [3].

The cage aquaculture unquestionably has impacted on lake water quality by enriching the lake with nutrients and organic carbon in the water column and by accumulating them in the sediment [2,7,9]. To what extend the excess of nutrients and carbon sources in the lake could affect the temporal changes of the trophic status and the thickness of anoxic hypolimnion layer where the anoxic water column is moving upward pushing up the epilimnion layer. As mostly on the deep tropical lakes, the stratification is permanent and the hypolimnion is permanently anoxic. The hypolimnetic hypoxia is common however if the magnitude of hypolimnetic hypoxia increases and the anoxic water column close to the water surface, with increasing the surface lake mixing by strong wind could initiate the upwelling that turns over the anoxic water column to the surface causing the entirely lake anoxic. Massive fish kill reported were mostly driven by this phenomena [3,10].

Although many studies have been done on the impact of cage culture on lake water quality in Lake Maninjau, trophic status even phytoplankton community but the studies were merely based on the short-term research [1,2,5,6]. We have reported the same type of study on nutrients dynamics from the long-term observation until 2011 however we did not discuss on the temporal changes on trophic status [2]. The study objectives are to evaluate the changes in water quality related to oxic-anoxic layer of water column and eutrophication and to observe and evaluate temporal changes of trophic status and secchi depth and to determine the key indicators for critical lake condition to develop possible early warning system. This study is based on the long-term monitoring data from on going research for about 8 years observation from 2005 to 2013.

2. Materials and Methods

2.1. Sampling Location

Lake Maninjau has surface area of 9737.5 Ha and maximum depth of 165m. The lake is located at 461.50 m above sea level in the Western area of Sumatra, Indonesia. The Lake Maninjau has several inlets of small streams and hot springs from the mountains. The lake has one outlet divided into a small surface outlet and the outflow from the hydropower turbine. Sampling station is located at S 0°22'33.0", E 100°11'35.1" which is a master station of the deepest area of the lake (Figure 1).

There were about 18630 units of floating net cage in all around the lake in 2013 reported in Lake Maninjau [1]. Water sampling was conducted at least 1 to 3 times each year during the day time. Deep profile water samples were collected by using 2 L Van Dorn water sampler. Water samples were transferred to HDPE bottles. All samples were acidified or preserved except the samples for sulfide analyses. Water samples for sulfide analyses were drawn by a syringe via Tygon tubing to maintain the reduced condition.

2.2. Physicochemical analyses

Physico-chemical characteristics such as temperature and pH was measured by using Water Quality Checker Horiba U-10, while dissolved oxygen (DO) measurement was done by using YSI 6000. Water transparency as secchi depth was measured by using a secchi disc. Water samples for total sulfide concentration were analyzed directly insitu by adding sulfide reagents to the water samples then measured by using HACH DR 2010 Spectrophotometer. Total Nitrogen (TN) and Phosphorus compound (TP) were analyzed by a colorimetric method using Spectrofotometer Shimadzu UV 2100 according to APPHA AWWA [11]. Prior to analyses, water samples for Chlorophyll-a measurement were added with saturated $MgCO_3$ then filtered by using Whatmann GF/F filter paper. The filtrates then were extracted using 90% aceton, subsequently measured at different wavelength using Spectrofotometer Shimadzu UV 2100 [11].

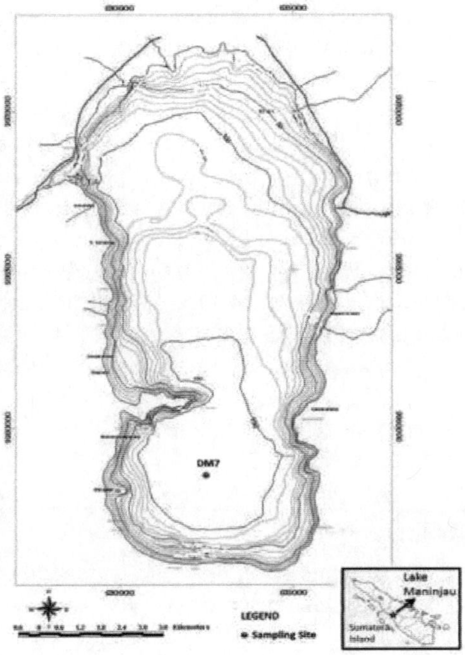

Figure1. Lake Maninjau and the master station.

2.3. Data Analyses

Trophic state index (TSI) of each lake was calculated based on chlorophyll-a, TP and secchi depth. The average of TSI scores is used to classify the trophic status of the lakes using the following equations from Carlson and Simpson [12].

$$\text{TSI(Chl)} = 9.81 \text{ LN [chlorophyll-a } (\mu g/L)] + 30.6 \quad (1)$$
$$\text{TSI(P)} = 14.42 \text{ LN [TP } (\mu g/L)] + 4.15 \quad (2)$$
$$\text{TSI(SD)} = 60 - 14.41 \text{ LN [secchi depth (m)]} \quad (3)$$
$$\text{TSI(Ave)} = [\text{TSI (Chl)} + \text{TSI(P)} + \text{TSI(SD)}]/3 \quad (4)$$

Optional TSI for total nitrogen was calculated using the following equation from Kratzer and Brezonik [13].

$$\text{TSI(N)} = 54.45 + \text{LN [TN (mg/L)]} + 4.15 \quad (5)$$

The trend of temporal changes of trophic status and secchi depth data are presented in simple radar chart. Simple correlation analysis is conducted using statistical analyses by SYSTAT.

3. Results and Discussion

3.1. Temperature and pH

Annual temperature and pH deep profiles of Lake Maninjau show slight change annually (Figure 2). Our previous study has shown the thermocline shifting slightly in 2009 and 2011 to the upper water column [2], but in 2013 it shifted back to lower water column at 10 – 20 m depth. The fluctuation of the annual temperatures are mostly influenced by climate variability. The lake water temperaturas are usually warmer in the hot season than those in the rainy season. As most of the deep lakes in tropical area, Lake Maninjau also experiences thermal stratification especially in the deepest area and the bottom water temperature is constantly colder (26.5 – 27 °C) than the temperature of the surface water

(28 – 30 °C). Similar pattern for annual pH profiles are observed where the chemocline layers also moved upward to the upper water column in 2011. Although the annual chemocline layer seems to fluctuate slightly the layer still falls within the range at 10 – 20 m depth. The chemocline pattern also likely is influenced by the climate variability. In the contrary to the temperature values, pH values of surface water gradually increase (9 – 11) and the bottom water pH values decrease (7-6) from 2005 to the recent years. High values of pH in the surface water indicate high rate of photosynthesis suggesting that the lake experiences eutrophication. Very high pH value of surface water was observed in 2008, 2009 and 2011. The water color was dark green indicating lake eutrophication. Green scum indicating the presence of toxic cyanobacteria (Microcystis) was also spotted in several areas. The same observation has been reported in previous studies [5, 6, 7]. Prolonged hot season stimulates the photosynthesis process which subsequently increses the water pH [14]. Lower pH values in the bottom water are usually resulted from the microbial degradation of complex organic carbon to simple organic carbon such us fatty acids and CO_2 [15]. Previous study by Lukman et al [1] and Henny [8] observed the increase in COD concentrations in the water column and volatile suspended solids in the sediments in Lake Maninjau by cage aquaculture activity.

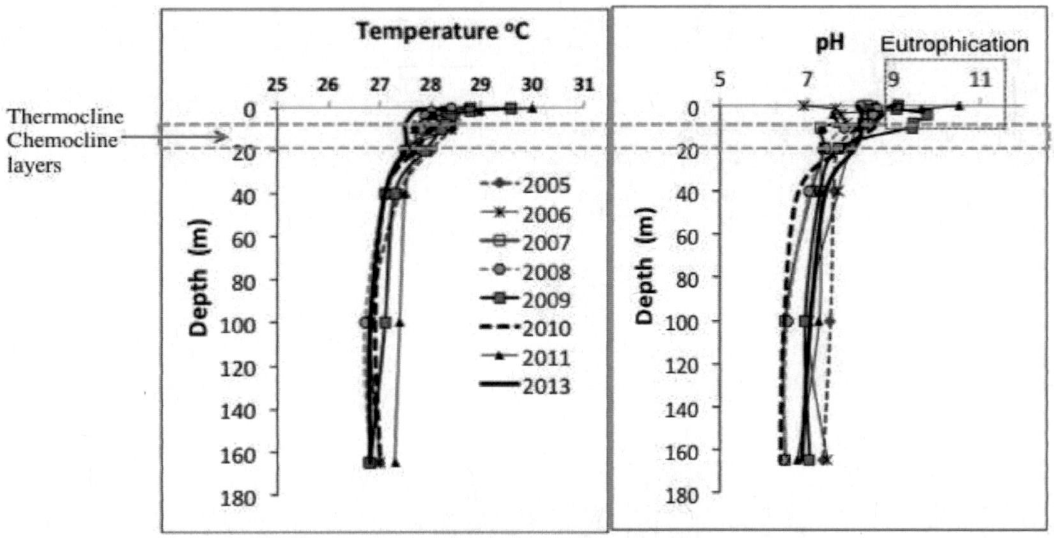

Figure 2. Deep profiles of temperature and pH

3.2. Dissolved oxygen (DO) and total sulfide

Sedimentation of feces and fish pellets from the cage aquaculture and algae drives DO depletion in the hypolimnion lakes by stimulating bacterial respiration and sulfide production from sulfate reducing bacteria (SRB) activity. DO and sulfide deep profiles show that the thickness of anoxic hypolimnion increases pushing up the oxic epilimnion (Figure 3). The results suggest that the anoxic water column is moving close to the surface water. The fluctuation is at the range of 10 - < 20 m. The anoxic line fell at depth of 30 m in 2005 and then moved upwards to at depth of 20 m in 2007 and even moved upward further to 8 m depth in 2011 and bounced back to at depth of 15-18 m in 2013. When we compare to annual sulfide profiles the same trends observed and the sulfide concentrations varied with the highest sulfide total concentrations of > 400 μg/L in 2009, however sulfide level is getting higher in the water column at depth of 20 m observed in the recent years. What the results suggest that fluctuation of oxic-anoxic layer is not only caused by sulfide diffusion or production itself but rather more due to organic carbon degradation by aerobic/facultative bacteria which consume up all the oxygen presence in that layer. Fermentative bacteria subsequently also degrade the excess organic

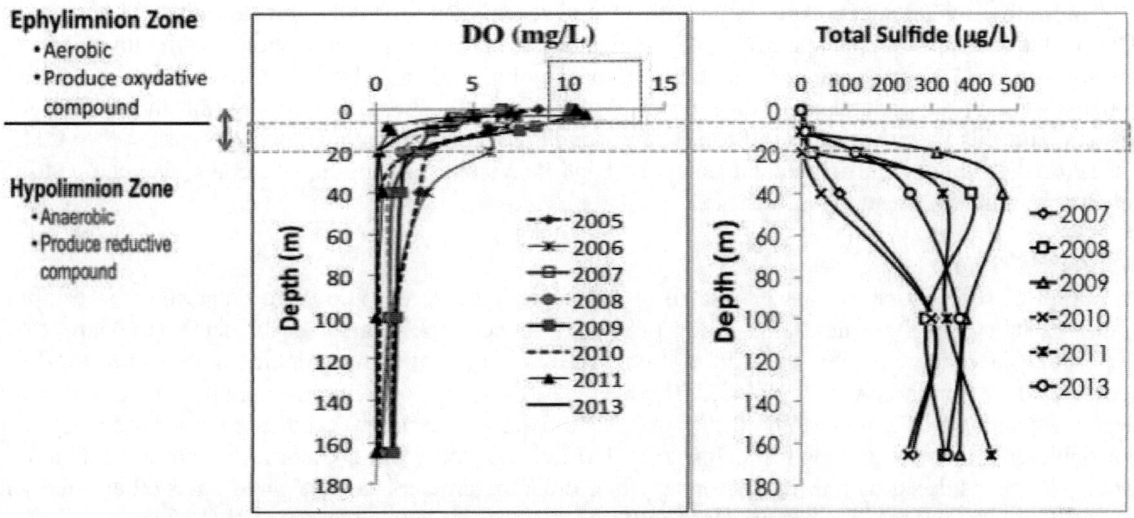

Figure 3. Deep profiles of dissolved oxygen (DO) and total sulfide

carbon producing more fatty acids and CO^2 which are subsequently utilized by other group bacteria such as denitrifiers, sulfate reduction bacteria (SRB), iron reducers and methanogenic bacteria. Excess organic carbon would trigger more microbial activity in the lake that can lead to oxygen deficit and elevated sulfide concentrations in lake. When the concentration of organic carbon exceeds the concentration of oxygen required, the oxygen depletion will occur [14,15]. DO and sulfide concentrations are in agreement indicating that when sulfide is presence DO is zero and vice versa. The annual changes of thermocline (temperature gradient), chemocline (pH gradient) and oxycline (oxygen gradient) are about the same pattern. These also suggest that most of mineralization and transformation of organics, nutrients and certain elements by microbial and chemical process both aerobically and anaerobically in Lake Maninjau occurs at this depth layer of 10 - < 20 m especially in the borderline of oxic –anoxic layer

Similar case with pH, DO values of surface water increased up to 9 - 11 mg/L in 2008 – 2011 indicating the lake eutrophication. Photosynthetic process during daytime will consume CO_2 and produce O_2 in the water subsequently increase pH and DO in the water [14]. Both pH >9 and DO >9 mg/L could be used as indicators for lake critical condition.

3.3. Trends of total nitrogen (TN), total phosphorus (TP), chlorophyll-a and water transparency

Trends of total nitrogen (TN), total phosphorus (TP), chlorophyll-a and water transparency for most area of the lake has been reported by our previous study [2]. More recent study has been reported on these parameters in the cage aquaculture area [1]. This study only shows the trends at the lake surface water of the deepest area. The results certainly show more elevated concentration of TN, TP and chlorophyll-a in the area near the fish cages when compared to those in the middle and the deepest area where the fish cages are more distance.

This study aims to show that even the water quality in the area far distance from the fish cages has been affected by the cage aquaculture activity. Due to lake stratification the observation in the deepest area can show more distinct changes in lake water quality.

Annual TN concentrations increase significantly from 2008 and so does the chlorophyll-a (Figure 5). The annual TP concentrations also increase gradually while the water transparency as secchi depth is in decrease trend. The results are in agreement with other study where the growing number of fish cage units every year indicates more nutrient inputs to the lake [1,3]. The water quality condition really depends on the climate variability during sampling time. More rainfall can dilute the pollutants and wash out the algae and lack of sunlight also limits the algal growth resulting in less photosynthesis process.

Chlorophyill-a has strong correlation with TP and secchi depth (r> 0.85) but less with TN (r =0.68). The results indicate that phosphorus is key indicator causing lake eutrophication and the lower water transparency is majorly caused by high chlorophyll-a and not by elevated suspended solids concentration. TN is abundant in lake water and it is usually phosphorus is limiting factor to algal growth. But this is not the case in Lake Maninjau. TP concentrations are > 50 μg/L in 2008 to 2013 exceeded the value of eutrophic condition [12]. High P concentration in the lake could also be due to P release from the sediment by sulfide [4].

3.4. Trophic status

Although eutrophication is closely related to chlorophyll-a concentrations, to determine the trophic status of lake using trophic status index (TSI) TP concentrations and secchi depth (SD) are also important. These three parameters are the key indicators to determine the trophic status. When the TSI scores of the three indicators are varied, the average TSI scores of TSI-chlorophyll-a, TP and secchi depth is usually used to determine the trophic status. TN is not considered as a strong indicator for eutrophication and not included in calculating TSI average score but sometimes it can be considered when N is considered a limiting factor or when other parameters do not agree each other such as condition when the lake also has problem with invasive macrophyte coverage [13]. TSI scores of 40 – 50 classify the mesotrophic condition, 50 – 70 classify the eutrophic condition and above 70 classify the hypereutrophic condition [12].

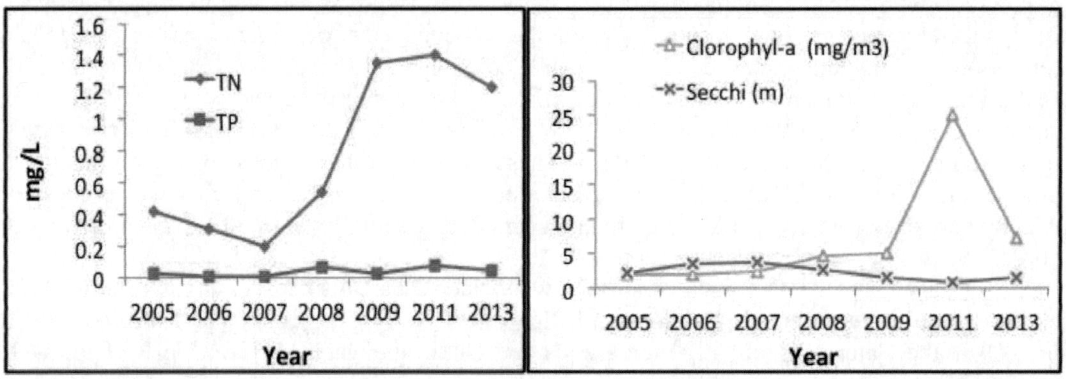

Figure 5. Trends of TN, TP, chlorophyll-a concentrations and water transparency as secchi depth

Based on the TSI scores for each indicator (chlorphyll-a, TP and SD) the trophic status of the lake could range from oligotrophic to hypereutrophic in which does not reflect the real condition. In this case the best predictor is using the average TSI scores of those three parameters. The trophic status of Lake Maninjau changes from mesotrophic in period of 2005-2007 to eutrophic in period of 2008 to 2013 based on the average TSI scores (Figure 6).

The temporal changes of trophic status of lake from 2005 to 2007 did not show any trend but from 2008 TSI scores for all three indicators and the average of three indicators show variability in which the months of July to September even up to December indicate the critical trophic condition (Figure 6). July 2011 was the record high for TSI score indicates highly eutrophic condition even to hypereutrophic level. The surface water pH and DO concentrations reflect the lake condition that year. Temporal changes of trophic status would likely be affected more by the climate variability such as monthly rainfall trend in a year. In the hot season the lake would likely experience more eutrophic condition than that in the rainy season. Rainfall will not dilute the pollutants and wash out the algae but also limit the algal growth due to less sunlight available.

Based on the TSI cores for each variable it shows that TSI(P)>TSI(chl)=TSI(SD). The results suggest that the zooplankton could probably feed more phytoplankton but still enough to cause the water turbid. The water turbid in Lake Maninjau is mostly caused by high chlorophyll-a concentration.

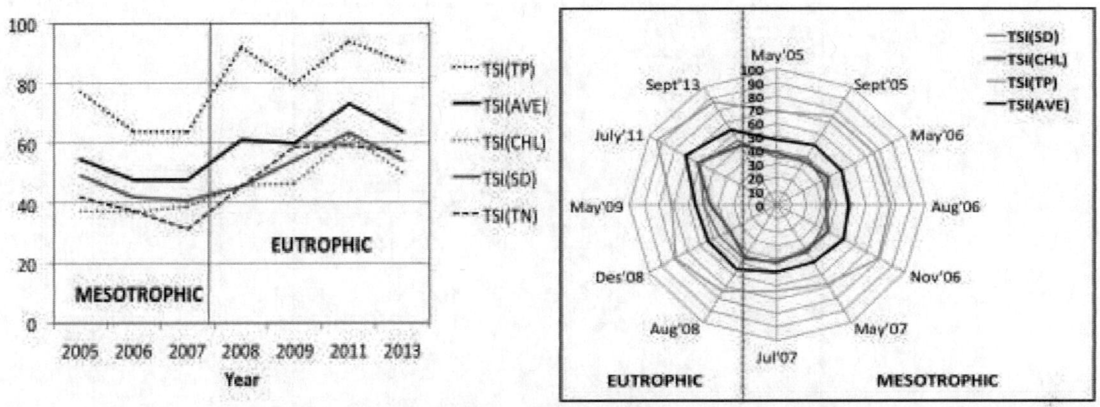

Figure 6. Annual and temporal changes of trophic status

3.5. Implication of indicators

This study also tries to indicate what the implication of indicators that can be used to determine the critical time of lake condition to help predicting when the incident of fish kill would likely to happen. The water quality of lake Maninjau is clearly degrading with the eutrophication problem and the anoxic water column layer is moving upward getting close to the surface. Accumulation of the nutrients and organic waste continuously triggers more microbial activity in the hypolimnion producing more sulfide and other toxic gas leading to oxygen depletion in the upper layer of hypolimnion. Both internal and external loading of phosphorus has prolonged Lake Maninjau in condition of eutrophic state. In long hot weather the algal bloom sometimes accompanied by the tocic algae of *Microcystis* which is also toxic to fish. Population of blue green algae counted more tan 40 up to 60% in almost all areas especially near the fish cages [6]. Prolonged to poor wtaer quality such as exposed to lower oxygen could effect ecosystem structure in lake [16]

The number of fish cage unit increase by double from 2005 to 2008. Poor water quality observed in 2008 to a low level leading to the incident of massive fish kill in February 2009. The incident of fish kill occurs more frequent every year and it happens 2 to 3 times a year [3]. Both eutrophication and oxygen depletion are the conditions that have lead to fish kill incident. Massive fish kill incident usually is also driven by strong wind that causes lake mixing and turns over the anoxic water column layer to the surface, but in a small case, fish kill usually occurs prior to highly eutrophic condition in the previous months. During cloudy or rainy condition with less sunlight during the day time the photosynthesis rate is lower and the respiration rate is higher at night subsequently the lake would likely experience the oxygen depletion and sulfide will diffuse up to the surface with faster rate taking up all the oxygen available and lead to fish kill incident.

It seems all understand what the cause of fish kill incident however no one knows when to take precautious prior to the incident. With climate variability sometimes the prediction can be obscure. However there is an indicator that can be used to develop early warning system. Online monitoring system also has been initiated by other study [17], however with the lake and weather conditions this system is still not working properly. Our evaluation of temporal changes of water transparency as secchi depth in 8 years clearly indicates that during hot season such as in July-September the lake water was more turbid than those in the early hot season such as in May (Figure 7). Monthly variability in secchi depth values also reflects on the condition during sampling. It is highlighted that there is strong correlation between cholorphyll-a concentration and secchi depth suggesting that the secchi depth could be used as the key indicator for early warning system of lake critical condition.

There were no fish kill incident reported in 2005 – 2008 when the lake water condition improved with secchi depth above 5 m. When the secchi depth decreased to < 3m in August till December 2008 in the following month of February 2009 the massive fish kill incident occurred.

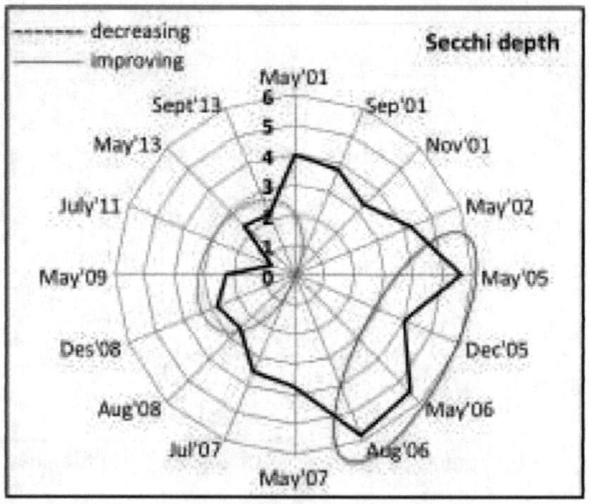

Figure 7. Temporal changes of water transparency as secchi depth

The eutrophication was also observed first in 2008-2009 with the worst condition observed was in 2011. The lake water turned to dark green containing high chlorophyll-a concentration with obvious presence of filamentous blue green algae of *Microcys* and with very low secchi depth reaching only 0.8 m. Fish kill incident has occurred almost every year since 2009. The incident was the biggest in 2009 and moderate in 2010-2012 and small in 2013 [3]. The pattern of fish kill incident seems to correlate with the prior conditions of lake water quality and the secchi depth. The water condition with secchi depth of <3m could likely still lead to fish kill incident. However this would need further study on monthly changes of trophic status and secchi depth and the fluctuation of oxic-anoxic layer depth. More thoroughly study on biogeochemical process in this oxic-anoxic layer could be needed to understand better the natural phenomenon in lake Maninjau.

4. Conclusions

The unsustainable cage aquaculture practice in Lake Maninjau unquestionably has been pressuring the lake by increasing the thickness of anoxic hypolimnion thereby moving upward the anoxic water column to the upper column and pushing up the epilimnion further up. The results suggest that although lake hypolimnion continuously generates sulfides, the anoxic layer moving upwards to the upper water column might also as results of other microbial process although it needs further investigation. Steadily increase in phosphorus concentrations in lake water suggests that the phosphorus is not a limiting factor for algal growth. The lake trophic condition has changed from mesotrophic to eutrophic after 2008 up to recent years. The secchi depth at below 3 m and pH>9 and DO> 9 mg/L during daytime could be used as early warning system for lake critical condition. The months of July to September with prolonged hot season could be the critical time for trophic condition for the lake. The results suggest that determination of these conditions further could help identify and predict the critical time for possibility of fish kill. However monthly temporal changes in trophic status and water transparency needs further observation to know when the peak of algal blooming and the lowest transparency and what duration of prolonged eutrophic condition coupled with weather condition would likely to cause the fish kill incident and subsequently early warning system for the lake can be developed.

Acknowledgement
The research has been funded by the Indonesian Government. Author would like to acknowledge Triyanto and Tri Suryono as the coordinator project of Lake Maninjau Research in 2005 to 2010 and the lab staffs of Research Lab of Limnology-LIPI who helped with the samples analyses, and Sutrisno, the staff of Lake Maninjau Research Station, who assisted with the field sampling.

References

[1] Lukman, Setyobudiandi I, Muchsin I and Hariyadi S 2015 Impact of Cage Aquaculture on Water Quality Condition in Lake Maninjau, West Sumatera Indonesia *International Journal of Sciences: Basic and Applied Research (IJSBAR,)* **23**(1), 120-137
[2] Henny, C and Nomosatryo S 2012 Nutrient Dynamicsin Lake Maninjau: Implication of fish cage fishery, *The Tenth International Symposium on South East Asia Water Environment,* November 7 – 11, Hanoi, Vietnam
[3] Henny C 2014 Ecotechnology and ecosystem based management tools as alternative solutions to reduce risks of ecosystem disaster and fish kill in Lake Maninjau, *Proc. Nat. Sem. Limnology* **8**, *September 16,* Cibinong
[4] Henny C and Nomosatryo S 2012 Sufide dynamics in Lake Maninjau: Implication on phosphate release hypolimnion, *LIMNOTEK,* **19**(1) 102-112
[5] Sulawesty F, Sutrisno, Hamdani A and Triyanto 2011 Kondisi Kualitas Air Beberapa daerah Pemeliharaan Ikan Karamba Karing Apung di Danau Maninjau, *LIMNOTEK,* **18**(1), 38-47
[6] Sulastri, Hartoto D I and Yuniarti I 2012 Environmental Condition, Fish Resources and Management of Maninjau Lake of West Sumatera, *Ind. Fish Res. J.* **18** (1*)* 130-145
[7] Sulastri 2001 *Lake Maninjau: Problems and Solutions*, Research Center for Limnology, Indonesian of Institute of Sciences, Cibinong, 70 p
[8] Henny C 2009 Dynamics of Biogeochemistry of Sulfur in Lake Maninjau, *LIMNOTEK,***16**(2), 34-35
[9] Garg J and Garg H K 2002 Nutrient loading and its consequences in a lake ecosystem, *Tropical Ecology* 43(2), 355-358
[10] Apip, Fakhrudin M, Sulastri, Subehi L and Ridwansyah I 2003 Assessing climate variables in physical process in Lake Maninjau, *LIMNOTEK,* **10**(1), 1-10
[11] APHA 2005 *Standard methos for the examination of water ans wastewater.* Washington D.C.
[12] Carlson R E and Simpson J 1996 *A coordinator's guide to volunteer lake monitoring methods,* North American Lake Management Society, 96p
[13] Kratzer C R and Brezonik P L 1981 A Carlson-type trophic state index for nitrogen in Florida lakes, *Wat. Res. Bull.* **17**,713-715
[14] Wetzel R G 2001 *Limnology: River and Lake Ecology*, Academic Press, San Diego.1006p.
[15] Widdel F 1988 *Microbiology and Ecology of Sulfate- and Sulfur-ReducingBacteria, In:* Zehnder A J B (ed.), Biology of Anaerobic Microorganisms. John Wiley & Sons, New York. 469-586
[16] Hackenson L 2005 Changes to lake ecosystem structure resulting from fish cage farm emissions, *Lakes and Reservoir Management,* **10**, 71-80
[17] Fachrudin M, Satya I A, Chrismada T, Hamid A, Foni A, Endra T and Kodarsyah 2012 Development of early warning system online monitoring system for environmental disaster: Case study of Lake Maninjau, *Proc. Nat. Sem. Limnology* **7**, 49-64

IOP Conf. Series: Earth and Environmental Science **31** (2016) 012028 doi:10.1088/1755-1315/31/1/012028

Activity and Stability of Biofilm Uricase of *Lactobacillus plantarum* for Uric Acid Biosensor

Dyah Iswantini*[1], Rescy Rachmatia [1], Novita Rose Diana [1], Novik Nurhidayat[2], Akhiruddin[3] and Deden Saprudin[1]

[1]Department of Chemistry, Faculty of Mathematics and Natural Sciences, Bogor Agricultural University, Bogor, 16680, Indonesia.
[2] Division of Microbiology R&D for Biology, The Indonesian Institute of Sciences, Bogor, Indonesia
[3]Department of Physics, Faculty of Mathematics and Natural Sciences, Bogor Agricultural University, Bogor, 16680, Indonesia.

E-mail : dyahprado@gmail.com

Abstract. Research of uric acid biosensor used a *Lactobacillus plantarum* was successfully conducted. *Lactobacillus plantarum* could produce uricase that could be used as uric acid biosensor. Therefore, lifetime of bacteria were quite short that caused the bacteria could not detect uric acid for a long time. To avoid this problem, development of biofilm for uric acid biosensor is important. Biofilms is a structured community of bacterial cells, stick together and are able to maintain a bacteria in an extreme environments. The purpose of present study was to determine and compare the activity of uricase produced by *L.plantarum,* deposited whithin biofilm and planktonic bacteria on glassy carbon electrode (GCEb & GCE), also to determine the stability of biofilm. The optimization process was conducted by using temperature, pH, and substrate concentration as the parameters. It showed that the activity of uricase within biofilm was able to increase the oxidation current. GCEb and GCE yielded the oxidation current in the amount of 47.24 µA and 23.04 µA, respectively, under the same condition. Results indicated that the optimum condition for uric acid biosensor using biofilm were pH 10, temperature of 40 °C , and uric acid concentration of 5 mM. The stability of GCEb decreased after 10 hours used, with decreasing percentage over 86.33%. This low stability probably caused by the unprotected active site of the enzyme that the enzyme is easier to experience the denaturation.

1. Introduction

Uric acid (2,6,8-trihidroxypurine) is the final product of purine metabolism in human body. The high concentration of uric acid in the blood caused hypertension and cardiovascular deseases [1]. Therefore, uric acid detection method is very important to develop. One method that is commonly and widely used to determine uric acid level is spectrophotometry method, but Spectrophotometry method has disadvantage such as: expensive, low sensitivity, and unstable to the light. Biosensor is an analytical tool to detect specific analyte, combining a physicochemical detector component and biological components [2].

Biosensor is a device that uses specific biochemical reactions mediated by isolated enzymes, immunosystems, tissues, organelles or whole cells to detect chemical compounds usually by electrical, termal or optical signal [3]. One of the enzymes used as a biological component in uric acid biosensor development is uricase. Uricase (urate oxidase) is an enzyme that participates in the purine breakdown pathway, catalyzing the oxidation of uric acid to allantoin and hydorgen peroxide in the presence of oxygen [4]. Generally, uric acid can be isolated from vertebrate animals. However, due to the complicated isolation procedure, limited materials and also high costs, then used alternatives sources such as molds, yeasts, and bacterias.[5].

To overcome the problem, research related to uric acid biosensor used *Lactobacillus plantarum* as uricase source was successfully conducted. *Lactobacillus plantarum* were known can produce enzymes, among of them are uricase for uric acid biosensor, pyruvate oxidase for pyruvate biosensor and lactic oxidase for lactic acid biosensor [6]. Lifetime of bacteria is quite short that makes uricase activity is not long enough to detect an uric acid. Biofilms are defined as communities of bacteria encased in a self-synthesized extracellular polymeric matrix that attaches to a biotic or abiotic surface [7]. This formation of biofilm can maintain the bacteria to survive in extreme environments, able to survive against an antibiotics and disinfectants. Jalilsood *et al* (2015) had researched that *L. plantarum* PA21 is a very potent biofilm produces provided specific local micro-enviroments which were favorable to some pathogen or spoilage microorganisms [8]. The aim of present study was to determine and compare the activity of uricase produced by *L.plantarum,* deposited within biofilm and planktonic bacteria on glassy carbon electrode (GCEb & GCE), also to determine the stability of biofilm. The ability of biofilm to detect the uric acid concentration also be measured.

2. Materials and Methods

Reagents

The uric acid used in this work is from nacalaiq tesque, Japan, $K_3Fe(CN)_6$ is from Merck, German, and the 3-dymetoxy-5-methyl-1,4-benzoquinnon (Q_0) is from Sigma, German.

Microorganism

The *Lactobacillus plantarum* strain used in this study was previously isolated and collected in microbiology laboratorium of LIPI Cibinong, Bogor, Indonesia.

Preparation of Cultures

The *L. plantarum* cells were cultured on Glucose Yeast Peptone (GYP) agar plates and were incubated in 37°C for 24 hours. Then *L. plantarum* cells were subcultured in 5 ml liquid GYP and were incubated in 37°C for 24 hours and stored as the starter. The starter was then mixed with the liquid GYP and incubated for 4 hours to reach the Optical Density 600 (OD_{600}) of 0,5-0,6. They were harvested by centrifugation at 10.000 rpm for 10 minutes at 4°C. The pellets were washed with physiological salt and were centrifugated in the same condition. Supernatant was removed and the boric buffer was added into the pellets and separated with the vortex.

The Formation of Biofilm

The biofilm was formed by using microtiter plate method with several modifications. The *L. plantarum* was inoculated onto GYP and was incubated in the shaker incubator at 37°C over night. The adsorbent measurement was conducted by using spectrophotometer to observe the amount of bacteria used. Then, the bacteria suspension was diluted in the GYP until reached the concentration of 10^6. The diluted cultur was inoculated into the microtiter plate with the liquid GYP as the negative control, then they were incubated for 72 hours. After that, they were washed with aquadest to remove the planktonic bacteria and were dried and colored. The coloration used violet crystal 1 % then was incubated in the room temperature for 15 minutes. After being washed and dried, the boric buffer was added. Then the absorbance was read by using microplate reader in the wavelength of 595 nm.

The Construction of Carbon Paste Electrode

Carbon paste electrode was constructed by mixing paraffin oil and graphit (1:2). Then the mixture was mixed with Q_0 which has dissolved into DMSO. The mixture was deposited in the glass electrode and the surface was leveled. The electrode was stored within 2 days before being used.

Electrochemical Measurement

Electrochemical measurement was conducted with cyclic voltametric by using eDAQ potensiostate (Ecorder 410) which was equipped with Echem software v 2.1.0. The electrodes used in the measurement were Ag/AgCl as the reference electrode, Pt as auxiliary electrode, and carbon paste as working electrode. Boric buffer solution was added into the measuring cell, and the current peak yielded was observed as the reference peak. Then the uric acid was added to observe the alteration of the current peak.

The Optimization of L. plantarum Uricase Activity

The optimization was conducted by varying the temperature (25-40) °C, pH (7.10-10.00), and uric acid concentration (1-5) mM by using *Response Surface Method.* After combining the factors of variables, then the study was conducted to gain the optimum activity.

The Determination of Electrode Stability

Electrode stability was determined from the uricase activity measurement after gaining the optimum condition. The electrode stabiliy can be observed by measuring the activity time to time. The decreasing of stability was confirmed by the decreasing of the current yielded in the measurement. The stability percentage was calculated with the equation (1).

$$\%\ stability = \left(100\% - \frac{[I_0 - I_t]}{I_0} \times 100\%\right) \quad \ldots (1)$$

where I_0 *is the* initial oxidation current and I_t is the oxidation current at certain time.

3. Results and Discussions

Characterization of Electrode

Before being used, the electrode was characterized with $K_3Fe(CN)_6$ to confirm that the electrode was proper to use. The proper electrode will result the voltamogram as showed in Figure 1.

IOP Conf. Series: Earth and Environmental Science **31** (2016) 012028 doi:10.1088/1755-1315/31/1/012028

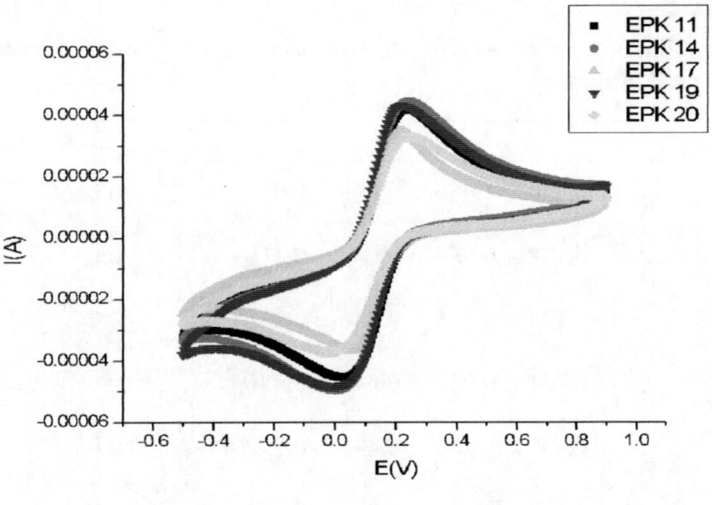

Figure 1. Voltamogram of carbon paste electrode characterization with $K_3Fe(CN)_6$

Formation of Biofilm of L. plantarum Cells

The confirmation of biofilm formation was conducted by using microplate reader Bio-Rad 680. Figure 2 showed the average OD at pH 7.10, 8.55, and 10.00. The average OD of pH 10.00 was lower than the average OD of pH 8.55, however the oxidation current of pH 10.00 ws higher than both pH 8.55 and 7.10. This may caused by the amount of enzyme excreted by the biofilm at pH 10.00 was much higher than at other pH. The cultures were inoculated onto wells 1,2,4,5,7,8, and 10, whereas the liquid GYP as negative control was inoculated onto well 12. Table 1 indicated that the average of OD of column 12 smaller than others column, it means that *Lactobacillus plantarum* is biofilm formation agent.

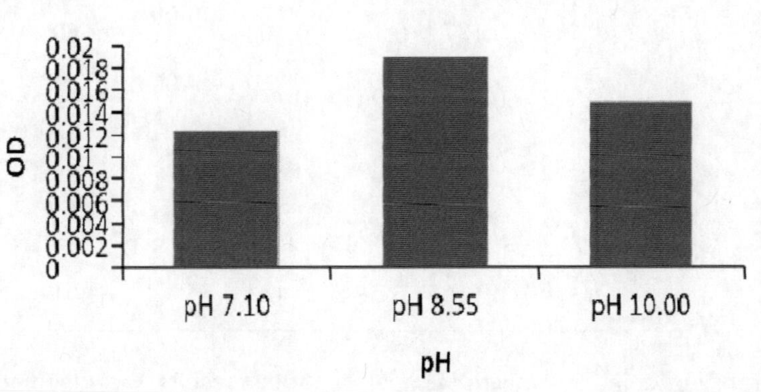

Figure 2. The comparation of average OD from biofilm extraction

IOP Conf. Series: Earth and Environmental Science **31** (2016) 012028 doi:10.1088/1755-1315/31/1/012028

Table 1. Result of biofilm of Optical Density (OD) measurement to confirm formation of *L. plantarum* cells biofilm

	1	2	3	4	5	6
A	0.160	0.059	0.045	0.093	0.133	0.046
B	0.138	0.084	0.045	0.116	0.077	0.048
C	0.154	0.125	0.046	0.141	0.094	0.049
D	0.111	0.056	0.042	0.086	0.131	0.056
E	0.139	0.205	0.045	0.092	0.081	0.051
F	0.128	0.097	0.039	0.104	0.100	0.055
G	0.172	0.076	0.044	0.146	0.106	0.051
H	0.138	0.107	0.042	0.084	0.107	0.049

	7	8	9	10	11	12
A	0.209	0.073	0.100	0.087	0.056	0.088
B	0.150	0.187	0.163	0.118	0.051	0.091
C	0.128	0.228	0.076	0.201	0.060	0.082
D	0.144	0.187	0.119	0.163	0.093	0.160
E	0.121	0.145	0.065	0.121	0.065	0.103
F	0.149	0.081	0.060	0.175	0.071	0.105
G	0.134	0.099	0.057	0.108	0.059	0.092
H	0.149	0.110	0.052	0.164	0.046	0.091

For the purpose of doing a comparative analysis of test results, the adherence capabilities of the test strains were classified into four categories (Table 2) [9].

Table 2 Classification of bacterial adherence by microtitre plate method

OD vaues of Tissue Culture Plate Method		
Mean OD value	Adherence	Biofilm formation
<0.062	Non	Non
0.062-0.124	Weak	Weak
0.124-0.248	Moderate	Moderate
>0.248	strong	Strong

(Saxena *et al.* 2014)

Result of formation of biofilm could detected uric acid and produced oxidation current higher than planktonic bacteria (Fig. 3). The oxidation current of *L. plantarum* biofilm and planktonic bacteria were of 47.24 and 23.04 μA, respectively, at the same pH, temperature, and uric acid concentration.

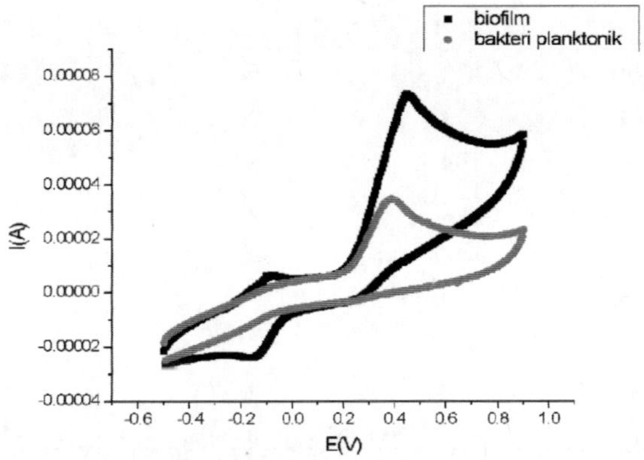

Figure 3 Voltammogram resulted by planktonic bacteria (-) and biofilm of *L. plantarum* (-)

Detection of uric acid using *Lactobacillus plantarum* biofilm could be performed at optimum condition of pH 10.00, temperature 40 ▯ , and uric acid concentration 5 mM. The stability of biofilm decreased after 10 hours used, with decreasing percentage over 86.33%. This low stability probably caused by the unprotected active site of the enzyme that the enzyme is easier to experience the denaturation.

The Optimization of Uricase Activity of Planktonic Bacteria

The combination of variables were varied by using Response Surface Method to result the variaton such as showed in table 2.

IOP Conf. Series: Earth and Environmental Science **31** (2016) 012028 doi:10.1088/1755-1315/31/1/012028

Table 2. The combination of optimization of *L. plantarum planktonic bacteria* uricase

pH	temp (°C)	[uric acid] mM	Oxidation Current (μA)
8.55	32.5	1	2.69
8.55	32.5	3	3.81
7.10	40.0	5	3.15
10.00	25.0	1	5.21
10.00	32.5	3	5.76
8.55	32.5	3	4.77
7.10	32.5	3	0.04
7.10	25.0	1	5.34
8.55	32.5	3	4.60
7.10	40.0	1	5.28
8.55	32.5	3	4.50
10.00	40.0	5	23.04
10.00	40.0	1	6.43
7.10	25.0	5	2.58
8.55	32.5	3	14.55
8.55	32.5	5	20.65
8.55	25.0	3	4.72
8.55	40.0	3	10.40
10.00	25.0	5	20.27
8.55	32.5	3	3.32

Figure 4 shows the voltamogram of uric acid measurement. There is the difference between the oxidation peaks of boric buffer as the reference and uric acid. The increasing of current peak was caused by the oxidation process of the uric acid into the allantoin, CO_2, and H_2O_2. The highest oxidation peaks of pH 7.10, 8.55, and 10.00 were 5.34 μA at 25°C, 20.65 μA at 32.5°C, and 23.04 μA at 40°C respectively.

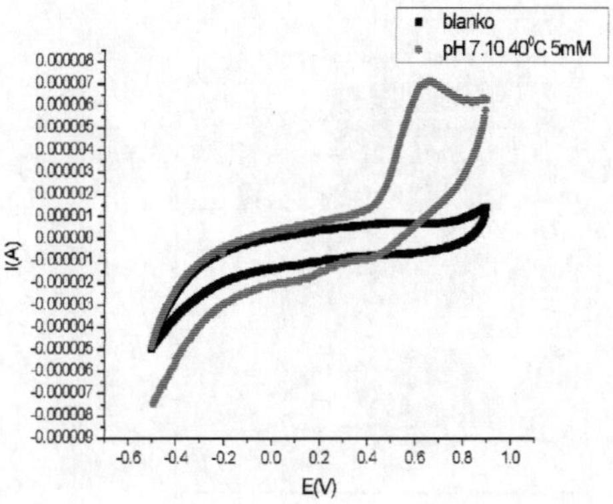

(a)

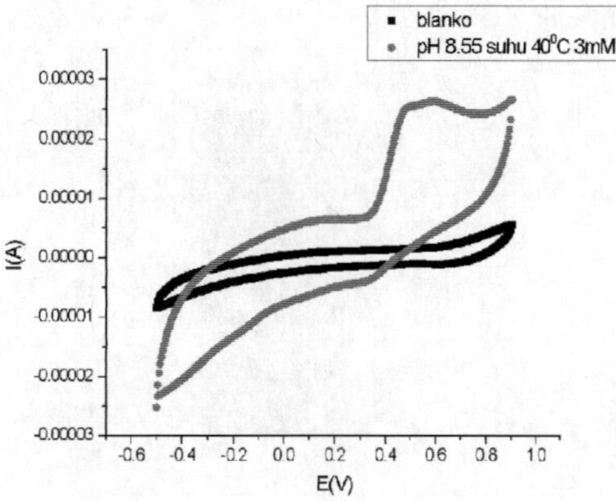

(b)

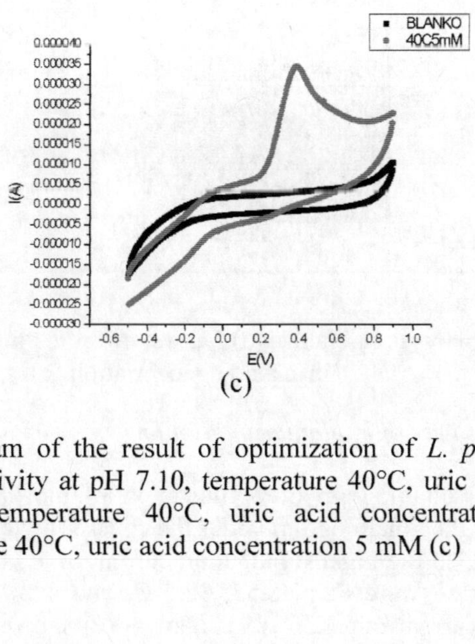

(c)

Figure 4 Voltamogram of the result of optimization of *L. plantarum planktonic bacteria* uricase activity at pH 7.10, temperature 40°C, uric acid concentration 1 mM (a), pH 8.5, temperature 40°C, uric acid concentration 3 mM (b), pH 10.00, temperature 40°C, uric acid concentration 5 mM (c)

To observe the interaction contour path of temperature vs pH, pH vs uric acid concenration, and temperature vs uric acid concentration, the plot contour analysis was performed. Optimum area was showed by the darker contour path and higher oxidation current. In figure 5, the interactions of pH, temperature, and uric acid concentration were showed. Figure 5 (a) showed the optimum area was at pH 10 and 40°C. Figure 5 (b) showed the optimum area was at pH 10 and the uric acid concentation was of 5 mM. Figure 5 (c) showed the optimum area was at 40°C and uric acid concentration was of 5 mM.

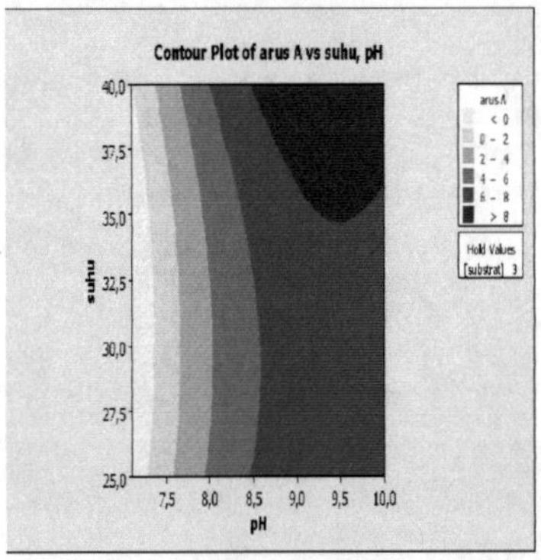

(b)

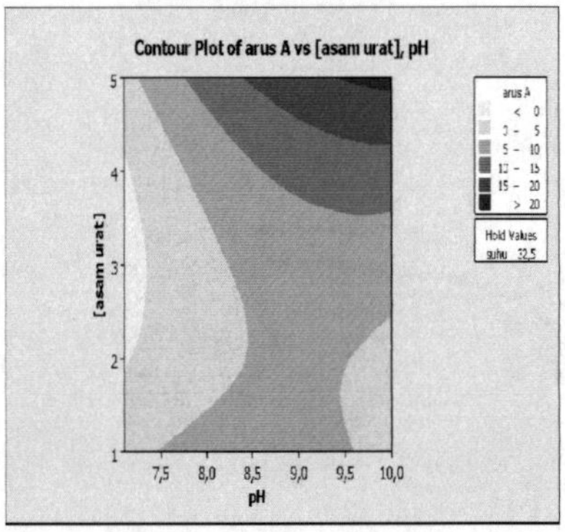

(a)

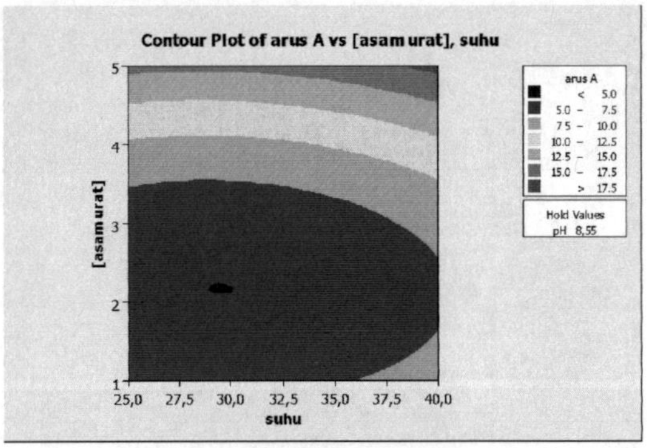

(c)

Figure 5. interaction contour: pH vs temperature (a), pH vs uric acid concentration (b), temperature vs uric acid concentration (c)

The Optimization of Uricase Activity of L. plantarum Biofilm

Figure 6 showed the voltamogram of optimization of biofilm uricase activity in various condition. Figure 6 (a) showed the voltamogram under the condition were of pH 7.10, 40°, and 1 mM of uric acid concentration, yielded the highest oxidation current of 1.74 μA. Figure 6 (b) showed the voltamogram under the condition were of pH8.55, 32.5°C, and 3 mM of uric acid concentration, yielded the highest oxidation current of 23.26 μA. Figure 6 (c) showed the voltamogram under the condition were of pH 10.00, 40°C, and 5 mM of uric acid concentration, yielded the oxidation current of 47.24 μA.

The oxidation current of biofilm uricase was higher than plantonic bacteria in pH 10.00. This indicates that biofilm uricase has far better activity than planktonic bacteria at that condition. The ability of biofilm which is able to survive in an extreem condition might caused the high of uricase's activity. However, in the neutral condition (pH 7.10) the biofilm uricase didn't show the good activity even no signal peaks yielded. This might caused by the biofilm bacteria couldn't optimized the condition over pH 7.

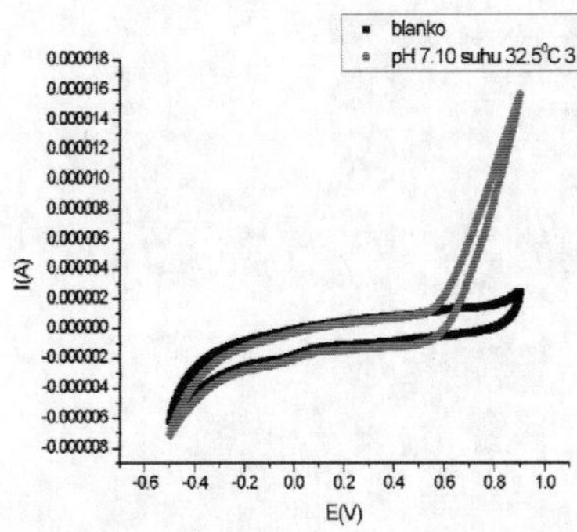

(b)

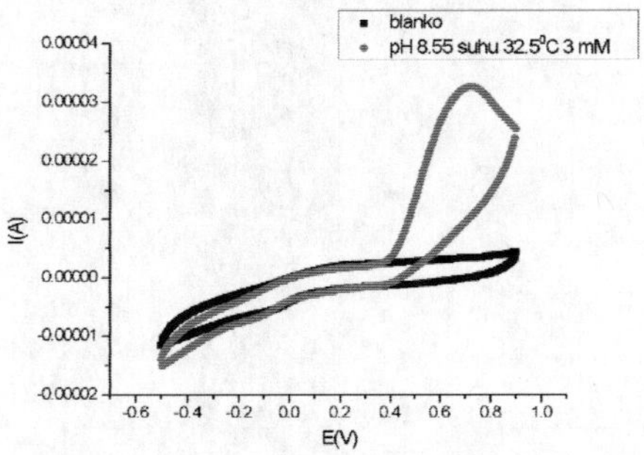

(a)

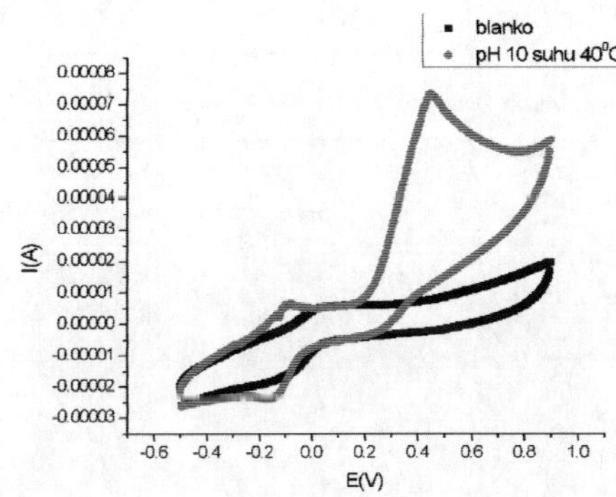

(c)

Figure. 6. Voltamogram of biofilm uricase activity at pH 7.10, 32.5°C, 3 mM of uric acid concentration (a), pH 8.55, 32.5°C, 3 mM of uric acid concentration, pH 10, 40°C, 5 mM of uric acid concentration

In figure 7, interactions beetwen temperature vs pH, pH vs uric acid concentration, and temperature vs uric acid concentration were observed by using contour plot. The darker area and higher oxidation current indicate the optimum area. Figure 7 (a) confirmed the highest oxidation current was at ph 10 and 40°C. At 32.5°C, the oxidation current yielded was very low, i.e. < 5μA. Figure 7 (b) confirmed the highest oxidatin current was at pH 10 and 5 mM of uric acid concentration. Uricase's activity wasn't optimum at pH < 7.5 which was showed by the low oxidation current. Figure 7 (c) confirmed the highest was at 40°C and uric acid concentration was of 5 mM.

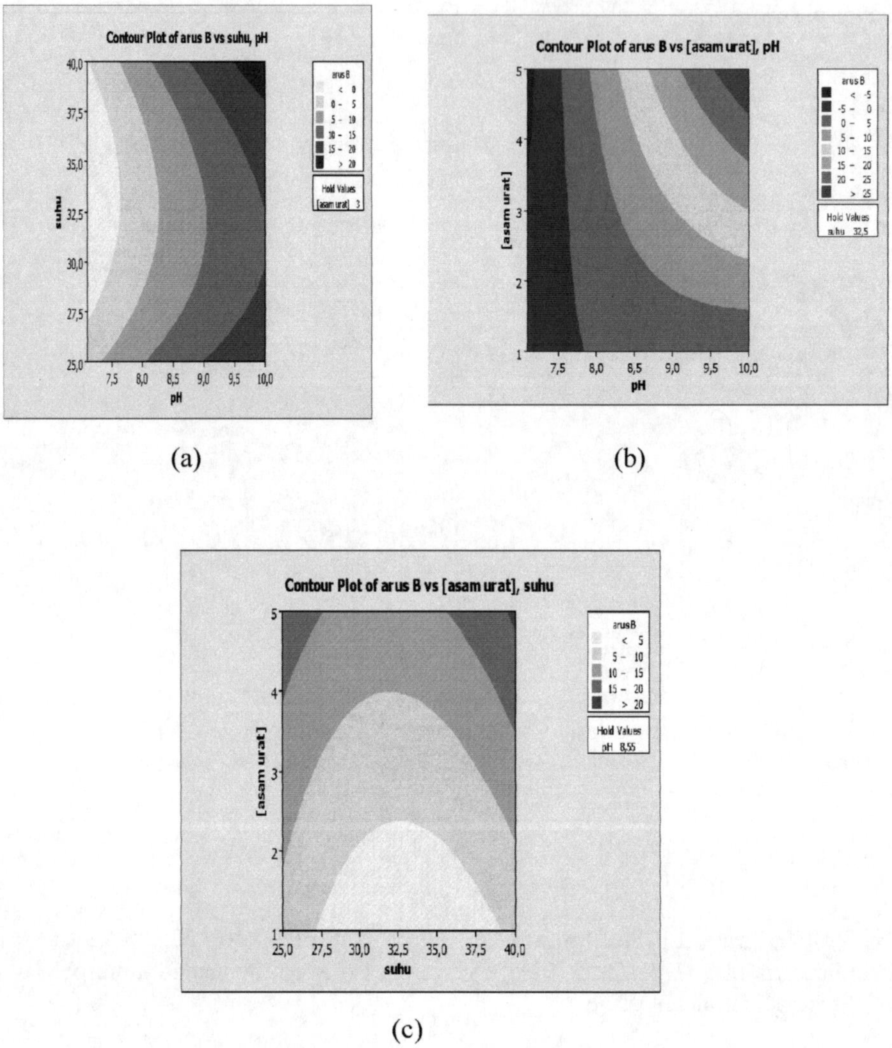

Figure 7. Interaction contour of pH vs temperature (a), pH vs uric acid concentration (b), temperature vs uric acid concentration (c)

Determination of the Stability of Electrode

Table 4 showed the stability of GCE and GCEb. The stability of GCE decreased over 28% after 4 hours. The GCEb uricase were more stable than another.

Table 4 The stability of GCE and GCEb

Time	Current (μA)	
	Planktonic	Biofilm
0	1.25	19.70
6	2.95	23.35
8	1.51	22.34
10	4.05	18.06

Figure 8 showed stability curve of two kinds of electrode. The stability of biofilm uricase contained electrodes increased after 6 hours, then significantly decreased after ten hours. The decreasing was over 86.33%. This low stability might be caused by the unprotected active site that the enzyme could be denaturated easily. The low stability could be addresed by immobilization method.

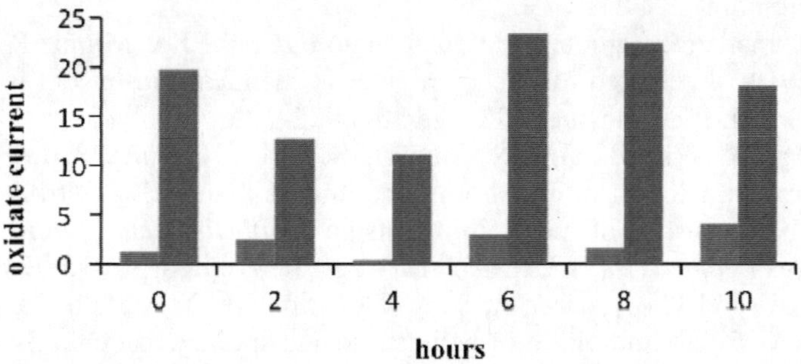

Figure 10. Stability curve of GCE (red) and GCEb (blue)

4. Conclusions

Lactobacillus plantarum could be functioned to form biofilm by incubating for 72 hours at 37°C. The uricase of *L. plantarum* biofilm is able to detect uric acid and yielded higher oxidation current than the uricase of planktonic bacteria. The optimum condition for biofilm uricase was at pH 10, temperature of 40°C, and uric acid concentration of 5 mM. The stability of *L. plantarum* biofilm uricase decreased after ten hours over 86.33%. This low stability might be caused by the unprotected enzyme's active site that the enzyme could be denaturated easily.

5. Acknowledgements

The authors would like to thank the Indonesian Government (IPTEK-DP2M DIKTI) who provided funding for this reasearch.

References

[1] Johnson,R.J, Kang, D.H, Feig D, Kivlighn,S Kanellis, J, Watanabe,S, Tuttle, K.R, Rodriguez- Iturbe,B, Herrera-Acosta J and Mazzali,M. 2003. Is there a pathogenetic role for uric acid in hypertension and cardiovascular and renal diseases. Hypertension; 41: 1183-1190.

[2] Arslan, F.,2008. An amperometric biosensor for uric acid determination prepared from uricase immobilized in polyaniline-polypyrrole film. Sensors Articels; 8:492-500.

[3] Monosik R, Stredansky M, Sturdik E. 2012. Biosensor – classification, characterization and new trends. *Acta Chimica Slovaca*. 5(1): 109-120. Doi: 10.2478/v10188-012-0017-z.

[4] Madga Aly, Sanaa Tork, Saleh Al-Garni, Reda Allam. 2013. Production and characterization of uricase from *Streptomyces exfoliates* UR10 isolated from farm wastes. *Turkish Journal of Biology*. 37: 520-529. doi: 10.3906/bty-1206-3.

[5] Atalla, M.M., Farag, R.H.; Eman, M.S., Abd-El-Lataif and E.A. Nehad, 2009. Optimum conditions uricase enzyme production by *Gliomastix gueg*. J Microbiolgy; 5:45-50.

[6] Gamella M, Campuzano S, Conzuelo F, Curiel J.A, Munoz R, Reviejo, Pingarron JM. 2010. Integrated multienzyme electrochemical biosensors for monitoring malolactic fermentation in wines. *Talanta.* 81: 925-933.

[7] Cha JO, Yoo JI, Yoo JS, Chung HS, Park SH, Kim HS, Lee YS, Chung GT. 2013. Investigation of biofilm formation and its association with the molecular and clinical characteristics of methicillin-resistan *Staphylococcus aureus*. *Osong Public Health Res Perspect*. 4(5): 225-232. doi.org/10.1016/j.phrp.2013.09.001.

[8] Jalilsood T, Baradaran A, Song AAL, Foo HL, Mustafa S, Saad WZ, Yusoff K, Rahim RA. 2015. Inhibition of pathogenic and spoilage bacteria by a novel biofilm-forming *Lactobacillus* isolate: a potential host for the expression of heterologous proteins. *Microb Cell Fact*. 14(96): 1-14. doi 10.1186/s12934-015-0283-8.

[9] Saxena S, Banerjee G, Garg R, Singh M. 2014. Comparative study of biofilm formation in *Pseudomonas aeruginosa* isolates from patients of lower respiratory tract infection. *Journal of Clinical and Diagnostic Research.* 8(5): 9-11. doi: 10.7860/JCDR/2014/7808.4330

Nanoparticles of *Selaginella doederleinii* leaf extract inhibit human lung cancer cells A549

Syaefudin[1], A Juniarti[1], L Rosiyana[1], A Setyani[1] and S Khodijah[1]

[1]Department of Biochemistry, Bogor Agricultural University, Bogor 16680, West Java, INDONESIA

Email: syaefudin01@apps.ipb.ac.id

Abstract. The aim of the present study is to evaluate cytotoxicity effect of nanoparticles of *Selaginella doederleinii* (*S. doederleinii*) leaves extract. *S. doederleinii* was extracted by maceration method using 70%(v/v) ethanol as solvent. Phytochemical content was analyzed qualitatively by using Harborne and Thin Layer Chromatography (TLC) methods. Nanoparticle extract was prepared by ionic gelation using chitosan as encapsulant agent. Anticancer activity was performed by using 3-(4,5-dimethylthiazol-2-yl)-2,5-diphenyltetrazolium bromide (MTT) assay. The results showed that *S. doederleinii* contains of flavonoids. Nanoparticle of *S. doederleinii* leaves extract greatly inhibited A549 cells growth (cancer cells), with IC_{50} of 3% or 1020 µg/ml. These nanoparticles extract also inhibited the growth of Chang cells (normal cells), with IC_{50} of 4% or 1442 µg/ml. The effective concentration of nanoparticles extract which inhibits cancer cells without harming the normal cells is 0.5% or 167 µg/ml. Further studies are needed to obtain the concentration of nanoparticles extract which can selectively suppress cancer cells.

1. Introduction

Lung cancer contributed 13% of the total number of new cancer cases diagnosed in the world [1]. Torre et al. [2] reported that number of death caused by lung cancer are greater than breast cancer. Currently, herbal medicine is one of the alternative drugs for curing the cancer, in addition to chemotherapy. One of the plants used as anticancer herbal medicine is *S. doederleinii*. Phytochemical screening of *S. doederleinii* discovers abundant of bioflavonoids. These compounds possess strong bioactivities, such as anti-inflammatory, antiviral, antioxidant, anticancer, and antitumor [3]. Anticancer research showed that essential oil from *S. doederleinii* could inhibit A549 and 7721 cell line [4].

Despite *S. doederleinii* has been widely used as anticancer agent, in-depth investigations are required to gain the information on anticancer ability of S. *doederleinii*, especially in its nanoparticle extract. Nanoparticle form of bioactive compounds expected to improve the substances distribution ability in human body and reach the targeted cells [5]. In this study, we evaluated the cytotoxicity effect of nanoparticle extract from *S. doederleinii* leaves on lung cancer cells A549.

2. Materials and methods

2.1. Plant collection and identification

S. doederleinii leaves were collected, identified, and authenticated by Biopharmaca Research Center, Bogor Agricultural University, Indonesia. The collected materials were washed in running tap water to eliminate clay and grimes followed by drying in the oven at 50°C, cut into small pieces and then powdered using mechanical blender and sieved with 40 mesh.

2.2. *Moisture content determination*
Determination of moisture content was done according to AOAC protocol [6].

2.3. *Plant extraction*
The powder of *S. doederleinii* were extracted using maceration method as described by Koirewoa [7]. Twenty grams of *S. doederleinii* powder was extracted in 200 ml of 70% ethanol for 24 h. The obtained extracts were subjected to vacuum evaporator (N-1100, Shanghai Eyela Co. Ltd., China) for excess solvent evaporation.

2.4. *Phytochemical identification*
The flavonoids, phytochemical compounds, were screened qualitatively using method described by Harborne [8]. *S. doederleinii* extract as much as the tip of a spatula is mixed with distilled water 10 ml. Drops of amyl alcohol, 5 drops of 1%(v/v) HCl and 70%(v/v) ethanol was added to this mixture. Afterwards, the solution was affixed with a little powder of magnesium (tip of a spatula). Yellow-orange-red appeared indicating the presence of flavonoids.

2.5. *Synthesis of nanoparticles*
S. doederleinii's nanoparticles were obtained by ionic gelation method. *S. doederleinii* dry extract was dissolved using a solvent and then stirred using the magnetic stirrer for 1 h. Then, the dissolved extracts were coated using chitosan with a ratio of 3.75:1 (v/v). A total of 12 ml of *S. doederleinii* extract was added to chitosan solution (45 ml). Magnetic stirring was conducted for 1 h for to ensure that crosslinking process was completed [9].

2.6. *Particle size analysis*
Particle size analysis of *S. doederleinii* nanoparticles was conducted by Particle Size Analyzer (PSA) using Beckman Coulter DelsaTM Nanoparticle Analyzer (USA) at 25°C [9].

2.7. *Zeta potential measurements*
Zeta potential measurements were conducted by micro-electrophoresis using a Malvern Zetasizer Nanoseries Nano ZS (Malvern Instruments, Herrenberg, Germany) [10].

2.8. *Cell culture*
Human lung cancer cell line A549 and normal cells (Chang cells) were obtained from ATCC (American Type Culture Collection). Cells were grown in Dulbecco's Modified Eagle's Medium (DMEM) medium supplemented with 10% Fetal Bovine Serum (FBS), 100 U/ml penicillin and 100 μg/ml streptomycin at 37°C in an incubator containing 5%(v/v) CO_2. All materials were purchased from Sigma-Aldrich, Germany.

2.9. *3-(4,5-dimethylthiazol-2-yl)-2,5-diphenyl tetrazolium bromide (MTT) assay*
MTT assay was performed as described by Wang et al. [11] with slight modification. Briefly, cells were seeded at a concentration of 5000 cells/100 μl DMSO (dimethyl sulfoxide) in a 96-well plate. After 24 h of incubation and cells' confluence has reached 50%, serial concentrations of nanoparticle extracts (0.06, 0.12, 0.25, 0.5, 1, 2, 4, 8, 12, and 16 %(w/v)), were added. Since *Curcuma zedoaria* Rosc. extracts (75 μg/ml) was added to drug treatment group, control group was added with DMSO. The use of *Curcuma zedoria* Rosc. as control is based on previous studies showing inhibition to various cancer cells [12, 13, 14]. Each concentration was repeated three times. These cells were incubated in a

humidified atmosphere with 5% CO_2 for 3 days. Then, 10 μl MTT (Sigma-Aldrich, Germany) solution (5 mg/ml) was added to each well and incubated at 37°C for 4 hours. In cells, dehydrogenase and reductase would reduce this solution into formazan, a purple color of artificial chromogenic products. The medium was removed and formazan was dissolved in 96%(v/v) ethanol 100 μl. The OD (optical density) was measured at 595 nm using a Bio-assay reader (Bio-Rad, USA). The growth inhibition was determined using following formula:

$$\%\ growth\ inhibition\ = \frac{OD\ control - OD\ sample}{OD\ control} \times 100\%$$

2.10. Statistical analysis
Cytotoxicity analysis was expressed as the mean ± SEM of pooled results obtained from at least three independent experiments. Statistical analysis was performed by one-way ANOVA and followed by Duncan test. Significance level was considered as $P < 0.05$.

3. Result and discussion

3.1. Extract and phytochemical content of S. doederleinii
The moisture content of *S. doederleinii* leaves powder was 7.62%(w/w). Moisture content described the content of organic compounds that can be isolated. Isolation method using ethanol as solvent yielded 10.87% (w/w) of extracts. Phytochemical screening showed that those extracts possessed flavonoids. It was proven with a yellow-orange appearance by Harborne' method. Previous phytochemical studies on the Selaginella species of Java led to discover many compounds, such as alkaloid, phenolic (flavonoid, tannin, saponin), and terpenoid (triterpene, steroid) [15].

3.2. Nanoparticles extract
Particle size analyzer resulted that the average nanoparticle of *S. doederleinii* were obtained 62.5 nm. This value was strengthened by previous research reported that nanoparticle encapsulated by chitosan resulted nanoparticle measurement at 40–100 nm [16]. *S. doederleinii* nanoparticle's extract was synthesized by physical modification with ionic gelation method. Crosslinking was made by adding sodium tripolyphosphate to extract. Toxicity effect after physical electrostatic interaction obtained by ionic gelation method less than that chemical process [17]. Nanoparticles from red betle extract were made by ionic gelation with chitosan and maltodextrin as chelating agent [18]. By controlling critical parameter in nanoparticle synthesis, such as chitosan concentration and ratio between chitosan and tripolyphosphate (TPP), there is spontaneous production of nanoparticle when TPP's were added at room temperature [19].

3.3. In vitro cytotoxicity using A549 and Chang cells
Figure 1 depicts the cytotoxicity activity of *S. doederleinii* nanoparticle extract. The results showed that this extract, at concentration of 0.6 to 1%, did not inhibit the growth of normal cells (Chang cells). In contrast, it could inhibit the growth of normal cells at concentration > 1%. Cytotoxicity analysis proved that this extract could inhibit cancer cells (A549 cells) at concentration > 0.25%. At concentration ≤ 0.25%, this extract has not been able to suppress the growth of cancer cells. This finding also indicated that *S. doederleinii* nanoparticle extract was toxic to cancer cells, but not to normal cells in particular concentrations, i.e. 0.5% and 1%.

The effective concentration of *S. doederleinii* nanoparticles extract required for 50% inhibition of the cell viability is expressed by IC_{50} [20]. Statistical analysis performed to determine the IC_{50} value. Anticancer assay towards normal cells showed that IC_{50} value was 4.33% or 1442 μg/ml, while IC_{50} for cancer cells was 3.06% or 1020 μg/ml. Wang et al. [21] reported that ethyl acetate extract from this plant had significantly exhibited anticancer activity with a low IC_{50} value. Other research stated that *S. doederleinii* had cytotoxic effect on murine cells [22].

Anticancer ability of this extract might be caused by its phytochemical compounds, especially flavonoids. Li et al. [23] reported that some of flavonoids, i.e. 2'',3''-dihydro-3',3'''-biapigenin and 3', 3'''-binaringenin from *S. doederleinii*, could inhibit human cancer cell. A combination of chromatographic techniques and cytotoxic assay also revealed that some natural secondary metabolites from the ethanolic extract of Selaginellaceae had ability to against cancer cells [22].

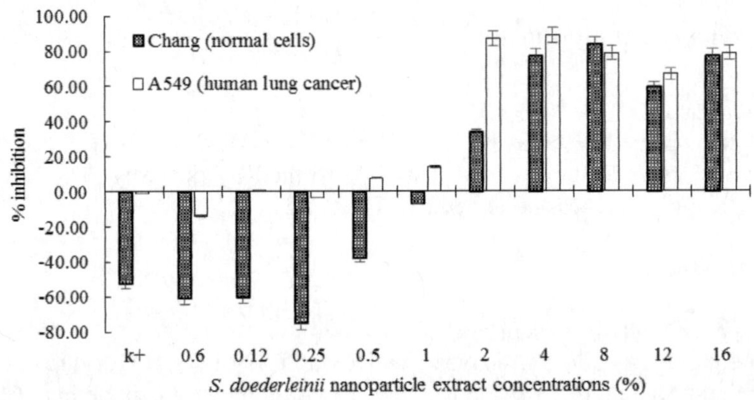

Figure 1. Cytotoxic activity of *S. doederleinii* nanoparticle extract.

4. Conclusion
We conclude that *S. doederleinii* extract contains flavonoids compounds. Its nanoparticle extract greatly inhibited A549 cell growth, with IC_{50} of 3% or 1020 μg/ml. At 0.5% or 167 μg/ml, the nanoparticles extract could inhibit lung cancer cells but not to normal cells. It suggests that this value indicated the effective concentration of nanoparticles extract suppressed cancer cells.

Acknowledgement
This study was supported by Indonesian Directorate General of Higher Education, Ministry of Research, Technology and Higher Education, Indonesia through the Program Kreativitas Mahasiswa 2015. The authors also appreciate Nanotech Herbal Indonesia, Indonesia, for its support in nanoparticle production.

References
[1] WCRF World Cancer Research Fund International Lung cancer statistics 2015 Available on: http://www.wcrf.org/int/cancer-facts-figures/data-specific-cancers/lung-cancer-statistics (8 July 2015)
[2] Torre L A, Bray F, Siegel R L, Ferlay J, Lortet-Tieulent J and Jemal A 2015 *Cancer J. Clin.* **65** 87-108
[3] Setyawan A D 2011 *Nus. Biosci.* **3** 44-58
[4] Gang W, Shun Y, Lei C, Yufen L and Hang S 2015 *Technol. Health Care.* **23** S21-S27.
[5] Brown S D, Nativo P, Smith J A, Stirling D, Edwards P R, Venugopal B, Flint D J, Plumb J A, Graham D and Wheat N J 2010 *J. Am. Chem. Soc.* **132** 4678-4684.
[6] AOAC 2006 Official methods of analysis of AOAC International 18th edn (Virginia: Association of Analytical Chemists)
[7] Koirewoa Y A, Fatimawali and Wiyono W 2012 *Pharmacon* **1** 47-52
[8] Harborne J B 1996 The flavonoids advance in research since 1986 (London: Chapman and Hall)
[9] Ikono R, Akwalia P R, Siswanto, Bambang W, Sukarto A and Rochman N T 2012 *Int. J. Engin. Tech.* **12** 5-9
[10] Müller R H 1996 Zetapotential und partikelladung in der laborpraxis – einführung in die theorie, praktische meßdurchführung, dateninterpretation (Stuttgart: Wissenschaftliche

Verlagsgesellschaft)
[11] Wang Z, Zhou J, Ju Y, Zhang H, Liu M and Li X 2001 *Biol. Pharm. Bull.* **24** 159-62
[12] Chen C C, Chen Y, Hsi Y T, Chang C S, Huang L F, Ho C T, Way T D and Kao J Y 2013, *J. Agric. Food Chem.* **61** 11418-27
[13] Shin Y and Lee Y 2013 *Toxicol. Res.* **29** 257-261
[14] Lakshmi S, Padmaja G and Remani P 2011 *Int. J. Med. Chem.* **2011** 1-13
[15] Chikmawati T, Setyawan A D and Miftahudin 2012 *Makara J. Sci.* **16** 129-133
[16] Qi L and Xu Z 2006 *Bioorg. Med. Chem. Lett.* **16** 4243-45
[17] Shu X Z and Zhu K J 2001 *J. Microencapsul.* **18** 237-45
[18] Dewandari K T, Yuliani S and Yasni S 2013 *J. Pascapanen* **10** 65-71
[19] Hu B, Pan C, Sun Y, Hou Z, Ye H, Hu B and Zeng X X 2008 *J. Agric. Food Chem.* **56** 7451-58
[20] Griffiths M and Sundaram H 2011 *Methods Mol. Biol.* **731** 451-465
[21] Wang J Z, Li J, Zhao P, Ma W T, Feng Z H and Chen K L 2014 *Evid. Based Complement. Alternat. Med.* **2015** 1-9
[22] Lin R C, Skaltsounis A L, Seguin E, Tillequin F and Koch M 1994 *Planta Med.* **60** 168-70
[23] Li S, Zhao M, Li Y, Sui Y, Yao H, Huang L and Xua L 2014 *Phytochem. Anal.* **25** 127-33

Modeling effectiveness of management practices for flood mitigation using GIS spatial analysis functions in Upper Cilliwung watershed

Suria Darma Tarigan[1]

[1]Department of Soil Sciences and Natural Resources Management, Bogor Agricultural University, Kampus IPB Dramaga, Bogor Indonesia.

E-mail: surya.tarigan@yahoo.com

Abstract. Flooding is caused by excessive rainfall flowing downstream as cumulative surface runoff. Flooding event is a result of complex interaction of natural system components such as rainfall events, land use, soil, topography and channel characteristics. Modeling flooding event as a result of interaction of those components is a central theme in watershed management. The model is usually used to test performance of various management practices in flood mitigation. There are various types of management practices for flood mitigation including vegetative and structural management practices. Existing hydrological model such as SWAT and HEC-HMS models have limitation to accommodate discrete management practices such as infiltration well, small farm reservoir, silt pits in its analysis due to the lumped structure of these models. Aim of this research is to use raster spatial analysis functions of Geo-Information System (RGIS-HM) to model flooding event in Ciliwung watershed and to simulate impact of discrete management practices on surface runoff reduction. The model was validated using flooding data event of Ciliwung watershed on 29 January 2004. The hourly hydrograph data and rainfall data were available during period of model validation. The model validation provided good result with Nash-Suthcliff efficiency of 0.8. We also compared the RGIS-HM with Netlogo Hydrological Model (NL-HM). The RGIS-HM has similar capability with NL-HM in simulating discrete management practices in watershed scale.

1. Introduction

Performances of discrete management practices such as infiltration well, retention ponds in reducing surface runoff had been investigated in plot scale. But, appropriate methodologies are still required to study the performance of the measures in a watershed scale. Many studies use catchment hydrological modeling like SWAT and HEC-HMS to investigate the performance of various management practices for flood management (Yustika, 2012, Budiyanto et al. 2015). SWAT model uses HRU (hydrologic response unit) as a smallest homogeneous unit to calculate hydrological processes (Arnold, 2012). Meanwhile, HEC-HMS uses sub-catchment as a smallest unit. In both cases, the surface area of discrete management practice like an infiltration well is too small to be represented by HRU or sub-catchment having an area 10 to 100 ha. One way to overcome this limitation is by using raster as a smallest unit analysis. One discrete management practice can be represented by one raster and using

GIS function the size of a raster can easily be adjusted according to the size of structural management practices. According to Arnold (2010), the raster delineation can simulate the impact of landscape position on management and runoff in explicit spatial detail.

The aim of this research is to use raster spatial analysis functions of Geo-Information System (RGIS-HM) to model flooding event in Ciliwung watershed and to simulate impact of discrete management practices on surface runoff reduction. The main advantage of the model lies in its ability to represent individual discrete management practices such as infiltration well in its single raster representation. The main equations used in the modeling process is Manning equation relating flow velocity with slope and surface roughness of the channel. In addition, SCS-CN equation was adopted to relate land use and the proportion of rainfall becoming surface runoff. The grid size is easily adapted to represent small size discrete management practice like infiltration well. Input of the model are topographic map (DEM), land use map, soil map and rainfall characteristics. GIS provide various hydrological functions and raster spatial analysis such as flow direction, flow accumulation functions, flow travel time and zonal statistic to depict flow hydrograph.

2. Methodology

The research was conducted in upper Ciliwung watershed situated in West Java, Indonesia (Figure 1). The outlet of the watershed situated in Katulampa. The watershed experience frequent big flooding in recent years. Effective management practices in upper watershed should be implemented to reduce peak discharge. To select effective management practices, a model should be constructed to test them at the watershed scale.

Surface water flow is the most important component in flood modeling. During rainfall event surface runoff flowing from the upper catchment as a uniform flow and it gradually accumulates forming channel flow as it move downwards of the catchment. At the upper catchment the water flow can be considered as normal flow (stationary flow), which can be approached by the Manning equation. This flow condition is a function of a resistance, slope and hydraulic radius. At the downstream, the flow tends to be non-stationery due to the backwater in the downstream channel. This study focuses on the flood wave generated by surface runoff from the upper catchment.

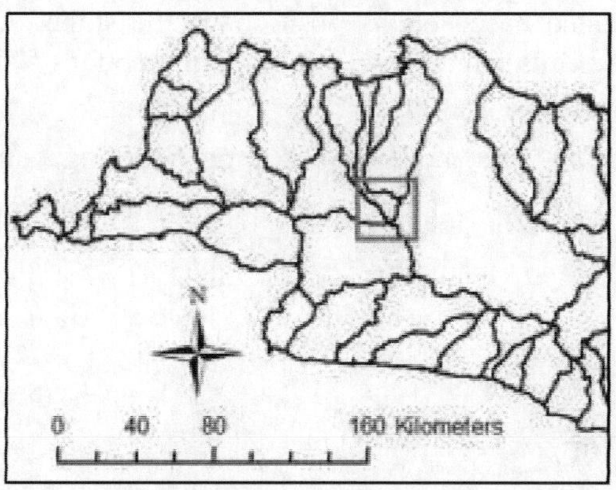

Source: BPDAS Citarum Ciliwung (2008)

Figure 1. Research location in Upper Ciliwung watershed, West Java, Indonesia

Runoff volume

Runoff volume calculation is the first step in every hydrological modeling. Runoff volume is primarily determined by the proportion of rainfall becoming surface runoff in particular watershed. Proportion of rainfall becoming surface runoff is determined by hydrologic soil groups (HSG) and land use characteristics. The combined impact of HSG and land use characteristics is expressed in term of CN

IOP Conf. Series: Earth and Environmental Science **31** (2016) 012030 doi:10.1088/1755-1315/31/1/012030

value. Based on Yustika et al. (2012), upper Ciliwung watershed dominated by HSG of type B. The CN values for HSG of type B are then differentiated primarily based on land use (Table 1). After determining the CN values, SCS equation is used to calculate proportion of rainfall becoming runoff using the following equation:

$S=(25400/CN)-254$ (1)
$Q= (P-0.2S)^2/(P+0.7)$ (2)
where,
Q = Runoff volume, mm
P = Rainfall, mm
S = Maximum recharge capacity of watershed
CN = Curve Number

The CN values and Manning coefficients based on dominant land use in upper Ciliwung are listed in Table 1.

Table 1. SCS curve number and Manning coefficients based on dominant land use in upper Ciliwung

Land use	Curve Number	Manning coefficient
Natural forest	0.55	0.6
Secondary forest	0.60	0.6
Mixed agriculture	0.66	0.5
Plantation	0.66	0.45
Settlements	0.85	0.012
Shrub	0.60	0.45
Bareland	0.86	0.018
Dryland farming	0.76	0.02

Source: Schwab et al. 1981

Flow velocity and travel time

Component of flow in an upper watershed consists of: a) overland flow, b) channel flow. Overland flow can be further differentiated based on its position on the slope. Channel cells are identified (using a GIS function) as those cells which have a minimum amount of >50 of upstream cells. In this model, baseflow is assumed constant.

At the upper catchment the flow was modeled based on Manning equation (DHV-MLD-Deltares 2011):

$V = \{s^{0.5} * (A/P)^{2/3}\}/n$ (3)

Where:

V = flow velocity in the cell (m3/s)
s = slope (m/m)
A = flow cross-sectional area (m^2)
P = wetted perimeter
n = Manning's coefficient

The slope is determined from DEM (Digital Elevation Model) available in grid size of 30m x 30m in the study area. The Manning's coefficient is determined according to the types the land-use for the particular cell (Table 1). Based on DHV-MLD-Deltares (2011) for overland flow where the depth relatively small compared to the width of the flow, the hydraulic radius (A/P) can be approximated by the depth of the flow in the cell. The Manning equation can be formulated as follows:

$V = \{s^{0.5} * d^{5/3}*W\}/n$ (4)

The flow travel time in the particular cell can be determined as follows.

$$T = x/V \qquad (5)$$

where:

V = flow velocity in the cell
s = slope (m/m)
d = depth of the flow
n = Manning's roughness coefficient
x = flow distance
W = flow width (for overland flow, it was estimated based on its position in the overland runoff process; for channel flow it was estimated based on observations in the catchment)

GIS functions were used to determine cumulative travel time from particular cell to the outlet based on the flow direction map obtained from GIS spatial analysis. By accumulating volume of excess rainfall in each of the cells along the flow direction and their respective time to reach the outlet, the flow hydrograph at the outlet of the watershed is determined using zonal statistics of GIS functions and plotting in a hydrograph.

Rainfall data

The model requires hourly rainfall data. Rainfall data used as an input for the model was obtained from Citeko meteorological station. Discharge data for model calibration was obtained from Katulampa hydrological station.

Model validation

The model was validated using flooding data events of Ciliwung watershed in 29 January 2014. The hourly hydrograph data and rainfall data were available during period of validation.

3. Results and Discussions

Model setup

Input of the model consisted of rainfall events, spatial data (soil, land use, topography). Soil and land use data were used to derive CN value. CN value determines partition of rainfall into surface runoff and infiltration (Figure 2).

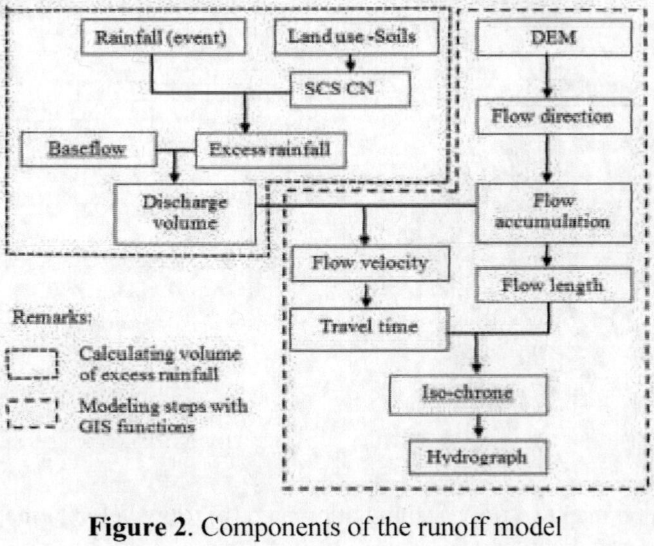

Figure 2. Components of the runoff model

Table 2. The hourly rainfall data from Citeko meteorological station and water level in Katulampa during flood event in 29January 2014.

Date	29 January 2014								30 January 2014						
Hour	17	18	19	20	21	22	23	24	01	02	03	04	055	06	077
Rainfall (mm)	0	0.1	2.7	5.2	45.5	63	25	1.3	0.2	2.8	0.1	0.1	2.5	8.5	3.5
Water level (m)	80	80	80	80	80	130	190	230	200	190	180	150	140	130	120

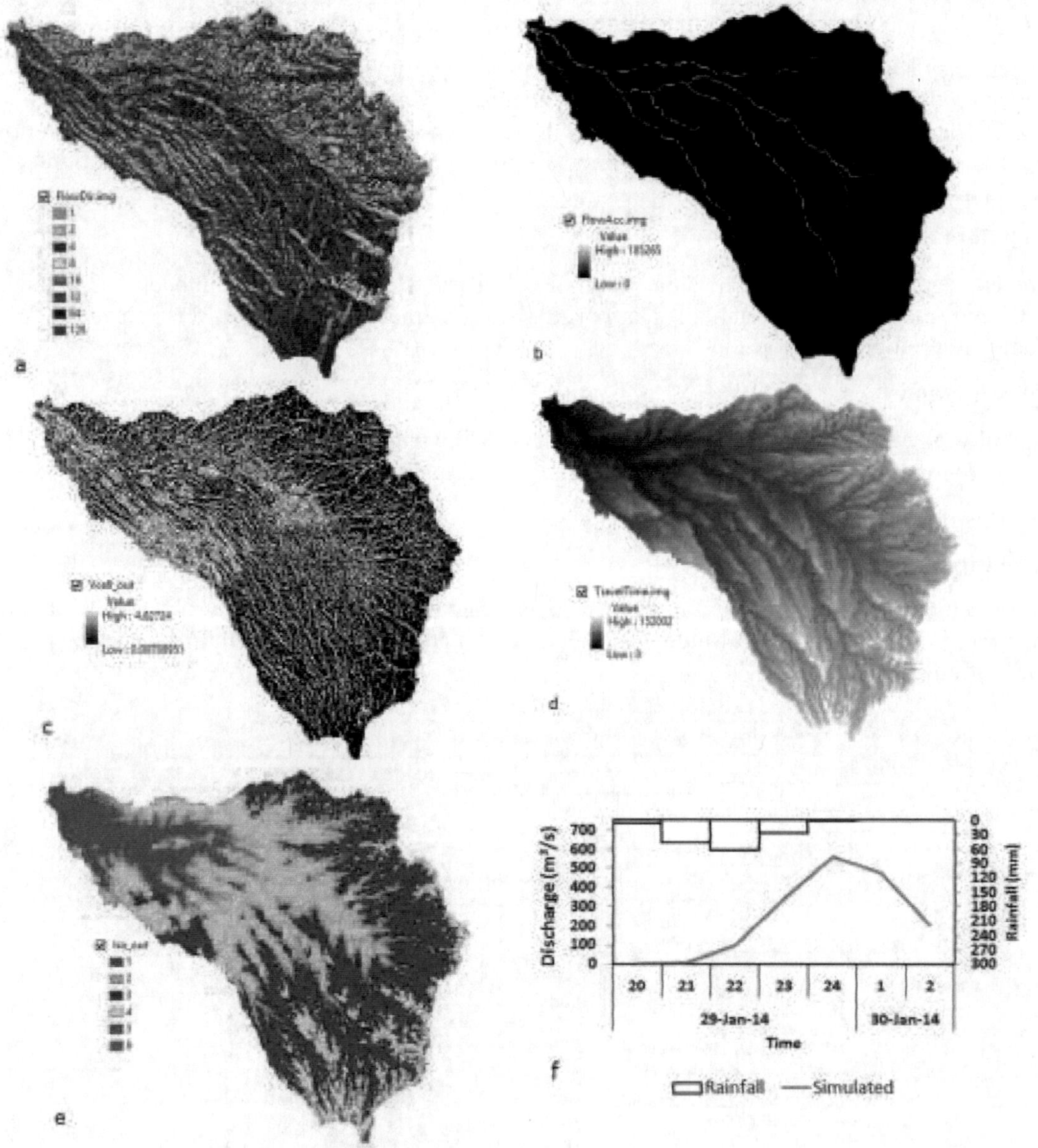

Figure 3. Flow direction map (a), flow accumulation map (b), flow velocity map (c), Travel time (d), Iso-chrone (e) and output hydrograph (f)

As the input for the model, we used hourly data available from Citeko rainfall station. We didn't use rainfall data from other two stations found in the watershed (Gunung Mas and Katulampa), since both stations had only daily rainfall data which couldn't be used to simulate hourly flood event. The hourly rainfall data from Citeko meteorological station in the upper Ciliwung and respective water level in Katulampa outlet during flooding event is shown in Table 2. The flood peak discharge occurred at 24.00 on 29 January 2014 generated by rainfall event started form 19.00 PM (Table 2).

Topography data or DEM was used to delineate the catchment and to derive flow direction (Flow_dir) and flow accumulation (Flow_acc) maps using GIS functions. Based on the Flow_dir, Flow_acc and rainfall data, the model calculated rainfall excess, flow velocity, flow volume excess, travel time, and Iso-chrone (Figure 3 a,b,c,d,e) using Equation 1 and 3. The final output of the model is discharge hydrograph at the outlet of the catchment at Katulampa (Figure 3 f).

Validation and Simulation

The flood event was initiated by single rainfall even started from 20.00 to 23.00 PM in 29 January 2014 with hourly amount of rainfall were 5, 46, 64, 25 mm consecutively. The flood started the rising limb 1.5 hour after rainfall initiation. The flood peak discharge (~525 $m^3 s^{-1}$) occurred at 23.30 AM on 29 January 2014 was accurately simulated by the model (Figure 4). The time period for rising limb and its trend were simulated by the model with good agreement with observed data. But, the magnitude of simulated discharge during the rising limb was underestimated by the model with decreasing trend toward the peak discharge. The decreasing limb one hour after the peak discharge was slightly overestimated by the model.

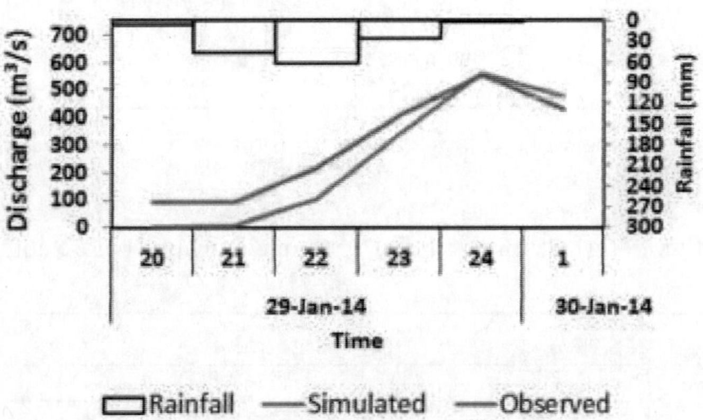

Figure 4. Simulated and observed peak discharge during flood event on 29 January 2014 in Katulampa outlet

The model was validated using observed hydrograph data recorded at Katulampa station during flood event in 29 January 2014 (Fig.1). The validation results were expressed as Nash-Sutcliff efficiency (NSE) (Nash and Sutcliff, 1970). In overall, the R^2 between observed and simulated discharge was 0.97 and the Nash-Suthclife Index amounted to 0.81. Both statistical indicators showed that the model can be used for prediction.

Impact of infiltration well in reducing flood peak discharge

After validation with good result, the model was used to simulate impact of infiltration in reducing flood peak discharge. Some 125 infiltration well were placed in agriculture land use such as mixed farming and dry land farming (Figure 5). Surface area of each infiltration well equals to single raster size which is 900 m^2. Total area of simulated infiltration well was 125 x 900 m^2 = 112,500 m^2 or equivalent to 11.25 ha which was on only 0.0008% of upper Ciliwung watershed. The simulated infiltration wells reduced 0.67% peak discharge (Table 3). The impact of those infiltration wells in reducing flood peak discharge was still minor, but the simulation showed that RGIS-HM was cable in integrating discrete management practices like infiltration well into the flow modeling. Result of this simulation showed that: a) discrete management practice like infiltration well can be represented by individual or combination of several rasters, b) this model can be used to determine numbers of infiltration wells required to reduce peak discharge into the targeted level. Raster size can easily be adjusted to accommodate the size of discrete management practice. This kind of adjustment is not possible using other hydrological models such as SWAT and HEC-HMS.

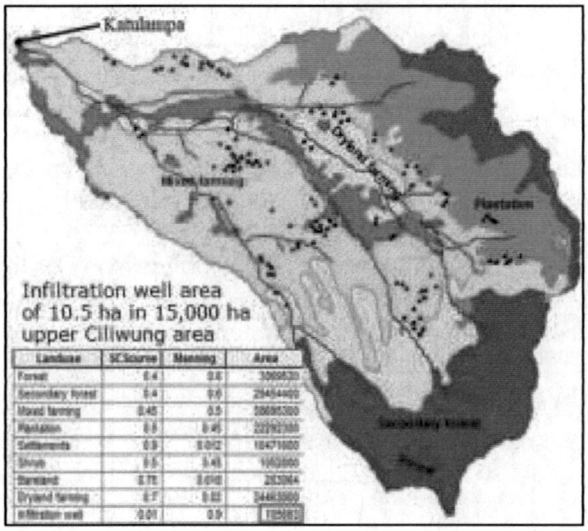

Figure 5. Simulated infiltration wells (black dots) with total surface area 10.5 ha in upper Ciliwung watershed with area of 15,000 ha.

Table 3. Impact of infiltration wells on flood discharge in Katulampa before and after infiltration well constructions

	Discharge (m3/s)		
Hour	Before	After	% Reduction
1	0.0	0.0	0.00
2	7.8	7.8	0.00
3	98.7	98.4	0.34
4	337.6	336.0	0.48
5	555.7	552.0	0.67
6	476.1	474.4	0.36
7	206.8	206.3	0.24
8	41.3	41.3	0.02
9	3.6	3.6	0.00

Netlogo Hydrological Model (NL-HM)

Netlogo is a multi-agent modeling environment for simulating natural and social phenomena. It is particularly suited for modeling complex systems evolving over time such as flood discharge hydrograph. The Netlogo world consists of agents. There are four types of agents: turtles, patches, links, and the observer. All of the agents can carry out their own activity simultaneously. The patches

IOP Conf. Series: Earth and Environmental Science **31** (2016) 012030 doi:10.1088/1755-1315/31/1/012030

are similar to the grid representation of the landscape and the water flow is being the turtles. Netlogo can import GIS data using GIS extension serving as the background of the patches including DEM data providing elevation for every patches. Therefore, Netlogo can be used to simulate surface water flow using GIS extension function of the Netlogo as follows:

```
extensions [ gis ]
globals [ road-dataset
          flow-dataset ]
patches-own [ p_catchment]
to setup
ca
  ;load all gis datasets
  set road-dataset gis:load-dataset "data/catchment.shp"
  set flow-dataset gis:load-dataset "data/riv1.shp"
  ;drawing polyline data from a shapefile
to display-watershed
  gis:set-drawing-color blue
  gis:draw road-dataset 1
end
;drawing polyline data from a shapefile
to display-flow
  gis:set-drawing-color blue
  gis:draw flow-dataset 1
end
```

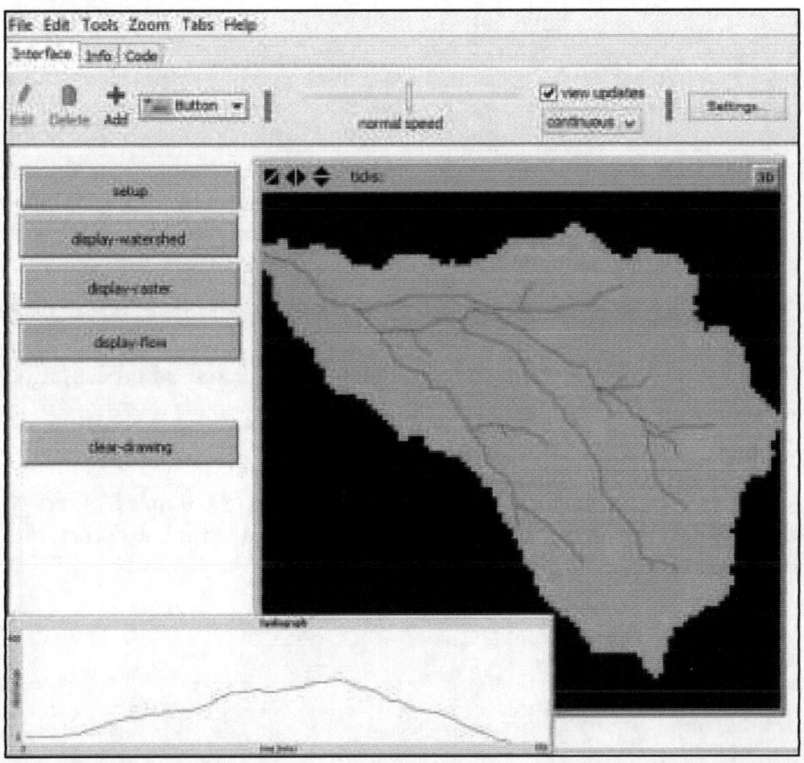

Figure 6. Example of Netlogo model and hydrograph for upper Ciliwung

To simulate water flow hydrograph, both equations mentioned in the methodology (Equation 1 and 3)

can be easily incorporated into the script below. Result of the script execution using Netlogo-interface is shown in Figure 6. The interface showed upper Ciliwung watershed boundary, the channel flow and resulting discharge hydrograph.

```
to go
; calculating runoff volume using SCS equation
 let cd gis:raster-sample channel-distance self
 ifelse cd < in-channel-dist
 [ set color blue ]
   if (landuse != 0)
     [ let cn curve-number landuse
       if (cn != 0)
        [ let s (25400 / cn) - 254
          let temp q rain- 0.2 * s
          set q (temp * temp) / (rain + 0.7) ] ] ] ]
………..
end
```

Using Netlogo script it is easy to assign infiltration well in to selected patches. It is also possible to differentiate resistance and storage condition between upper and lower patches of the watershed. Other strength of NL-HM compares to RGIS-HM is its ability to perform dynamic simulation of flood hydrograph (Fig. 6). The only disadvantages of NL-HM is its longer processing time compared to that of RGIS-HM for larger watershed.

4. Conclusions

Modeling flood discharge using GIS spatial function (RGIS-HM) showed good Nash-Suthcliff index of 0.8. Despite its underestimation during the rising and decreasing limbs, the model was very successful in determining the flood peak discharge. The main advantage of the model compared to existing model such as SWAT and HEC-HMS models is its ability to simulate small and discrete management practices like infiltration well having size smaller than 1 ha by assigning its attributes to a single raster. This capability enables RGIS-HM to be used to test performance of discrete management practices like infiltration well in watershed scale.

Netlogo modeling environment can also be used to model and to simulate peak flood discharge and the impact of discrete management practices. Its strength lies in its simple and easily constructed script and also its dynamic visualization capability. The disadvantage is the increased processing time for a large watershed.

5. Acknowledgements

This study was conducted in the framework of the joint Bogor Agriculture University-Monash University research collaboration in Water Infrastructure and was funded by the Directorate General of Higher Education (DIKTI), Indonesia

References

[1] Arnold JG, Kiniry JR, Srinivasan R, Williams JR, Haney EB, Neitsch SL. 2012. Soil Water Assessment Tools: Input/Output Documentation Version 2012. Texas Water Resources Institute

[2] Arnold JG, Allen PM, Volk M, Williams JR, Bosch DD. 2010. Assessment of different representations of spatial variability on swat model performance. Transactions of the ASABE 53(5): 1433-1443. ISSN 2151-0032.

[3] BPDAS Citarum Ciliwung. 2008. Pengelolaan DAS terpadu DAS Citarum. Ditjen Rehabilitasi Lahan dan Perhutanan Sosial. Ministry of Forestry Indonesia.

[4] Budiyanto, S., Tarigan S., Sinukaban N., Murtilaksono K.. 2015. The Impact of Land Use on Hydrological Characteristics in Kaligarang Watershed. 8(2):125-130 Intl. J Sciences and Engineering. DOI: 10.12777/ijse.8.2.125-130

[5] DHV-MLD-Deltares. 2011. Modeling for peak runoff zoning. 6Ci's River Basin Territory Project - B2 report.

[6] Nash JE, Sutcliffe JV. 1970. River flow forecasting through conceptual models, 1, A discussion of principles, J Hydrol 10:282–290.

[7] Schwab GO, Frevert RK, Edminster TW, Barnes KK. 1981. *Soil and Water Conservation Engineering.* Third Edition. United States of America (US): The Ferguson Foundation Agricultural Engineering Series.

[8] Yustika RD, Tarigan SD, Hidayat Y, Sudadi U. 2012. Simulation of Landuse Management System Using ArcSWAT Model in Ciliwung Hulu Watershed. Informatika Pertanian 21(2):71-79. ISSN 0852-1743.

IOP Conf. Series: Earth and Environmental Science **31** (2016) 012031 doi:10.1088/1755-1315/31/1/012031

Study of the Effectivity of Several Tree Canopy Types on Roadside Green Belt in Influencing The Distribution Vertically and Horizontally of CO gas Emitted from Transportation Activities to Vicinity of The Road

Bambang Sulistyantara[1]*, Nizar Nasrullah[1], Indung Sitti Fatimah[1], Prita Indah Pratiwi[1]

[1,] Department of Landscape Architecture, Bogor Agricultural University, Kampus IPB Dramaga, Bogor, Indonesia

E-mail: bbsulistyantara@yahoo.co.id, nizarnasrullah@yahoo.com, indungsittifatimah@gmail.com, pritaindahpratiwi@ymail.com

Abstract. High volume of vehicle leads to the increase of emission of pollutants level in major cities of Indonesia. Carbon monoxide (CO) is categorized as the main gas pollutants from transportation that are harmful to human health. Plants could be used as roadside green belt to reduce the level of pollutants emitted from the transportation. The purpose of this research is to determine tree canopy type that effectively reduce CO gas concentration, to determine the relation between tree canopy types and pollutant distribution vertically and horizontally. The research was conducted on roadside green belt of Jagorawi Highway, especially on the plot of glodogan (*Polyanthea fragrans*), plot of mahogany (*Swietenia macrophylla*) and a control plot (open filed). Air sampling was conducted to measure the concentration of CO at three elevation 1.5 m, 5 m, and 10 m at each distance 0 m, 10 m, and 30 m. Concentration of CO was analysed using Iodometri method. Vertical distribution of CO gas shows that the concentration increases with the increasing of sampling elevation on the plot of *Polyalthia fragrans* and *Swietenia mahogany*, but the control plot shows the opposite. Horizontal distribution shows that the concentration decreases at the distance 10 m on the plot of *Polyalthia fragrans* and *Swietenia mahogany*, but the concentration increases again at the distance 30 m. At the distance 10 m and an elevation 1.5 m, the highest decline percentage of CO occurs on the plot of *Swietenia mahogany* (45.1%), on the plot of *Polyalthia fragrans* is just 22.2%, while in control plot, it increases by 2.2%. At the distance 30 m and elevation 1.5 m, the concentration increased again on all of the plots. Thus roadside green belt with a thickness 10 m is not effective in reducing the concentration of CO at the distance 30 m or in residential areas.

1. Introduction

The increase of population resulted in high demand for means of transportation as migration tool of both people and goods from one to another place. Based on the data of Badan Pusat Statistik Kota Bogor (2013), the population of Bogor city in 2012 increased by 3.87% from the previous year. While the number of public transportation in Bogor city has reached 4,644 units in 2012, it increases by 3.61% from the previous year. The number of vehicle certificate issued by the Bogor Police is 17.112

* Please direct further correspondence with this author

for passenger cars, 2.935 for freight cars, 142 for buses and 55.444 for motorcycles. High volume of vehicle, the high pollutants released so that the air pollution level in major cities of Indonesia, including Bogor continues to rise.

Pollutants could be defined as chemicals which are added or inserted into the atmosphere through human activity causing the increase of substances concentration (Krupa 1997). Pollutants could be classified into primary pollutant and secondary pollutant. The primary pollutant is a pollutant directly released into environment from natural processes and human-related activities. Primary pollutants could be deposited onto the surface of the ground or carried by the wind and fused with other primary pollutants to form secondary pollutants. Pollutants could be in the form of gas and solid particles. The negative impact of air pollution is harmful for human health, vegetation, ecology and proverty. Therefore, air pollution should be controlled in an integrated manner from a variety of approaches in order to be fulfilled healthy air quality.

Based on Government Regulation Number 41 Year 1999 which establishes air quality standards, types of air pollutants quality standard has been set including SO_2, CO, NO_2, oxidants (O_3), hydrocarbons, PM10, PM22, Pb and Dust. To achieve the quality standard of air quality, emission limitation and reduction of pollutants from the air is one of effective approach. Pollutants that have been released into the atmosphere could only be reduced by vegetation in green open spaces, and reduced by natural processes such as rain washing pollutants in the atmosphere or the flow of wind diffusing pollutants into the wider atmosphere so that the concentration of pollutants could be decreased.

Pollutants produced from combustion of fossil fuels are gas emissions of vehicles such as CO, hydrocarbons (HC), SOx, NOx and particles. The most harmful pollutants are the particles, NOx, SOx, hydrocarbons, and the lowest is carbon monoxide (CO) (Darmono 2001). Even though CO is less toxic than other pollutants, CO is categorized as the main gas pollutants which reach almost half of all existing air pollutants. CO is a colorless, odorless and tasteless gas emitted from combustion process, so that its existence is not recognized by humans. CO could be harmful to health because it could combine with haemoglobin in the blood to form COHb (Krupa 1997).

One of the most important green spaces in the city is roadside green belt. Roadside green belt consists of vegetation planted outside the roadside or in median of the road. Vegetation on roadside green belt not only provides the aesthetics and comfort, but also it reduces the negative impacts of transportation activity such as air pollution and noise into the environment (Gary 2014).

Naturally diverse forms of tree canopy consist of rounded, cylindrical, umbrella, palm, heart, and fan shape (Cheng, 2011). Moreover, the density of the leaves will determine the shade of the tree canopy. In planning roadside green belt, it is necessary to apply the form of trees which effectively reduces air pollution emitted from the highway. Therefore, the effectivity of various tree canopy types on roadside green belt in reducing the concentration of air pollution should be investigated, especially CO gas in the vicinity of the road by examining the distribution of pollutant concentrations horizontally and vertically.

2. Literature Review

Pollutants produced from a varied source. Krupa (1997) suggests that the pollutants could be produced from static or dynamic sources. Static source is a source of pollutants that has a stationary position, such as a volcano or a factory chimney. Dynamic source is a source of pollutants that could move the position, for example a motor vehicle. Pollutants generated from the motor vehicle have multiple mechanisms of distribution, which spread to sideways and upwards (diffusion).

Pollutants produced from combustion of fossil fuels are emissions of motor such as CO, hydrocarbons (HC), SOx, NOx and particles. The toxicity of the pollutants is different. The most harmful pollutants to health are the particles, followed by NOx, SOx, hydrocarbons, and the lowest is carbon monoxide (Darmono 2001). Even though they are less toxic than other pollutants, but carbon monoxide (CO) is categorized as the main pollutants that reach almost half of all existing air pollutants. CO is colorless, odorless and tasteless gas, so that its existence is not recognized by humans. CO tends to diffuse from the near-surface soil and fills an enclosed space which can be very harmful to humans. CO could be harmful to health because it could combine with haemoglobin in the blood to form carboxyhaemoblobin (COHb). CO is highly toxic at concentrations greater than 1000 ppm, it could cause asphyxiation. It is the condition when body and brain in shortage of O_2 because blood has bind CO (Krupa 1997). In addition, CO could affect the nervous and cardiovascular systems such as hardening of the arteries. This can lead to disruption of the blood supply to parts of the body such as the heart muscle that cause chest pain until death.

To reduce the level of pollutants in the air, several ways could be applied, either reduce pollutants at source (effluent) or reduce pollutants in ambient air. In order to reduce pollutants at source, some ways could be applied are improve the combustion process, select raw materials that do not produce high levels of pollutants, and use better technology. In order to reduce the level of pollutants at the source, the biologic mechanisms, namely the use of plants and vegetation is generally applied.

Plants as landscape elements of the road, either tree, bush or shrub has the potential and important role in reducing air pollutants. Plants have the potential morphologies such as leaf shape, leaf thickness, number of stomata, trichomes and others which support the potential absorption of air pollutants. The process of gas absorption by plants occurs mainly on leaves. Most of the gas exchange occurs in the leaf through the stomata (Gardner et al 1991). The ability of plants to absorb the gas from the air especially could be known from plant metabolic processes such as photosynthesis, respiration, transpiration, and stomata conductivity. According to Grey and Deneke (1978), species with a moderate to high stomata density has potential of good absorber of pollutant.

The streets in the city or on the highway have been completed by the roadside green belt, either at the roadside or median of the road. A variety of plants that exist on roadside green belt could be optimized to reduce the levels of pollutants generated by motor vehicles. An adequate existences of green open space with the type of vegetation that can absorb pollutants effectively becomes a very important requirement for the health and comfort to urban community.

3. Objectives of Research

This research has a common goal to study differences in the ability of different tree canopy to absorb CO gas pollutant, while the specific objectives are:

- to determine tree canopy type that effectively reduce CO gas
- to determine the relation between tree canopy types and pollutant distribution vertically and horizontally

3. 1. Expected Output

The results of this study are expected to give an idea of the importance of plant selection with tree canopy which is effective in reducing pollutants on roadside green belt, and it could be useful for guidelines in the selection of plants, especially tree, in the preparation of planting plan on roadside green belt.

3.2. Activity Overview

Similar study has been conducted is the research related to the function of vegetation in reducing the levels of pollutants in ambient air. Aspects that have been studied include species with high capacity to adsorb pollutant particles of lead (Pb), nitrogen oxides (NOx) as well as the width of the green belt relationships which are effective in reducing pollutants. The new aspect had been examined is the relationship between the type of tree canopy and its effectivity in reducing CO gas. The result of this research activity is expected to complete the preparation of guidelines for plant selection, especially trees that effectively reduce CO gas for roadside green belt.

4. Research Methods

4.1. Location and Time of Research
To conduct the research regarding the relation between canopy shapes and reduction of pollutants' concentration in vicinity of the road had been conducted in the end of June 2015 on roadside green belt of Jagorawi Highway in Cibubur from km 8 - 12.5 on Jakarta – Bogor track. Air sampling was conducted for measurement of CO gas on June 11, 2015.

Preparation of tools and materials for CO measurement of air quality parameters as well as the analysis was conducted in the laboratory of Environmental Research Centre (PPLH) LPPM IPB.

Table 1. Canopy shape of green belt and location at Jagorawi Highway

Canopy shape	Tree species	Position (km)	Sampling time/ Replication
Rounded	Mahoni (*Swietenia macrophylla*)	8.800	10 - 11 am, 01 - 02 pm, 04-05 pm
Cylindrical	Glodogan bulat (*Polyathia fragrans*)	9.800	10 - 11 am, 01 - 02 pm, 04-05 pm
Control	Open space	12.500	10 - 11 am, 01 - 02 pm, 04 - 05 pm

4.2 Implementation of Research

4.2.1. Measurement of CO Gas Concentration
This activity was conducted with air sampling on plot as following:
(1) Plot of green belt planted with rounded Mahogany (Swietenia macrophylla) (2) Plot of green belt planted with cylindrical Glodogan (Polyalthea fragrans), and (3) control plot in the area without green belt, open space). In each plot, sampling was conducted at the distances 0, 10, 30 m. Distance 0 m is the edge of road side green belt that close to road curb or at point 5 m from the edge of the road (asphalt). At each point, sampling was conducted at the elevation 1.5, 5, 10 m above soil surface. Air sampling was conducted using Impinger Air Sampler.

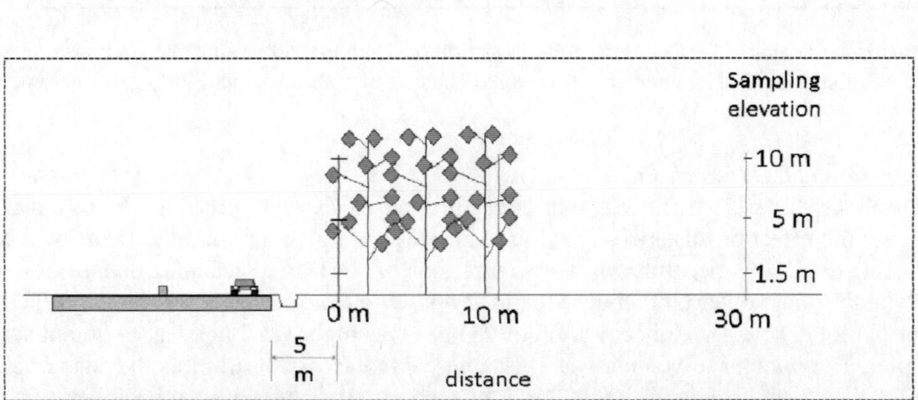

Figure 1. Sampling elevation of air quality measurement

IOP Conf. Series: Earth and Environmental Science **31** (2016) 012031 doi:10.1088/1755-1315/31/1/012031

4.2.2. Measurement of CO Level in the Laboratory

Measurement of CO was conducted following the method of Iodometri in PPLH IPB Laboratory.

4.2.3. Data Analysis

The effectivity of the reduction of CO gas concentration by each type of canopy will be found by comparing air quality from the results of measurement on green belt with air quality on control plot. The effect of roadside green belt to concentration reduction of pollutant would be formulated through experimental design with three factors: type of canopy of green belt, distance from the road, and the elevation above soil surface. Furthermore, to determine the influence of the experiment factors, the data of CO gas concentration was analysed using ANOVA. If the experiment factor was significant, test of Duncant Multiple Range at the level of $\alpha \leq 5\%$ would be applied. To assess which canopy type better in reducing concentration, the value of concentration reduction rate are compared between plots.

4.2.4. Formulation of Recommendation

Recommendations for the selection of plant type which could be used for planting plan for roadside green belt based on canopy types of plant which has the highest effectivity of pollutant reduction and the relation between the shape of the canopy and reduction of pollutant concentration.

5. Results And Discussions

5.1. Source of Pollutant

Vehicles passing on Jagorawi Highway are categorized as dynamic source of pollutant. At the time of CO gas sampling, the volume of vehicles reached 10.163-13.035 vehicles per hour (Table 2). Table 2 shows the emissions on June 11, 2015 decreased at noon, and then increased again in the afternoon.

Table 2. Time of air sampling, total vehicle passed the Highway, and estimation of CO gas emitted by vehicle

Time of sampling	Total Vehicles	Estimation of CO emission / km (g)	Wind velocity (m/second)		
			Polyalthia	Mahagony	Control
10 – 11 am	13.035	689.163	0.3- 0.5	1-1.5	0.5
01 – 02 pm	13.270	617.360	0.6-0.7	1.5-1.7	0.5-0.9
04 - 05pm	10.163	769.077	1.2 – 1.7	2-3.1	0.5 – 0.6

Description:
Emission of gasoline vehicle 60 g CO / km, emission of diesel vehicle 0.69 gram CO / km based on Strauss and Mainwaring in Sulistjorini (2009). The vehicle assumed consists of 10% diesel and 90% gasoline type vehicle.

5.2. Concentration of CO in each Plot

Results of analysis of variance showed that concentration of CO was very significanly differ among plots. However the effect of distance from the road, elevation of sampling, and their interraction was not significant to CO concentration. Concentration of CO in *Swietenia mahagony* plot was significantly higher than other plots, but value in *Polyalthia fragrans* plot was not significantly differ against control plot (Table 3). Concentration of CO gas was higher at 0 m or at pollutant source point than other plot. The number of vehicle was the same passed across of all plots, but wind velocity was the fastest in *Swietenia mahagony plot* (1-3.1 m/second). It is probably the higher wind velocity enhaced more CO pollutant distribution to *Swietenia mahagony* plot.

Table 3. Average of CO concentration in the plot of air sampling in several distance from the road.

Distance (m)	Plot			Average
	Polyalthia fragrans	*Swietenia mahagony*	Control/ open field	
0	2896.1	7620.2	3074.4	4530.3
10	2573.0	3581.6	2919.0	3024.9
30	2964.6	9065.0	3341.3	5123.6
Average	2811.5[b]	6755.7[a]	3111.7[b]	

Description. Values which it is followed by different alphabet character is significantly different by Duncant Multiple Range Test at the level of $\alpha \leq 1\%$.

5.3. The Distribution of CO Gas Concentration

Distribution of CO gas concentration on plot of *Polyalthia fragrans* is presented in Figure 2. Analysis of variance performed effect of sampling distance and elevation and their intreraction to CO concentraion were not significant. Vertical distribution shows at the distance 0 m, the concentration tend to increases with the increasing of sampling elevation, at the distance 10 m the concentration at the elevation 10 m is higher than elevation 0 m, while at the distance 30 m, the concentration slightly decreases with the increasing of sampling elevation.

Horizontal distribution at the elevation 1.5 m shows that concentration of CO tend to increases with the increasing of distance up to 30 m. At the elevation 5 m, the concentration decreases at the distance 10, then increases again at the distance 30 m. While, at the elevation 10 m, the concentration decreases until at the distance 30 m.

If the concentration at the distance 0 m as a reference, so that on plot of Polyalthia with the elevation 5 m and 10 m, the concentration decreases at the distance 10 m by 22.2% and 21.6%, and at the distance 30 m and elevation 10 m, concentration decreased by 22.3% (Table 4).

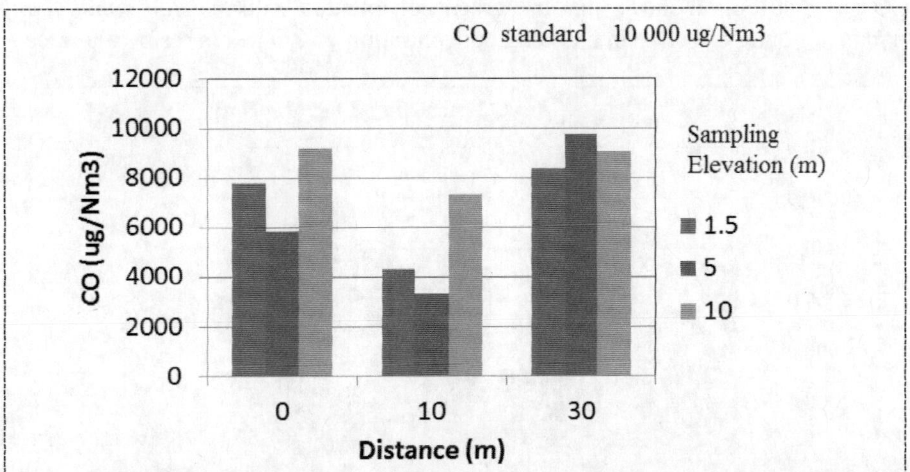

Figure 2. The concentration of CO gas on plot of Glodogan

Distribution of CO gas on plot of *Swietenia mahogany* is shown in Figure 3. Analysis of variance performed effect of sampling distance and elevation and their intreraction to CO concentraion in Swietenia mahagony were also not significant.Vertical distribution of CO gas shows generally at the distance 0 and 10 m, the concentration tends to decrease at the elevation 5 m and then increase again at the elevation 10 m, while at the distance 30 m, the concentration increases with the increasing of elevation sampling point.

Horizontal distribution of CO gas on plot of *Swietenia mahogany* shows that at the elevation 1.5, 5, 10 m, the concentration decreases at the distance 5 m, and then increases again at the distance 30 m. If the concentration at the distance 0 m as a reference, at the elevation 1.5, 5, and 10 m, the concentration at the distance 10 m declines by 45.2%, 43% and 20.4% respectively, while at the distance 30 m and elevation 10 m , the concentration decreased by 1.4% (Table 4).

Table 4. Concentration in distance 10 m and 30 m if concentration at the distance 0 used as reference in each plot of sampling

Elevation (m)	Distance (m)						
	0	10			30		
		Mahagony	Polyathia	control	Mahagony	Polyathia	control
	%						
1.5	100	54.9	119.2	102.2	107.4	134.9	105.4
5	100	56.9	77.8	52.8	166.6	108.2	82.8
10	100	79.6	78.4	248.2	98.6	77.7	171.2

In each plot, thickness of the trees is 10 m, the land use from distance 10 m to 30 m is open space. Especially on plot of *Swiethenia mahogany*, the land use at the distance 30 m is residential area. This shows that CO gas concentration decreases with the existence of vegetation. At the distance 30 m, concentration increases, it might be caused by additional CO gas flowing from elevation 10 m. This pattern of CO gas distribution shows that the green belt planted with *Swietenia mahogany* with the thickness 10 m is not quite effective in reducing concentration of CO gas in residential area.

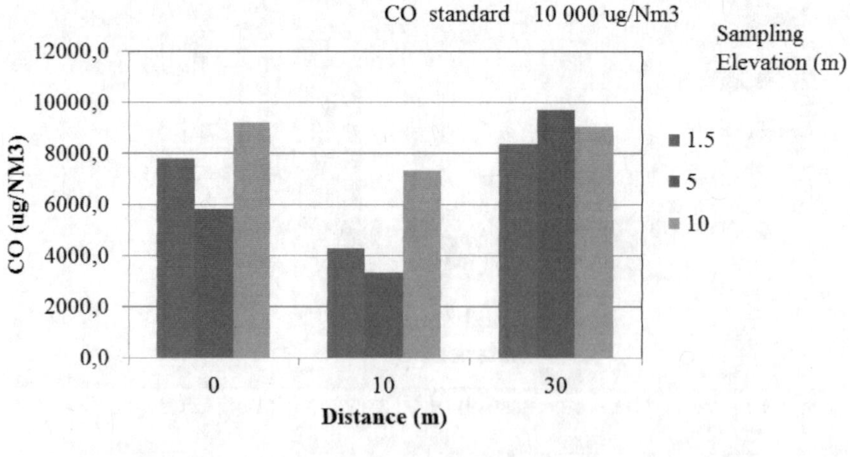

Figure 3. The concentration of CO gas on plot of Mahogany

The distribution of CO gas on control plot is presented in Figure 4. Vertical distribution shows that the concentration tends to decrease with the increasing of sampling elevation.

Horizontal distribution shows that at the elevation 1.5 m, concentration is not reduced by the addition of distance up to 30 m. At the elevation 5 m, the concentration tends to decrease with increasing of distance up to 30 m. At the elevation 10 m, the concentration increases at the distance 10 m, then declines at the distance 30 m. If the concentration at the distance 0 as a reference, then the control plots at the elevation 5 m, the concentration decreased by 47% at the distance 10 m, and decreased by 17% at the distance 30 m (Table 4).

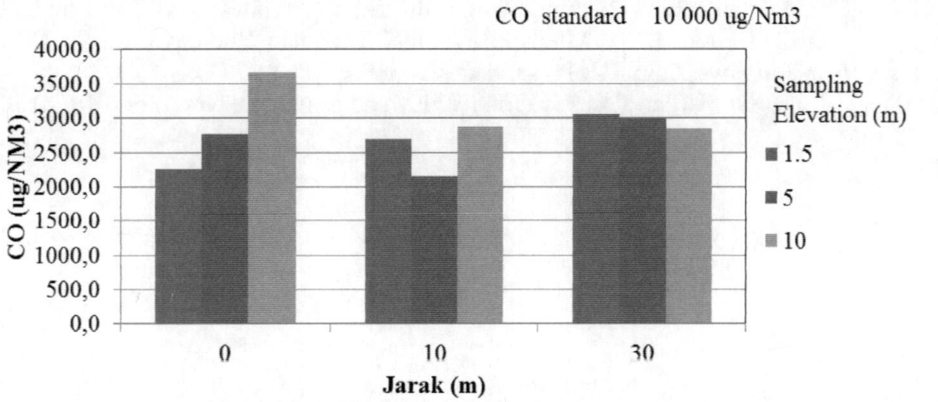

Figure 4. Concentration of CO gas on control plot

If it is compared to the control plot, road side green belt of *Swiethenia mahogany* could reduce the concentration at the elevation 1.5 m and distance 10 m with a higher percentage than *Polyalthia fragrans* and control.

6. Conclusions and Recommendations

Distribution of CO gas concentration increases with the increasing of sampling elevation on the plot of *Polyalthia fragrans* and *Swietenia macrophylla*, but the contrary trend is found in the plot of control.

Roadside green belt of *Swietenia macrophylla* with rounded canopy type is more effective than *Polyalthia fragrans* with cylindrical canopy type in reducing CO concentration at the distance 10 m, but concentration rebound at the distance 30 m. Therefore roadside green belt with the width 10 m is not yet effective to reduce CO concentration in residential area.

It is suggested to study CO gas distribution in wider roadside green belt in order to know the minimum width that effective to protect environment at the vicinity of road.

Acknowledgement

We would like to thank to DIKTI (Direktorat Jenderal Pendidikan Tinggi) who support this research through BOPTN (Bantuan Operasional Pendidikan Tinggi Negeri). We also would like to thank to PT JASA MARGA who has permitted us to conduct this research.

References

[1] A. Gary, Green Infrastructure for Landscape Planning (Routledge, New York, 2014).
[2] Badan Pusat Statistik Kota Bogor, Bogor dalam Angka (BPS, Bogor, 2013).
[3] Darmono, Lingkungan Hidup dan Pencemaran: Hubungannya dengan Toksikologi Senyawa Logam (UI Press, Jakarta, 2001).
[4] S. V. Krupa, Air Pollution, People, and Plants (The American Phytopathological Society, Minnesota, 1997.
[5] Chen G, Landscape Architecture: Planting Design Illustrated (ArchiteG. Inc, California, 2011).
[6] Gardner F. P., R. B. Pearce, R. L. Mitchell, Fisiologi Tanaman Budidaya, Susilo H, translator, translation from: Crop Physiology (UI Press, Jakarta, 1991).
[7] Grey G. W., F. J. Deneke, Urban Forestry (John Wiley and Sons. Inc, New York, 1978)

IOP Conf. Series: Earth and Environmental Science **31** (2016) 012032 doi:10.1088/1755-1315/31/1/012032

Analysis of the Equatorial Lower Stratosphere Quasi-Biennial Oscillation (QBO) Using ECMWF-Interim Reanalysis Data Set

Givo Alsepan[1]*, Sandro Wellyanto Lubis[2], Sonni Setiawan[1]

[1] Department of Geophysics and Meteorology, Bogor Agricultural University (IPB), Indonesia
[2] Physics of the Atmosphere, GEOMAR Helmoltz Centre for Ocean Research Kiel, Germany

E-mail: givoalsepan@yahoo.com

Abstract. The ERA-Interim data set from Europe Center for Medium Range Weather Forecasting (ECMWF) was used to quantitatively analyze the characteristic of equatorial quasi-biennial oscillation (QBO). Analysis of spatial and temporal of the data showed that the zonally symmetric easterly and westerly phase of QBO regimes alternate with period of ~27.7 months. Based on Equivalent QBO Amplitude (EQA) method, the maximum amplitudes in zonal mean zonal wind (u), temperature (T), vertical shear (du/dz) and quadratic vertical shear (d^2u/dz^2) are ~28.3 m/s , ~3.4 K, ~4.8 m/s/km, and ~1.0 m/s/km^2 respectively. The amplitudes decay exponentially with a Gaussian distribution in latitude. The twofold-structure of QBO descends downward at rate of ~1 km/month. The temperature anomaly can be used to analyze the characteristic of QBO which satisfies the thermal wind balance relation in the lower-stratosphere due to very small contribution of the mean meridional and vertical motion. Moreover, the concentration of the total column ozone (TCO) in the tropics is significantly influenced by QBO. During the westerly phase of QBO, the TCO is relatively increased in the lower-stratosphere, but decreased during the opposite phase.

1. Introduction

The stratospheric region in the equator is dominated by descending downward of westerly and easterly wind which was first introduced by Ebdon (1960) and Reed *et al* (1961), and then it was called as Quasi-Biennial Oscillation (Angell and Korshover 1964). QBO is an oscillation of westerly and easterly wind descending downward in the lower-stratosphere of the equatorial region (~16 – 50 km) with period of 28 months in average (Baldwin *et al* 2001). The easterly phase of QBO has greater maximum amplitude than the westerly phase, and the amplitudes have symmetrical pattern in the equator with a Gaussian distribution (Baldwin *et al* 2001; Pascoe *et al* 2005).

Mechanism of QBO was first shown by Holton and Lindzen (1972), where QBO is a result of the interaction between mean flow and the equatorial planetary waves. Holton and Lindzen were the first

*Corresponding author

people who conducted the modeling of QBO based on the propagation of Kelvin waves and Rosby-gravity waves. Holton (2004) explained that QBO is only occurred in the equator because Kelvin waves and Rosby-gravity waves are only formed in a region around 12^0 N – 12^0 S. The equatorial region provides a great energy originated from releasing of laten heat in macro scale of convective clouds for creating the planetary equatorial waves and forcing them to the lower-stratosphere. QBO is not occurred in high latitude regions because there is no enough energy to force the equatorial planetary waves to the lower-stratosphere, although there is a formation of clouds and "front" area in high latitude.

In addition to discuss about zonal wind oscillation, QBO also influences the chemical and physical process in the lower-stratosphere. Holton (2004) has proved that QBO satisfies the thermal wind equation if wind shear of zonal wind is associated with temperature in the lower-stratosphere. Meanwhile, Baldwin *et al* (2001) showed that temperature and ozone data sets can be used to analyze QBO.

This study uses ERA-Interim data sets from ECMWF (Europe Center for Medium Range Weather Forecasting). ERA-Interim is the last atmospheric reanalysis data sets produced by ECMWF after ERA-40. These data sets are available from January 1979 to present and has the pressure levels from 1000 to 1 hPa (Dee *et al* 2011).

The aims of this study are to quantify the characteristics of QBO (period, oscillation amplitude, propagation rate, Gaussian distribution of amplitude), to prove the thermal wind equation, and to show the relationship between QBO and total column ozone (TCO) using zonal wind, temperature, and TCO data sets obtained from ECMWF-Interim Reanalysis.

2. Methods

ERA-Interim data sets from ECMWF (Europe Center for Medium Range Weather Forecasting) during 30 years (January 1981 to December 2010) were used in this study, and the spatial resolution of the data is 220 km x 220 km. The data are monthly data sets from 1 hPa (~49 km) to 1000 hPa consisting of zonal wind (u), temperature (T), and total colum ozone (TCO). Temperature and zonal wind data sets are used to quantify QBO and to analyze the thermal wind balance relation, and total column ozone data sets are used to analyzed the relationship of QBO and the fluctuation of ozone in the lower-stratosphere.

2.1. Equivalent QBO Amplitude (EQA)

EQA was first introduced by Randel *et al* (2004). This method is developed to calculate the amplitude of QBO by determining the standard deviation of zonal wind anomaly, then the standard deviation is multiplied by $\sqrt{2}$. The mathematical formula can be formulated as follows:

$$EQA = s_{anomaly}\sqrt{2} \quad (1)$$

EQA method can also be used to investigate the characteristic of QBO using other parameters such as temperature, wind shear, and quadratic vertical shear.

2.2. Spectral Analysis

Spectral analysis is used to calculate a hidden period or to estimate a spectral density of time series data. This method is a modification of Fourier analysis. Fourier analysis can be represented by using Fourier series, where every periodic function can be represented by unlimited sum of sine and cosine function. Mathematical function of Fourier series can be formulated as follows:

$$f(t) = \frac{a_0}{2} + \sum_{i=1}^{\infty}\left(a_n \cos\frac{n\pi t}{1} + b_n \sin\frac{n\pi t}{1} \right)$$

$$a_n = \frac{1}{1}\int_{-1}^{1} f(t)\cos\frac{n\pi t}{1}\,dt$$

$$b_n = \frac{1}{1}\int_{-1}^{1} f(t)\sin\frac{n\pi t}{1}\,dt \quad (2)$$

Spectrum $f(\omega)$ is a Fourier transformation of autocovariance function which can be formulated as follows:

$$f(\omega) = \frac{1}{\pi}\sum_{k=-\infty}^{\infty} \gamma(k)\exp^{-i\omega k} \quad (3)$$

2.3. Cross-Correlation

Cross-correlation is used to explain an association between two non-linear randomly variables. In analysis of time series, cross-correlation between two time series data describes cross-correlation coefficient $r_{xy}(k)$ which is calculated by using a simple formula of Box and Jenkins (1976) as follows:

$$r_{xy}(k) = \frac{c_{xy}(k)}{s_x s_y} \quad (4)$$

Where k is a *lag* with $k = 0, \pm 1, \pm 2, ..., n$ and $s_x s_y$ is a standard deviation of series x and y respectively. Meanwhile, $c_{xy}(k)$ is a cross-covariance function at $lag(k)$ which can be obtained from:

$$c_{xy}(k) = \frac{1}{N}\sum (x_t - \bar{x})(y_{t+k} - \bar{y}); t = 1; N-1; k \geq 0 \quad (5)$$

Where $\bar{x}$ and $\bar{y}$ are the average of respective variables and t is time. Cross-correlation is generally used to determine *lag* value of two non-liniear time series data.

3. Results

3.1. Characteristics of QBO

Time-height section of monthly zonal mean zonal wind from ERA-Interim data sets during 1981 to 2010 was shown in Figure 1, and the result is appropriate with the time-height section found by Naujokat (1986) using radiosonde data sets and Gray et al (2001b) using rokcetsonde data sets. Oscillation of westerly and easterly zonal wind is dominantly occurred in the lower-stratosphere about ~100 – 5 hPa (~16 – 35 km), meanwhile the atmospheric layer above 5 hPa is dominated by Semiannual Oscillation (SAO) which is appropriate with the result of Pascoe *et al* (2005). Generally, easterly zonal wind is stronger than westerly zonal wind.

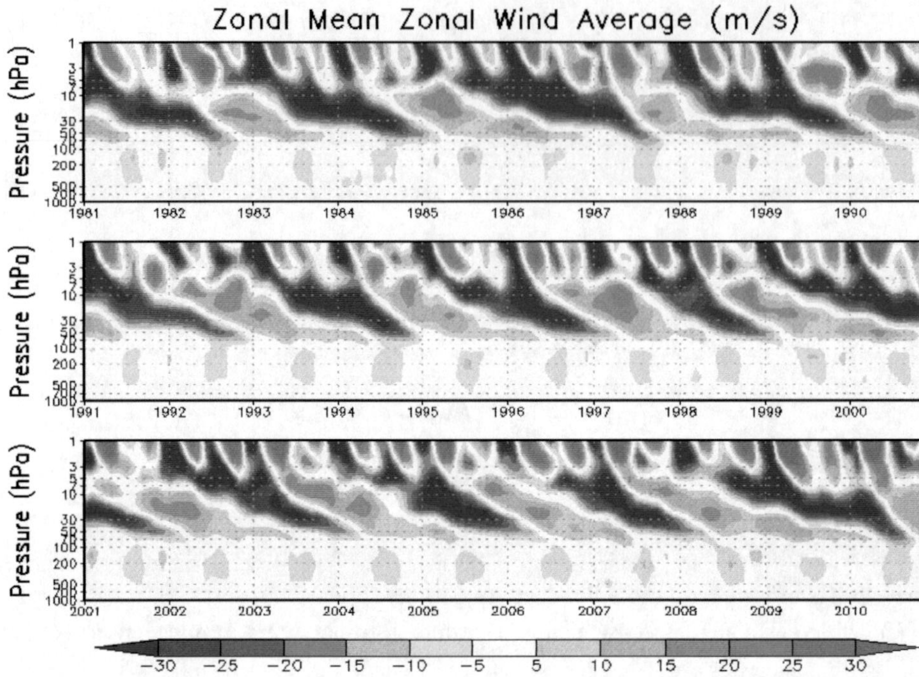

Figure 1. Monthly zonal mean zonal wind using ERA-Interim data sets (1981 to 2010) in a function of time and pressure. Blue color indicates easterly zonal wind and red color indicates westerly zonal wind.

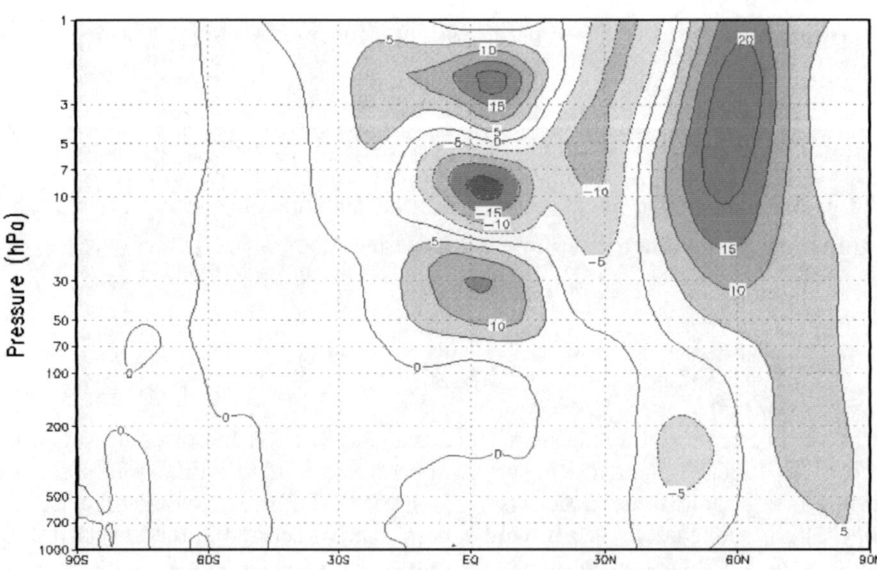

Figure 2. Latitude-height section of zonal mean zonal wind anomaly based on ERA-Interim data sets in January 1981. Blue color indicates easterly zonal wind and red color indicates westerly zonal wind.

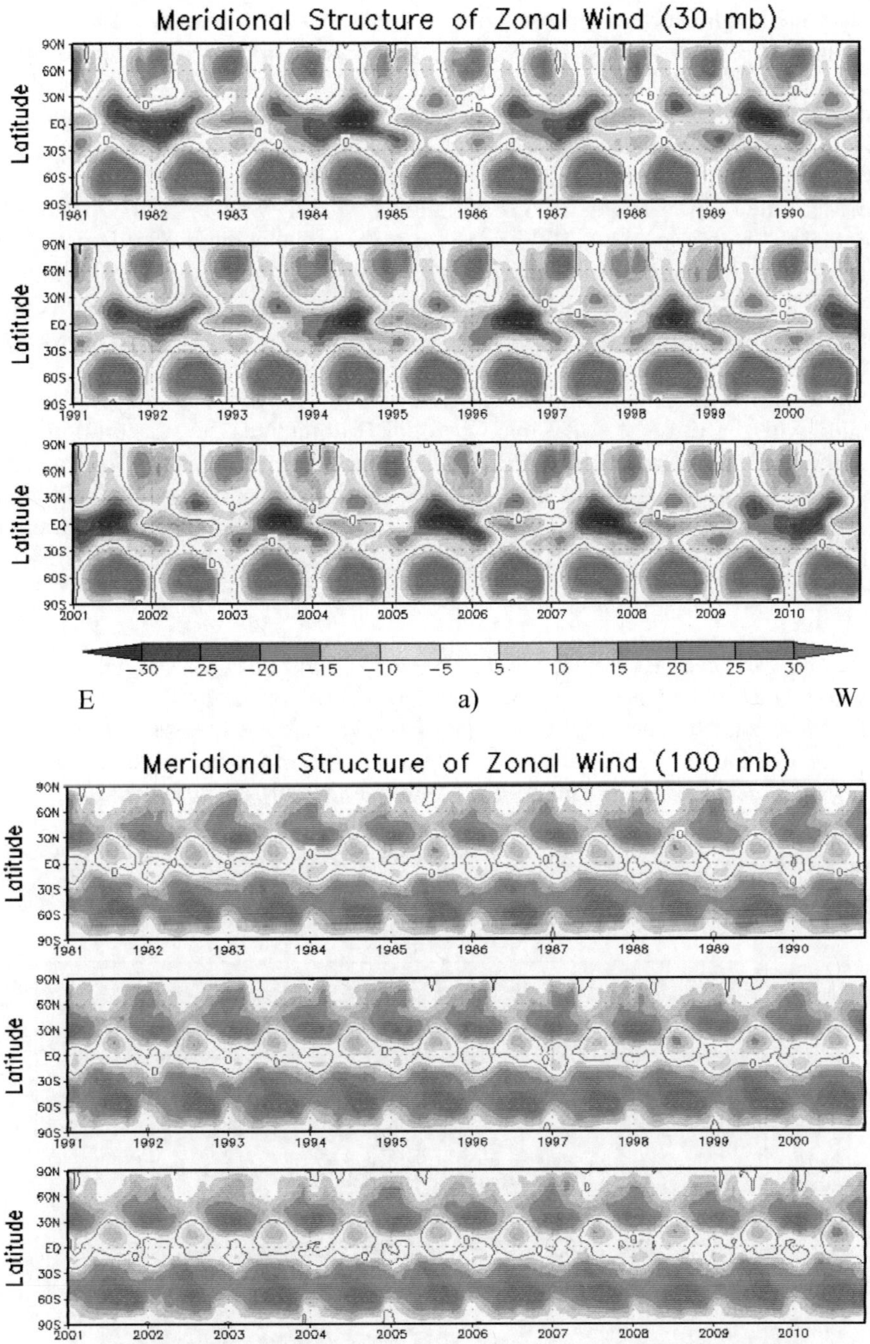

Figure 3. Meridional structure of zonal mean zonal wind. a) at 30 hPa, b) at 100 hPa.

Figure 2 is latitude-height section of zonal mean zonal wind in January 1981 which shows the threefold-structure. The easterly phase (E) is above the westerly phase (W) of zonal wind at ~100 – 5 hPa in the equator, so it is called as the twofold-structure of QBO. Another structure above ~5 hPa in the upper-stratosphere is indicated as SAO. The structures are symmetry in latitude, where the amplitude is increased in the equator and decreased in the high latitude.

Based on time-latitude section in Figure 3, oscillation of zonal wind is clearly occurred at ~30 hPa (Figure 3a) compared with a level at ~100 hPa (Figure 3b) in the equator. Oscillation of westerly and easterly zonal wind at ~100 hPa can not be seen clearly, so it indicates that oscillation of QBO is stronger at ~30 hPa than at ~100 hPa. In Figure 3, we can also see that oscillation of QBO ascends in the equator and descends in the higher latitude.

Figure 4a shows that the maximum amplitude of QBO is at ~20 hPa (~27 km) above the surface of equatorial region with the value of ~28.3 m/s where QBO dominates the variability of zonal wind. In the other side, the maximum amplitude of QBO in temperature is occurred at ~30 hPa (~25 km) above the surface with the value of ~3.4 K (Figure 4b). The amplitudes of zonal wind and temperature are increased in the equator, but they are exponentially decreased with a Gaussian distribution in latitude (Figure 5). Those conditions indicate that QBO is only occurred in the equatorial area. According to Holton (2004), QBO is only strongly formed in the area around 12^0 N – 12^0 S.

Derivative of zonal wind can also be used to analyze the structure of QBO. Wind shear (du/dz) is the first derivative of zonal wind. The result shows that QBO in the lower-stratosphere with the maximum amplitude of ~4.8 m/s/km is occurred at ~30 hPa (~25 km) above the surface (Figure 6a). Second derivative of zonal wind is a quadratic vertical shear (d^2u/dz^2) which shows that the maximum amplitude of QBO is occurred at ~50 hPa (~21 km) above the surface with the value of ~1.0 m/s/km^2 (Figure 6b). The amplitudes of QBO in wind shear and quadratic vertical shear are increased in the equator, but they are exponentially decreased with a Gaussian distribution in latitute (Figure 7).

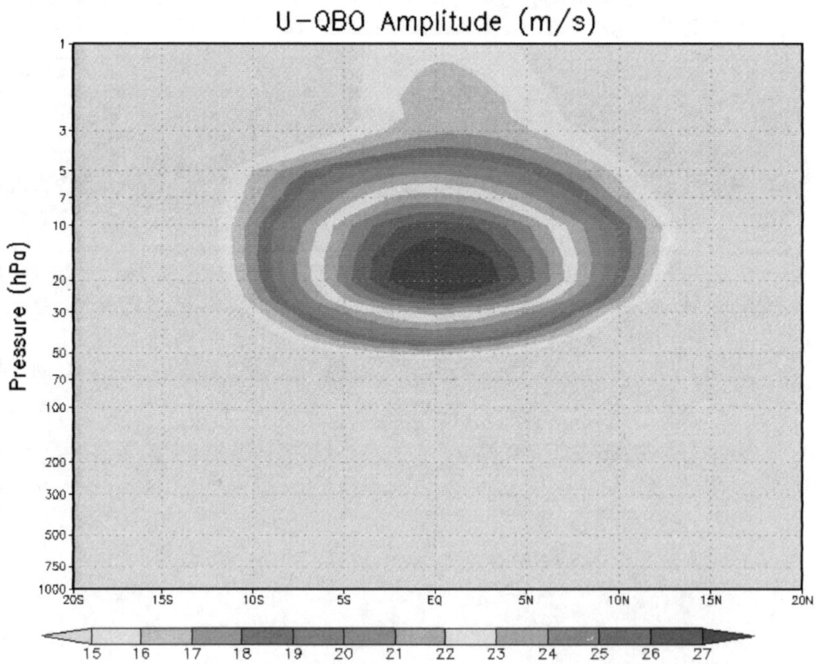

a)

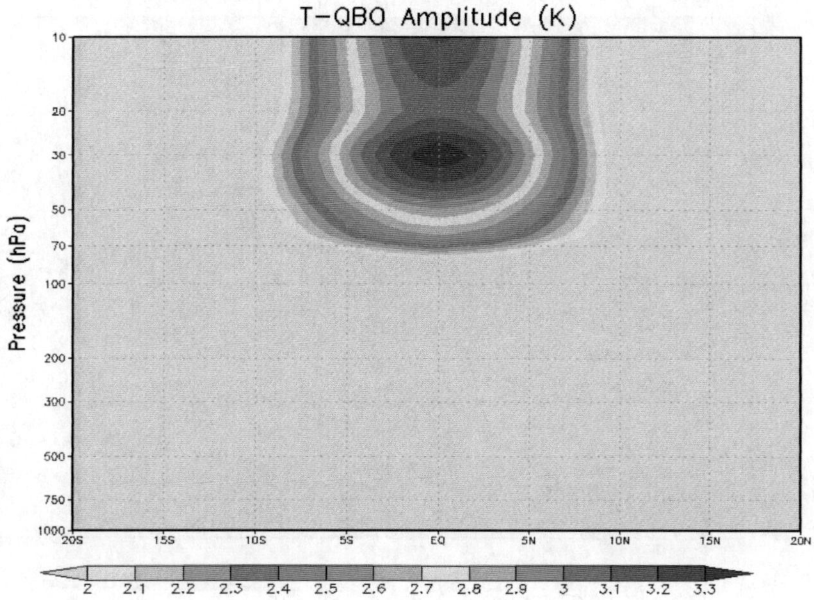

b)

Figure 4. Amplitudes of QBO using EQA method. a) U-QBO amplitude, b) T-QBO amplitude.

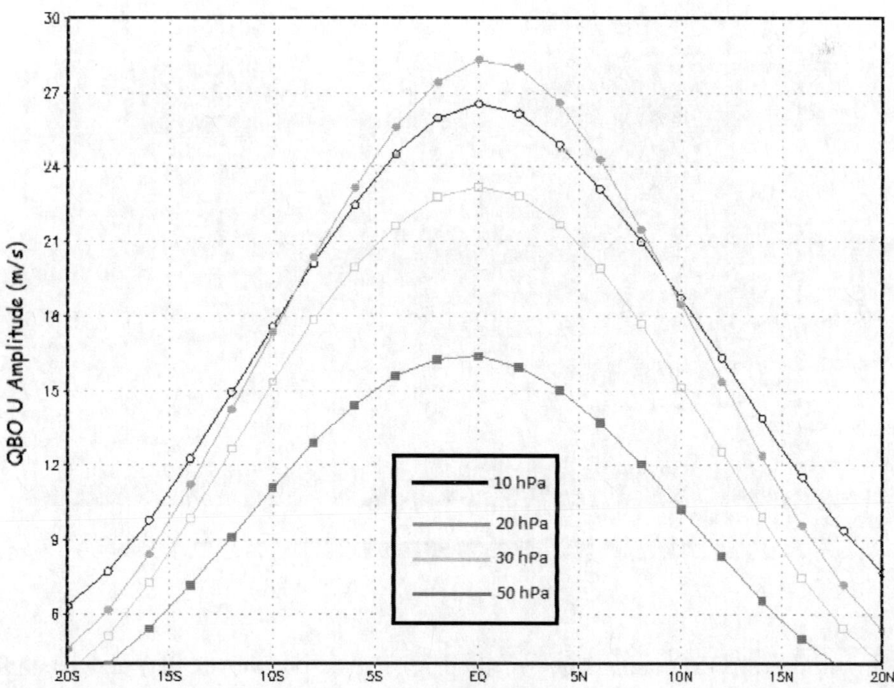

Figure 5. Meridional profile of U-QBO with a Gaussian distribution.

IOP Conf. Series: Earth and Environmental Science 31 (2016) 012032 doi:10.1088/1755-1315/31/1/012032

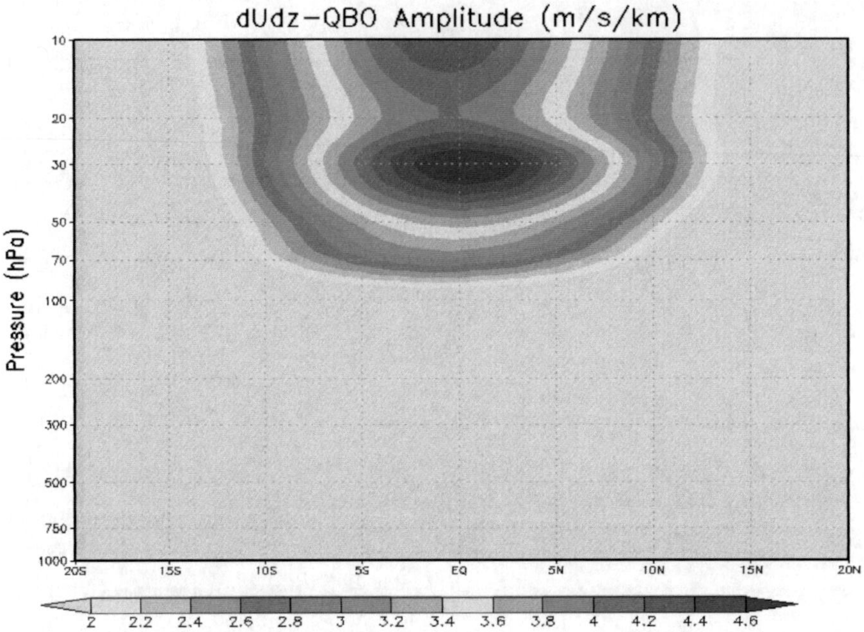

a)

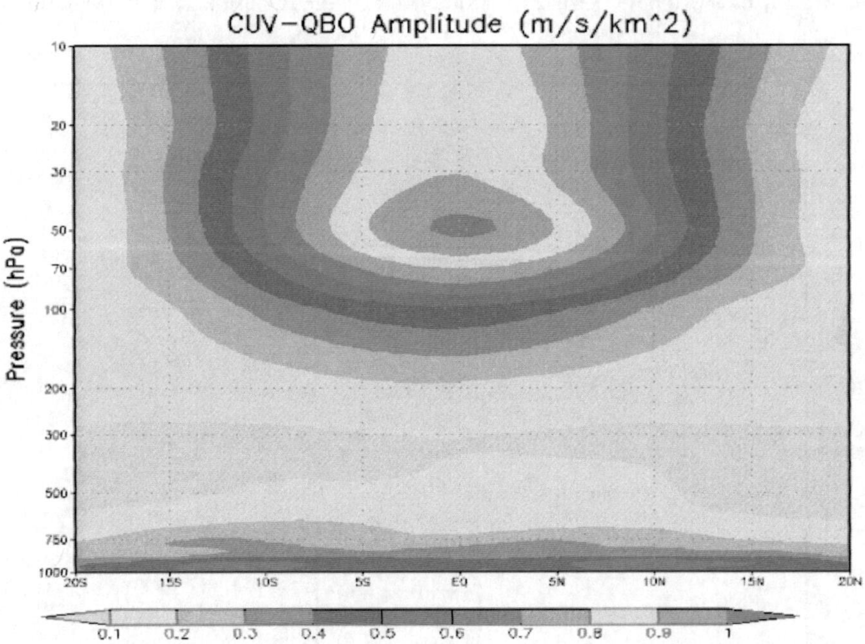

b)

Figure 6. Amplitudes of QBO using EQA method in first derivative (dU/dz) and second derivative (d^2U/dz^2) of zonal wind. a) dU/dz-QBO amplitude, b) d^2U/dz^2-QBO amplitude.

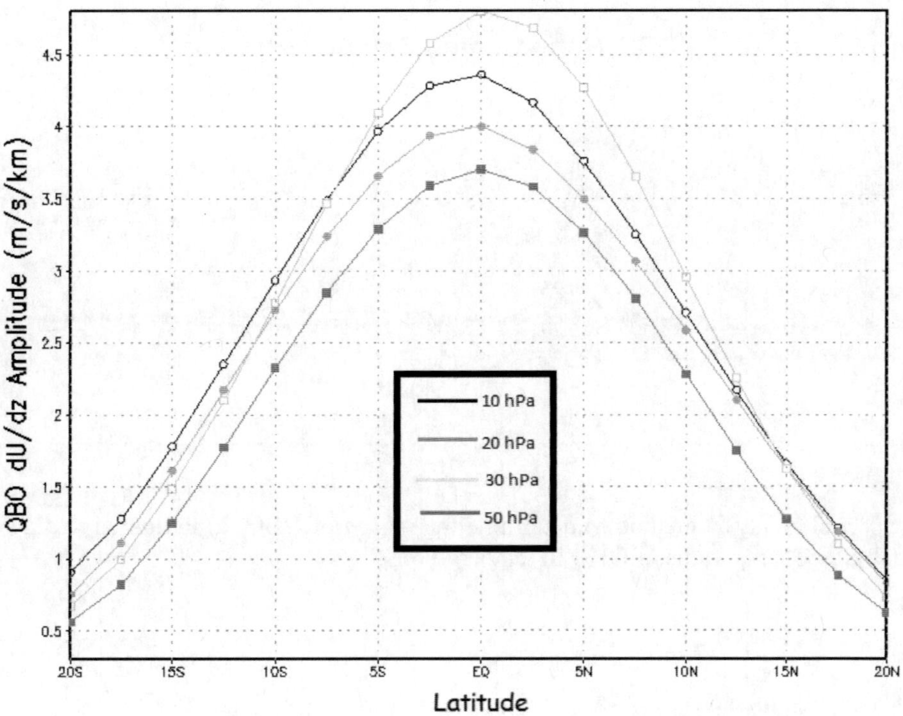

Figure 7. Meridional profile of dU/dz-QBO with a Gaussian distribution.

Based on spectral analysis, the oscillation of zonal wind shows a period of ~27.7 months (Figure 8a) at ~30 hPa above the surface. This result is marginally shorter than ~28.4 and ~28.2 months as measured by Baldwin *et al* (2001; observation) and Pascoe *et al* (2005; ERA-40) respectively. The similar period of oscillation is also occurred in temperature (Figure 8b).

Using cross-correlation method between zonal wind at 10 hPa and 30 hPa (Figure 9a) and at 10 hPa and 50 hPa (Figure 9b) above the surface, it can be calculated that QBO descends downward at rate of ~1.06 km/month. The result agrees with the propagation rate of ~1 km/month measured by Holton (2004).

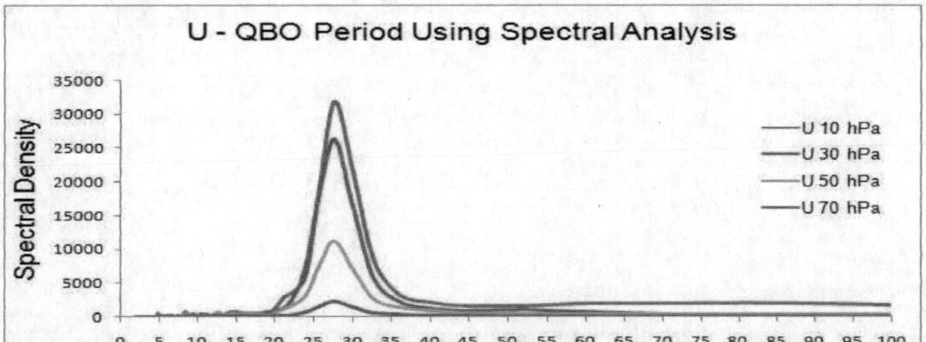

a)

IOP Conf. Series: Earth and Environmental Science **31** (2016) 012032 doi:10.1088/1755-1315/31/1/012032

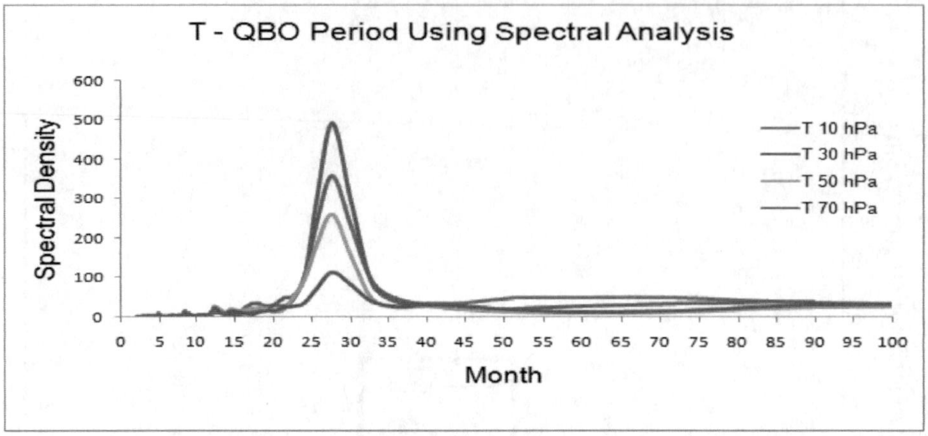

b)

Figure 8. Spectral Analysis method to determine the period of QBO. a) Periodicity value of QBO in zonal wind, b) Periodicity value of QBO in temperature.

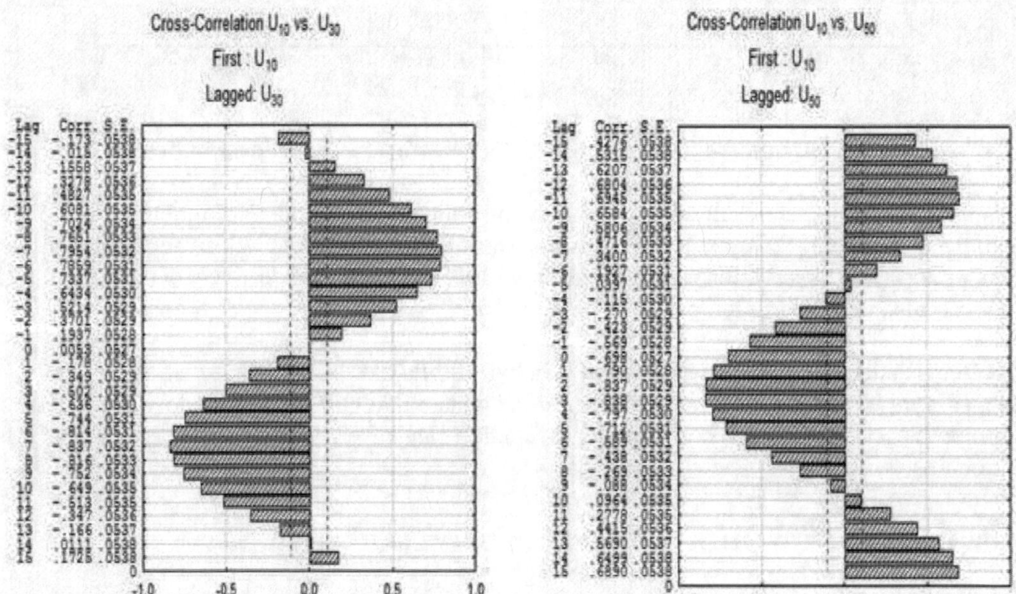

a) b)

Figure 9. Cross-correlation of zonal wind anomaly. a) Cross-correlation between zonal wind anomaly at 10 hPa and 30 hPa, b) Cross-correlation between zonal wind anomaly at 10 hPa and 50 hPa.

3.2. Relationship of Temperature and Wind Shear
Relationship of QBO-dU/dz (wind shear anomaly) and QBO-T (temperature anomaly) satisfies the thermal wind equation due to very small contribution of the mean meridional and vertical motion (Holton 2004). Anomaly of westerly wind shear associates with anomaly of warm temperature in the equator, and anomaly of easterly wind shear associates with anomaly of cold temperature (Pascoe *et al* 2005). Figure 10 shows the relationship between zonal mean of temperature anomaly at 30 hPa and zonal mean of wind shear anomaly at 50 – 30 hPa above the surface. Anomaly of temperature in QBO is consistent at ±4 K at 30 hPa (Baldwin *et al* 2001).

QBO is symmetry to the equator and the contribution of mean vertical and meridional motion is very small to the oscillation, so the relationship of wind shear and temperature in QBO satisfies the thermal wind equation. The equation can be formulated as follows :

$$\beta y \frac{\partial \overline{u}}{\partial z} = -\frac{R}{H}\frac{\partial \overline{T}}{\partial y} \tag{6}$$

In the equator, $\frac{\partial \overline{T}}{\partial y} = 0$ due to $y = 0$. By using L'Hopital rule, the thermal wind equation in the equator can be formulated as follows :

$$\frac{\partial \overline{u}}{\partial z} = -\frac{R}{H\beta}\frac{\partial^2 \overline{T}}{\partial y^2} \tag{7}$$

Where u is zonal wind, T is temperature, z is log-pressure height, y is latitude, R is gas costant of dry air, $H \approx 7$ km, and β is variation of Coriolis parameter in latitude (Andrews *et al* 1987).

The equation explains the meaning of Figure 10. Second derivative of temperature has an apposite sign with temperature in the equator, so wind shear and positive anomaly of temperature (warm) are in a same phase during the westerly phase of QBO. During the easterly phase of QBO, wind shear is in a same phase with negative anomaly of temperature (cold).

An exception is occurred in 1991, where temperature is suddenly increased (Figure 10 and 12). This event is caused by the eruption of Mt. Pinatubo (Philippine) on 14th – 15th of June 1991. The eruption created a giant cloud of aerosol in the stratosphere surrounding the earth during 2 weeks and spreading to the tropical region around 30^0 N – 20^0 S (McCormick and Veiga 1992). Sulphate aerosol of the eruption heated the stratosphere by absorbing the infrared radiation from the earth (Robock 2000). By absorbing the radiation, temperature in the stratosphere is significantly increased.

3.3. Association of QBO and Total Column Ozone (TCO) in the Stratosphere
Concentration of ozone in the stratosphere is resulted by a combination of transport and photochemistry process at ~30 – 20 hPa above the surface (Brasseur *et al* 1999). Figure 11 shows that there is a defference of ozone concentration in the westerly and easterly phase of QBO, where ozone concentration is relatively increased during the westerly phase of QBO, but decreased during the opposite phase. It indicates the transport process of zonal wind oscillation in QBO to the fluctuation of ozone in the lower-stratosphere.

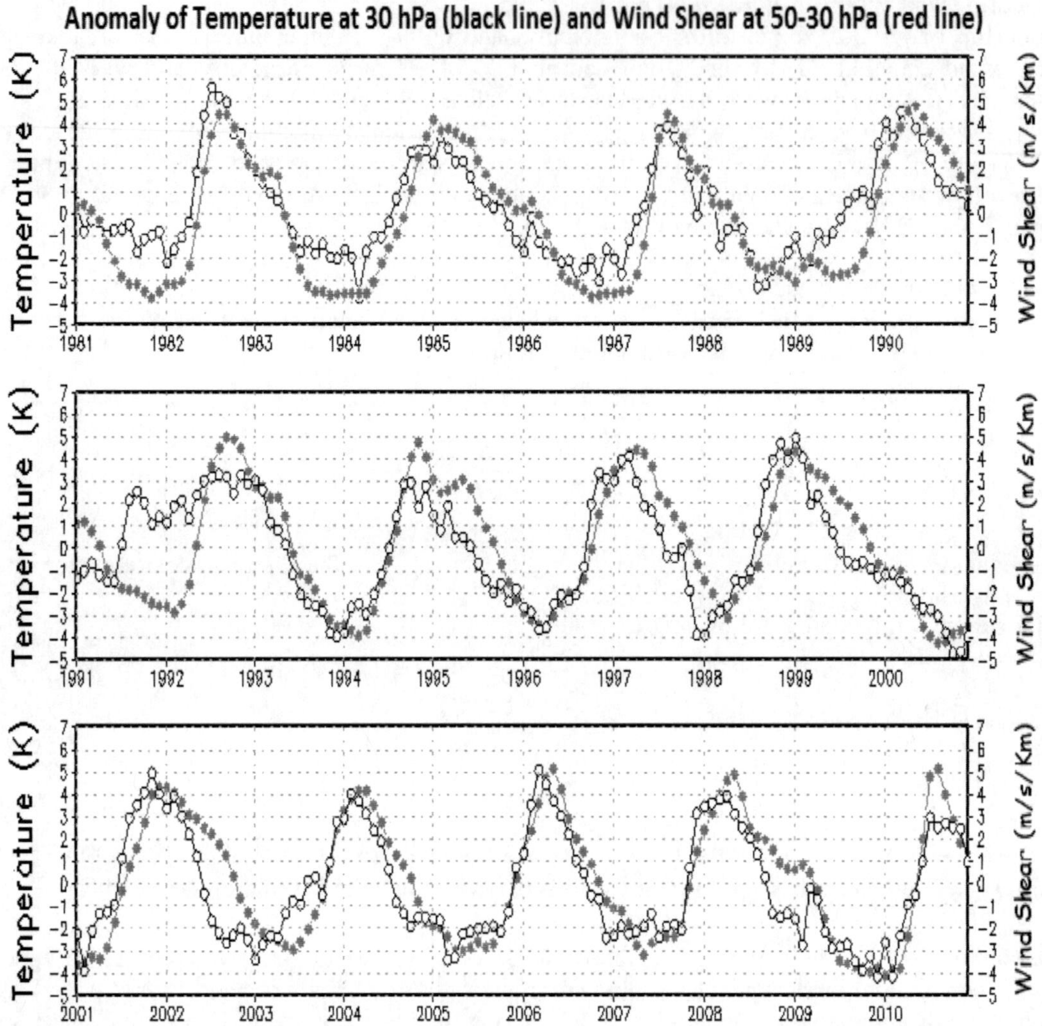

Figure 10. Relationship of temperature anomaly (T) at 30 hPa and wind shear anomaly (dU/dz) at 50-30 hPa above the surface.

Ozone depletion is not only related to the depletion of ozone substances, but also related to temperature in the stratosphere. Negative anomaly of temperature in QBO (Figure 12) shows that lower temperature in the easterly phase of QBO is associated with the depletion rate of ozone causing the ozone depletion. In the other side, negative anomaly of TCO is generally associated with negative anomaly of temperature in QBO, because ozone absorbs the sun radiation and infrared radiation. The lower concentration of ozone in the stratosphere causes a lower absorbtion of radiation, so temperature is decreased in the lower-stratosphere. During the westerly phase of QBO, anomaly of temperature will be increased as long as TCO is increased.

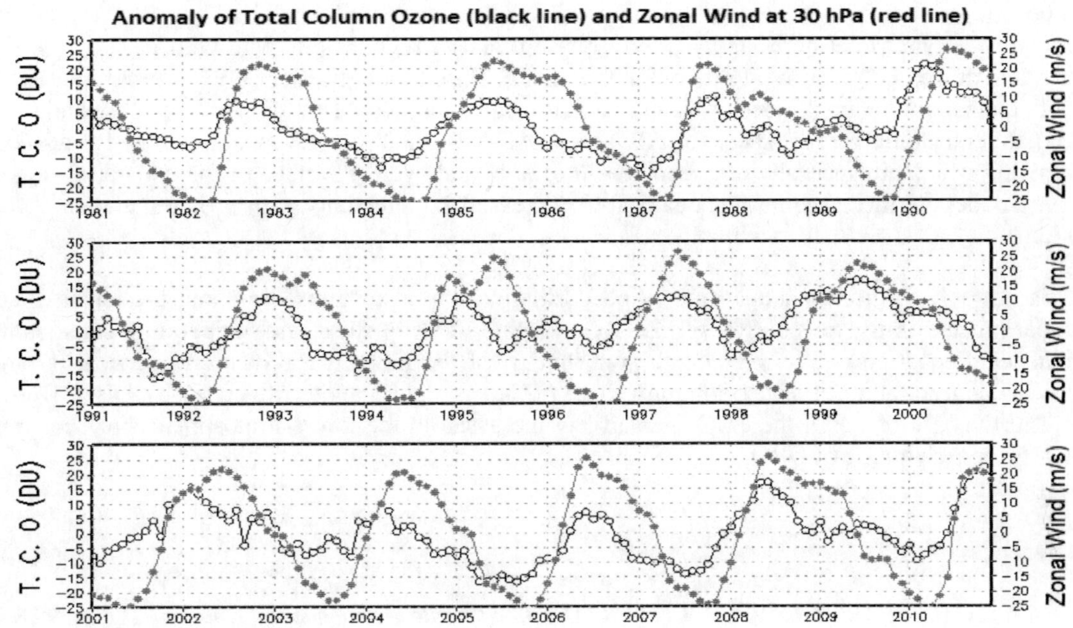

Figure 11. Relationship of Total Column Ozone anomaly (TCO) and zonal wind anomaly (U) in QBO at 30 hPa above the surface.

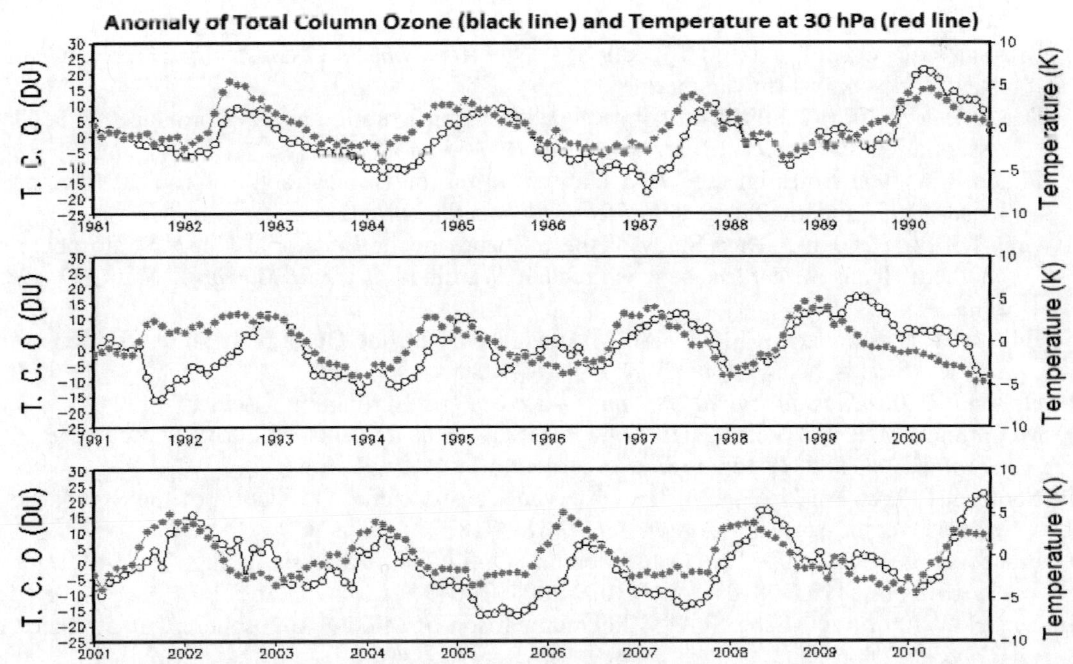

Figure 12. Relationship of Total Column Ozone anomaly (TCO) and temperature anomaly (T) in QBO at 30 hPa above the surface.

4. Conclusion

The zonally symmetric easterly and westerly phase of QBO regimes alternate with period of ~27.7 months and descend downward at rate of ~1 km/month. Based on Equivalent QBO Amplitude (EQA) method, the maximum amplitudes of QBO in zonal mean zonal wind (u), temperature (T), wind shear (du/dz), and quadratic vertical shear (d^2u/dz^2) are ~28.3 m/s, ~3.4 K, ~4.8 m/s/km, and ~1.0 m/s/km^2 respectively which are generally occurred at ~30 – 20 hPa above the surface of the equatorial region. The amplitudes are increased in the equator, but decreased exponentially with a Gaussian distribution in latitude.

QBO is not only to discuss about zonal wind oscillation, but also to discuss about the physical and chemical process in the lower-stratosphere. Relationship of wind shear and temperature satisfies the thermal wind equation due to very small contribution of the mean meridional and vertical motion. Moreover, concentration of Total Column Ozone (TCO) is significantly influenced by QBO. During the westerly phase of QBO, the TCO is relatively increased in the lower-stratosphere, but decreased during the easterly phase of QBO.

5. References

[1] Andrews G A, Holton J R and Leovy C B 1987 *Middle Atmosphere Dynamics* (Orlando, FL: Academic Press)
[2] Angell J K and Korshover J 1964 Quasi-biennial Variations in Temperature, Total Ozone, and Tropopause Height *J. Atmos. Sci.* **21** 479-429
[3] Baldwin M P *et al* 2001 The Quasi-biennial Oscillation *Rev. Geophys.* **39** 179-229
[4] Box E P and Jenkins G M 1976 *Time Series Analysis: Forecasting and Control* (Michigan: Michigan University)
[5] Brasseur G P, Orlando J J and Tyndall G S 1999 *Atmospheric Chemistry and Global Change* (New York: Oxford University Press)
[6] Dee D P *et al* 2011 The ERA-Interim Reanalysis: Configuration and Performance of the Data Assimilation System *Q. J. R. Meteorol. Soc.* **137** 553-597
[7] Ebdon R A 1960 Notes on the Wind Flow at 50 mb in Tropical and Subtropical Regions in January 1957 and in 1958 *Q. J. R. Meteorol. Soc.* **86** 540-542
[8] Gray L J *et al* 2001b A Data Study of the Influence of the Equatorial Upper Stratosphere on Northern-Hemisphere Stratospheric Sudden Warnings *Q. J. R. Meteorol. Soc.* **127** 1985-2003
[9] Holton J R and Lindzen R S 1972 An Updated Theory for the Quasi-biennial Oscillation of the Tropical Stratosphere *J. Atmos. Sci.* **29** 1076-1080
[10] Holton J R 2004 *Introduction to Dynamics Meteorology* (Burlington: Elsevier)
[11] McCormick M P and Veiga R E 1992 SAGE II Measurements of Early Pinatubo Aerosols *Geophys. Res. Lett.* **19** 155-158
[12] Naujokat B 1986 An Updated of the Observed Quasi-biennial Oscillation of the Stratospheric Winds Over the Tropics *J. Atmos. Sci.* **43** 1873-1877
[13] Pascoe C L *et al* 2005 The Quasi-biennial Oscillation: Analysis Using ERA-40 Data *J. Geophys. Res.* **110** D08105 doi:10.1029/2004JD004941
[14] Randel W J *et al* 2004 The SPARC Intercomparison of Middle Atmosphere Climatologies, *J. Climate.* **17** 986-1003
[15] Reed R J, Campbell W J, Rasmussen L A and Rogers R G 1961 Evidence of a Downward Propagating Annual Wind Reversal in the Equatorial Stratosphere *J. Geophyis. Res.* **66** 813-818
[16] Robock A 2000 Volcanic Eruptions and Climate *Rev. Geophys.* **38** 191-219

IOP Conf. Series: Earth and Environmental Science **31** (2016) 012033 doi:10.1088/1755-1315/31/1/012033

The use of Moringa Oleifera Seed Powder as Coagulant to Improve the Quality of Wastewater and Ground Water

Hendrawati,[1,2] Indra Rani Yuliastri,[2] Nurhasni[2] , Eti Rohaeti[3], Hefni Effendi[3], Latifah K Darusman[3]
[1]Doctoral Program in Environmental and Natural Resources Management (ENRM), IPB
Email: hendrawati@uinjkt.ac.id
[2]Department of Chemistry Faculty of Science and Technology State Islamic University Jakarta
[3]Bogor Agricultural University IPB

E-mail: hendrawati@uinjkt.ac.id

Abstract. Wastewater and ground water treatment are mostly using Polyaluminum Chloride (PAC), a synthetic coagulant, which possess health risk and require expensive cost. This research was carried out to observe the effect of *Moringa oleifera* seed as natural coagulant to replace synthetic coagulant. *M. oleifera* reduced 98.6% turbidity of wastewater, 10.8 % of its conductivity, 11.7% of its BOD and removed its metal contents (Cd, Cr, Mn). When applied to ground water, *M. oleifera* removed the turbidity of ground water as much as 97.5%, while reduced the conductivity and BOD of ground water 53.4 % and 18%, respectively. The use of *M. oleifera* also reduced total number of coliform. The advantage of using *M. oleifera* is that it does not reduce pH as PAC, hence does not require further treatment to adjust pH of the treated water.

1. Introduction

Water is an important resource for life. Entire living organisms on earth need water for life. However, water can be problematic if it is not available in the right conditions. Water is used by human for various purposes, therefore the cleanliness of water consumed is very important since water is known to affect the health. Today, the quality of water becomes a major problem that needs serious attention. Good quality water has become an expensive item, because many water sources has been polluted by waste coming from the various human activities. This leads to declining quantity of water sources that could not meet the ever growing need.

In the provision of clean drinking water, besides the quantity and continuity, the quality must meet the applied standards. The ideal water should have some characteristics such as clear, colorless, tasteless, odorless, phatogen-free, harmful chemical-free and non-corrosive. Water is also expected not to leave sediment in all distribution organs. This standard was set to prevent the occurrence and the spread of waterborne diseases.

To achieve this standard, there is one common technique applied in water treatment process, which is coagulation-flocculation. Coagulation is the process of coagulating colloidal particles due to the addition of synthetic materials to neutralize charged particles thus forming a precipitate due to the force of gravity. Coagulant can be synthetic materials such as ferrous sulfate ($Fe(SO_4)$), aluminum sulfate or alum ($Al_2(SO_4)_3$), and Poly Aluminum Chloride (PAC) ($(Al_2(OH)_3Cl_3)_{10}$).

Coagulation is one of the most common ways to reduce the pollutant contents in the water body that are present as turbidity, color and organic matters. Coagulation is also used to reduce the metal ion content in water. Separation of these colloids can be done by the addition of synthetic coagulant or biocoagulant followed by slow agitation (flocculation) that causes coagulation of colloidal particles so they can be separated by sedimentation [1].

The common methods of water purification using synthetic materials such as aluminum sulfate (alum) and calcium hypochlorite are not efficient, because these materials are imported and thus make the water cost becomes relatively expensive in most economically developed countries and is not affordable for most rural population. Therefore, some people try to get the water source from dams, mining, small streams, rivers, and lakes. Water from these sources is usually turbid and contaminated with microorganisms that may cause various diseases.

Several findings from previous research in Postnote (2002) demonstrated the use of synthetic materials for water purification can be severely hazardous to health if something goes wrong in their treatment during processing [2]. The report considered the high level of aluminum in the brain is a risk factor causing Alzheimer's disease. Other studies have raised doubt about the feasibility of inserting aluminum into the environment by the use of aluminum sulfate as a coagulant continuously in the water treatment process. However, Davis (2006) found no conclusive evidence about the correlation between aluminum and Alzheimer's disease [3].

Besides synthetic chemicals, there are natural ingredients that can be derived from tropical plants which can be used as coagulants, including moringa seeds (*Moringa oleifera*). The use of natural ingredients from local indigenous plants to clear muddy water is not a new idea [4]. From existing reports, there were allegations that the powder of Moringa seeds has antimicrobial properties. Previous research found that Moringa is not toxic [5] and recommended for use as a coagulant in developing countries. Various studies have been conducted and showed that moringa seeds are effective as biocoagulant to improve physico-chemical properties of contaminated water. *M. oleifera* functions as coagulant trough adsorption and neutralization mechanisms [6]. *Moringa oleifera* is potential as organic pollutant absorber in simulation solution [7]. *M. oleifera* is reported able to eliminate the turbidity and dissolved organic matters of river water [8]. Damayanti *et al.* (2011) made a membrane consisted of *M. oleifera*, PAC and zeolite for palm oil effluent treatment [9].

Indonesia is rich in biodiversity, and moringa tree can grow well, easy to find and easy to cultivate in various regions. Therefore it is not difficult to use Moringa seeds as a natural coagulant or biocoagulant for water clarifying process. The use of natural coagulants in water treatment process is expected to provide more advantages than the use of synthetic materials because they are natural and reported as safe to be consumed. The cost of using natural coagulants will be less expensive than that of alum. Effectiveness of natural coagulant for water purification will be tested also in the wastewater treatment process. Therefore, research should be conducted to find out the effectiveness of moringa seed in improving water quality. Water quality parameters that need to be investigated are including turbidity, electrical conductivity, pH and temperature, metal absorbing capability, and ability to decrease microbial content.

2. Methods

2.1. Sample Collection

Wastewater samples were taken from a textile industry in Karawang. Wastewater sampling was conducted in March, April and May 20??, where the company is in normal production period. Wastewater samples were taken from the equalization basin. Groundwater samples were taken from a well in Pamulang. Analysis was conducted at the Research Laboratory, Center for Integrated Laboratory, Syarif Hidayatullah State Islamic University Jakarta.

2.2. Materials and equipment

Materials used in this study were the test substance (sample) and chemicals. Wastewater test material was taken from a textile industry in Karawang and ground water was taken from the

Pamulang, *Moringa oleifera* seeds were taken in March 20??, in Pasawahan, Purwakarta. The chemicals used were Poly Aluminum Chloride (PAC) brand Kuriflock concentration of 100 mg/L, Single Strength Lactose Broth (Merck), Double Strength Lactose Broth (Merck), oxygen-alkali iodide azide reagent Sodium Iodide (NaI) as an oxidant, sodium thiosulfate ($Na_2S_2O_3$) 0.025 N (Merck), sulfuric acid (H_2SO_4) 6 N (Merck), manganese sulfate ($MnSO_4$) 4 M (Merck), and starch indicator.

The instruments used were pH meter (Myron L ARH1), digital thermometer, portable conductometer (Myron L ARH1), portable turbidity meter (HANNA Instrument), Atomic Adsorption Spectrophotometer (Perkin Elmer), magnetic stirrer (Cymarec*2), cuvet, durham tube and glasswares

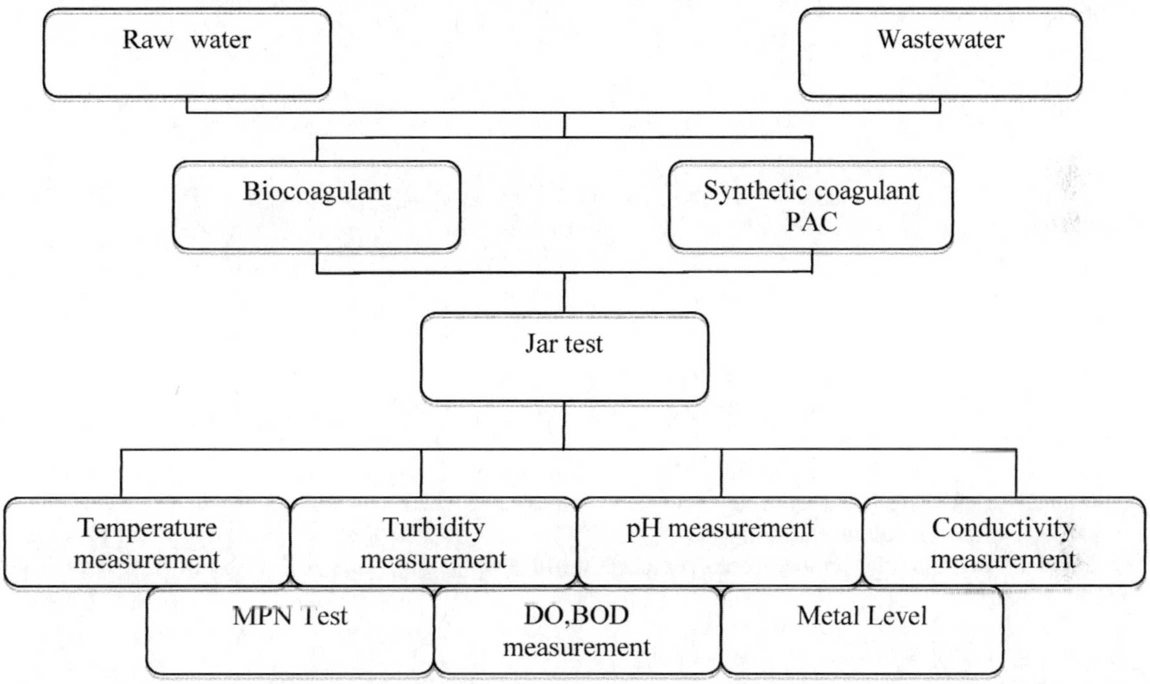

Figure 1. Research Flowchart

3. Results and Discussion

In the wastewater and groundwater treatment process using *M. oleifera* as coagulant, the optimum dose required to treat the wastewater and ground water is different. This can be seen in Table 1, a dose of 100 mg/L gave the best result for the wastewater as the treated water showed the lowest value of turbidity, closest pH value to 7, and the lowest value of conductivity. While the ground water needed 80 mg/L of coagulant. This showed that the optimum dose of *M. oleifera* as coagulant was influenced by the level of the initial turbidity of the water to be processed. The optimum dose to treat water with initial turbidity around 300 FTU (wastewater) is 100 mg/L or higher. The lower the initial turbidity, the lower the optimum dose. Excessive coagulant addition will not cause turbidity to disappear until the lower limit. Determination of optimum dose can be done by the lab-scale Jar Test method.

Table 1. Effect of coagulant addition to the tested parameters of the sample

Tested Parameter	Wastewater				Ground Water			
	Control	PAC (100 mg/L)	*M. oleifera* 80 mg/L	*M. oleifera* 100 mg/L	Control	PAC (100 mg/L)	*M. oleifera* 80 mg/L	*M. oleifera* 100 mg/L
Temperature (°C)	28.27	28.63	28.66	28.83	28.40	28.90	28.90	29.00
Turbidity (FTU)	85.79	32.19 (89.6% reduction)	6.75 (97.9% reduction)	4.913 (98.6% reduction)	49.92	21.68 (89.4% reduction)	5.44 (97.5% reduction)	6 (97.4% reduction)
pH	5.10	4.83	5.67	6.20	6.96	5.21	7.38	7.42
Conductivity (µS)	1109.70	1906.30	1052.50	1004.65 (10.8% reduction)	227	238.70	219 (53.4% reduction)	216.35

3.1. The effect coagulant addition to water temperature

The addition of powdered moringa seed (*M. oleifera*) and the PAC as a coagulant in wastewater treatment processes and water purification did not significantly affect the temperature. The initial temperature of the wastewater samples was 28.27 °C and the highest temperature after the addition of coagulant was 29 °C. While the initial temperature of ground water sample was 28.40 °C and the highest temperature after coagulant addition was 29.00 °C. The use of coagulant in water treatment process did not drastically change the temperature. Temperature of each sample was still in the normal temperature range for water.

3.2. The effect of the use of M. oleifera to water turbidity

Table 1 shows that the optimum concentration of *M. oleifera* coagulant to reduce wastewater turbidity was 100 mg/L, while the ground water only need 80 mg/L of *Moringa oleifera*, which reduced 97.9% and 97.5% turbidity, respectively. *M. oleifera* coagulant has better coagulation capability to reduce water turbidity compared with PAC at concentration of 100 mg/L that was able to reduce turbidity by 89.6% in wastewater and 89.4% in ground water.

The highest turbidity value in wastewater after coagulation process using *M. oleifera* was 6.75 FTU and the lowest turbidity value was 4.913 FTU. Meanwhile, the highest turbidity value in ground water after coagulation process using *M. oleifera* was 5.44 FTU and the lowest was 6 FTU. When *M. oleifera* concentration exceeded the optimum dosage, turbidity raised up because all colloids have been neutralized and precipitated with an optimum dosage, so the excess coagulants will cause turbidity in water as they did not interact with oppositely charged colloidal particles.

Turbidity in the water is caused by suspended solids, both organic and inorganic substances. Inorganic substances include crack of rock, sand, mud, and dissolved metals. Organic matters originating from domestic and industrial waste could serve as good environment for bacteria to grow. Besides microorganisms, algae and plankton can also cause cloudiness in the water.

When the coagulant was added to the sample and followed by rapid stirring, the resulting cationic protein from *M. oleifera* was distributed to all parts of the liquid and then interacted with the negatively charged particles that caused dispersed turbidity. Such interactions disturb the force that stabilize the particles, so that it can bind to small particulates to form precipitate. This process is called coagulation. An additional advantage in this case is, that all the mud that comes from grain coagulation *M. oleifera* is biodegradable and is an organic material. Unlike alum, coagulation activity is strongly influenced by the natural alkalinity of the water itself. So it is necessary to add material, such as lime, to increase alkalinity or pH of the water that will be coagulated using alum. As a result, the sludge that is produced through this process has a larger volume than the one produced by the coagulant *M. oleifera* seeds.

3.3. The effect of M. oleifera usage to water pH

The degree of acidity (pH) is one of the most important factors affecting the coagulation process. When the coagulation process is not carried out at optimum pH range, it could lead to failure of the floc formation and poor quality of water produced. The optimum pH for each coagulant is different. Specific coagulant will not work optimally in more acidic or more alkaline environment than the optimum pH value. Based on the analysis, the optimum pH obtained using *M. oleifera* biocoagulant is at pH 6-8. At this value, amino acids ionize to produce carboxylate ion and proton, proton charge attracts electrons (colloids) to form neutral group and then produce floc [10].

Recommended pH range for water by WHO (2006) is between 6.0 and 8.0 [11]. The treatment was performed in the range of 4.83 to 7.42 which increased by increasing the coagulant dosage. The pH of 100 mL of distilled water increased from 7.6 to 8.28 mg after the addition of *M. oleifera* powder. This can be explained by the fact that the solution becomes more alkaline due to the ability of *M. oleifera* as a coagulant lies in the presence of cationic water-soluble protein found in its skin and seeds. This causes the acceptance of protons in water by the alkaline amino acids present in *M. oleifera* protein that results in the release of hydroxyl groups which make the solution becomes alkaline [12].

In the sample that used the PAC as coagulant, the pH will be more acidic with the addition of coagulant dosage. During water treatment process, alum produces acid which in turn lower the pH. Increased acidity may occur due to the aluminum trivalent cation that act as Lewis acid and accept lone electron pairs [4] In the synthetic coagulant (PAC), a decrease of pH was caused by the presence of free hydrogen ions (H^+) generated from the hydrolysis reaction, which is when the coagulant reacts with water. Generally, the higher the amount of coagulant, the higher the decrease of pH.

3.4. The Effect of M. oleifera Usage on Conductivity

Conductivity in water is vary widely. The solubility of minerals in different geographic regions also varies. There is no standard value but high level of conductivity is not allowed in drinking water for consumers [12]. Table 1 shows that the optimum dosage of *M. oleifera* to decrease conductivity was at 100 mg/L. At this concentration, *M. oleifera* reduced 10.8% conductivity of wastewater and 53% conductivity of ground water. Compared with the initial conductivity, the addition of a coagulant in wastewater and ground water made the conductivity value down to 1109.70 μS/cm and 227 μS/cm, respectively. The decline of conductivity value was affected by the initial conductivity value. Initial low conductivity value decline larger than the higher initial conductivity value. However, the excess use of *M. oleifera* beyond optimum dosage made conductivity value rise again due to the presence of unbound ions.

Table 1 shows increasing conductivity value with the addition of coagulant PAC, from 1109.7 μS/cm to 1906.3 μS/cm. Whereas initial conductivity in the control of ground water was 227 μS/cm and after the addition of PAC was 238.6 μS/cm. High conductivity value was determined by the presence of mineral ions and dissolved inorganic compounds. The addition of coagulant *M. oleifera* may result in the dispersion of some mineral ions and inorganic compounds into a floc which will then be precipitated and separated from the solution. This caused the reduction of electrical conductivity.

The addition of coagulant PAC increased the conductivity of water due to the reaction of water with acidic or alkaline metals. Water can also react with salt which will cause the rise of conductivity value. Another reason is the inorganic compounds that dissociate in water, which contribute to water's ability to conduct very large electric current. Conductivity or electrical conductivity of water depends on the concentration of ions in water. In the coagulation process, moringa seeds give little effect on the degree of acidity and conductivity. *M. oleifera* seed solution reacted as positively charged natural polymer coagulant.

3.5. The Effect of M. oleifera Usage in Metal Level

There have been many efforts conducted by scientists to reduce the concentration of dissolved metals in the environment. One of them is coagulation process. The observed metals in this study were Cd, Cr, and Mn. Cd and Cr metals components of dying agents used in textile industries, while Mn are found in the groundwater. Determination of metal content in this study was carried out by using Atomic Absorption Spectroscopy. The result can be seen in Table 2.

Table 2. The effect of coagulant addition on the metal level

Tested Parameter	Wastewater				Ground Water			
	Control	PAC (100 mg/L)	*M. oleifera* 80 mg/L	*M. oleifera* 100 mg/L	Control	PAC (100 mg/L)	*M. oleifera* 80 mg/L	*M. oleifera* 100 mg/L
Mn (mg/L)	6	0.092	-	-	0.594	0.265	-	-
Cd (mg/L)	6	0.024	-	-	-	-	-	-
Cr (mg/L)	6	-	-	-	-	-	-	-

Tabel 2 shows that the addition of coagulants can reduce the level of metals in wastewater significantly. The addition of *M. oleifera* coagulant decreased the Cd metal content from 6 ppm to undetectable, whereas PAC coagulant decreased the Cd content from 6 ppm to 0.024 ppm (99% reduction). Cr content was decreased from 6 ppm to undetectable with the addition of *Moringa oleifera* and PAC. The addition of *Moringa oleifera* lowered the Mn content from 6 ppm to undetectable, whereas the addition of PAC lowered the Mn content from 6 ppm to 0.092 ppm (98% reduction). The presence of dissolved Cd and Cr metals was not detected in tested ground water. Dissolved Mn amount in ground water is 0.594 mg/L. After the addition of *M. oleifera* coagulant, it became undetectable. Meanwhile, Mn level decreased to 0.265 after the addition of PAC (55.4% reduction).

This study proved that the coagulant may reduce level of metals in the wastewater. This could be due to the addition of a coagulant that will form flocs and pull those metals into the flocs. Chemical wastewater treatment is usually performed to remove particles that are not easily precipitated (colloidal), including heavy metals. With the addition of coagulant, removal of such materials in principle takes place through changes in the properties of these materials, which can be precipitated from not easily deposited (coagulation-flocculation), either with or without oxidation-reduction reactions, and also takes place as a result of the oxidation reaction. Decreased levels of these metals may also occur because of *Moringa* amphoteric protein binds to the oppositely charged metal ions-binding compounds, which causes the metal ions to precipitate. Alkaline pH generated by the addition of *Moringa* also allows the positively charged metal ions to precipitate as insoluble metal hydroxides due to the release of OH groups of *M. oleifera*. This is supported by the fact that in the treatment without coagulant, to precipitate the metal is done by applying a solution of alkali (eg. lime) to form hydroxide precipitate of the metals. Precipitated metal will be more stable if the pH of water is above 10.5. This of course is not effective in the treatment of wastewater and ground water because it will require additional process to lower the pH value.

3.6. The Effect M. oleifera usage on Total Coliform

More than 80% of diseases in tropical areas are caused by the use of unclean water. The presence of coliform in surface water, such as ground water, comes from poor sanitation [13], while the presence of microbes in the wastewater comes from biological treatment of wastewater treatment processes. This process involves decomposing bacteria to decompose organic compounds in wastewater. The use of water for consumption without prior treatment can cause serious illness that spread through the water (serious water-borne disease).

Processing the water by coagulation using *M. oleifera* as natural coagulant showed that the treatment with *M. oleifera* provided additional advantage of reduced total microbes (Table 3). *M. oleifera* treatment reduced e-coli values by 80% in wastewater treatment. While the use of *M. oleifera* in ground water treatment reduced microbial content by 45%.

With proper mixing, the moving particles will enlarge and form flocs that tend to fall to the bottom of the vessel due to gravity. This confirms the effectiveness of *Moringa* as coagulant for the purification of dirty water. Furthermore, the decrease in total coliform number was also affected by alkaline condition generated by *Moringa*. Most microorganisms grow well at pH 6.0-8.0, but some of them can grow well at pH 3 (acidophiles) and at pH 10.5 (alkaliphiles). Coliform bacteria are facultative anaerobic microorganisms that can grow in aerobic environments and in fermentation condition that produces lactic acid. Therefore these bacteria can still grow at low pH environment, coliform bacteria can still grow, but they can not survive alkaline pH [14]. The addition of *Moringa* as caogulant affect the increase in pH which in turn stop them from growing..

M. oleifera seeds also have bactericidal activity, which have been proved by Oluduro and Aderiye (2007) in their research. Bacterial species *S. faecalis* and P. *aerugenosa* which were cultured in water, stop growing back after *M. oleifera* seeds were added. When the seeds of *M. oleifera* are crushed and dissolved into the water, protein produces a positive charge that acts like a magnet and attracts dominant negatively charged particles such as clay, silk, and other toxic particles. This is in accordance with the invention of Schwarz (2000) that the flocculation process removes about 90-99% of bacteria that are usually attached to solid particles, so the bacteria will be aggregated together to form flocs and can be removed from the water [19].This suggests that bacteria in water were not only inactivated in a dormant state, but also were killed. *M. oleifera* removes both gram negative and gram positive bacteria [20-21].

Table 3. The effect coagulant addition on total coliform, DO and BOD of Wastewater and ground water.

Tested Parameter	**Wastewater**			**Ground Water**		
	Control	**PAC (100 mg/L)**	***M. oleifera* 100 mg/L**	**Control**	**PAC (100 mg/L)**	***M. oleifera* 100 mg/L**
Total Coliform (MPN/100ml)	>1100	1100	210 (80% reduction)	28	20	11 (45% reduction)
DO (mg/L)	14	12.4	11.2	10.8	10	8.4
BOD (mg/L)	6.8	6.3	6	6.4	6.2	5.2

3.7. The Effect of M. oleifera Usage on Dissolved Oxygen

Contaminants in industrial wastewater of textiles consist of organic and inorganic materials that are dissolved or dispersed in water and coarse solids, like the residue of the fiber and yarn. Textile dyes from organic or inorganic compounds can also cause an increase in the BOD. Solute removal originated from textile dyes in wastewater will reduce its BOD.

Table 3 shows that the addition of coagulant can affect the decline of dissolved oxygen in wastewater and groundwater. The addition of 100 mg/L *M. oleifera* in control wasterwater reduced

20% DO from 14 mg/L to 11.2 mg/L. The addition of PAC decreased the DO content by 11.5%. In groundwater, *M. oleifera* lowered 22% DO value from 10.8 mg/L to 8.4 mg/L. PAC only decreased 8% of DO value. The level of biological oxygen demand (BOD) also declined 11.7% from 6.8 mg/L to 6 mg/L in the control wastewater after the addition of *M. oleifera.* Meanwhile, PAC decreased the BOD level by 7%. The addition of *M. oleifera* as coagulant into ground water decreased 18% of BOD level from 6.4 mg/L to 5.2 mg/L, whereas the addition of PAC was only able to decrease the BOD level by 3%. The addition of coagulant decreased the DO level due to the increase in inorganic (PAC) and organic (*M. oleifera*) substances which cause increased oxygen needs to oxidize these substances.

3.8. General Discussion

In the wastewater and groundwater treatment process using *M. oleifera* as coagulant, the optimum dose required to treat the wastewater and ground water is different. This can be seen in Table 1, a dose of 100 mg/L gave the best result for the wastewater as the treated water showed the lowest value of turbidity, closest pH value to 7, and the lowest value of conductivity; while the ground water needed 80 mg/L of coagulant. This showed that the optimum dose of *M. oleifera* as coagulant was influenced by the level of the initial turbidity of the water to be processed. The optimum dose to treat water with initial turbidity around 300 FTU (wastewater) is 100 mg/L or higher. The lower the initial turbidity, the lower the optimum dose. Excessive coagulant addition will not cause turbidity disappear until the lower limit. Determination of optimum dose can be done by the lab-scale Jar Test method.

Compared with Indonesian regulation number KEP-51/MENLH/10/1995 on Standard Quality of Liquid Waste for Industrial Activities, the result achieved by this research is safe to be use. The standard quality for pH ranges from 6.0 to 9.0. The pH score after the addition of coagulant were 6.2 and 7.38. The standard quality of BOD is 60 mg/L, while the coagulation process gave 5.2 mg/L and 6.2 mg/L. Maximum levels of Cr, Cd, and Mn permitted in the industrial waste are 0.5 mg/L, 2 mg/L, and 2 mg L, respectively, whereas the result of using *M. oleifera* treatment was undetected.

Several scientific studies suggested that *Moringa* seeds could serve as coagulant because it contains low molecular weight water soluble protein [15-16].) Protein will be positively charged when dissolved in water [17]. Protein will act as positively charged synthetic materials [18] and can be used as synthetic polymer coagulant. Therefore, *Moringa* can be called as a coagulant. Since this coagulant is derived from plants and without any synthetic process, it is also called natural coagulant or biocoagulant. The most likely mechanism that occurs in the coagulation process is the adsorption and neutralization of the voltage or adsorption and bonding between unstable particles. It is difficult to determine which mechanism that occurs because both mechanisms may occur simultaneously. But, the most common mechanisms of coagulation that involves *Moringa* seeds are the adsorption and voltage neutralization [19]. When *Moringa* seeds that have been processed (powder) were poured into the dirty water, the protein in the seeds will bind the negatively charged particulates that cause turbidity, such as clay, bacteria, dust, and others. Thus, the particulates are collected and agglomerated into larger molecules, which will settle to the bottom, then it will be easy to separate the water and contaminant.

4. Conclusion

Moringa oleifera was proven to be able to be used for textile wastewater treatment and purification of groundwater. *M. oleifera* can be used in the coagulation process because it has properties as a natural coagulant. Our result suggested that *M. oleifera* is effective at the concentration of 80 to 100 mg/L as a coagulant to treat wastewater from textile industry and ground water. *M. oleifera* optimum coagulant dose is also influenced by the initial state of the sample to be coagulated. The heavier the burden of pollution, the higher the optimum dose that is needed.

5. Acknowledgment
The authors gratefully acknowledge to Department of Chemistry, Faculty of Science and Technology, State Islamic University Syarif Hidayatullah Jakarta

References

[1] Tebbut, T.H.Y. 1982. *Principles of Water Quality Control*, Translate edition, Mohajit. Bandung: ITB. Indonesia.

[2] Postnote. 2002. *Access to Water in Developing Countries. No.178.*

[3] Davis, K. L. 2007. *Alzheimer's Disease*. Redmond, WA: Microsoft Corporation.

[4] Sutherland J.P., G.K. Folkard, M.A. Mtawali and W.D. Grant. 1994. *Moringa oleifera* as Natural Coagulant. *Journal of WEDC Conference*. University of Leicester, UK.

[5] Grabow WOK, Slabert JL, Morgan WSG, Jahn SAA. 1985. Toxicity and Mutagenicity Evaluation of Water Coagulated with *Moringa oleifera* Seed Preparations Using Fish, Protozoan, Bacterial, Enzyme, and Ames *Salmonella* Assays. http://www.h2ou.com/h2wtrqual.htm#References

[6] Bhatia, S., Othman, Z., & Ahmad, A. L. (2007). Coagulation–flocculation process for POME treatment using Moringa oleifera seeds extract: Optimization studies. *Chemical Engineering Journal*, *133*(1-3), 205–212. doi:10.1016/j.cej.2007.01.034

[7] Akhtar, M., Moosa Hasany, S., Bhanger, M. I., & Iqbal, S. (2007). Sorption potential of Moringa oleifera pods for the removal of organic pollutants from aqueous solutions. *Journal of Hazardous Materials*, *141*(3), 546–56. doi:10.1016/j.jhazmat.2006.07.016

[8] Sánchez-martín, J., Ghebremichael, K., & Beltrán-heredia, J. (2010). Bioresource Technology Comparison of single-step and two-step purified coagulants from Moringa oleifera seed for turbidity and DOC removal. *Bioresource Technology*, *101*(15), 6259–6261. doi:10.1016/j.biortech.2010.02.072

[9] Damayanti, a, Ujang, Z., & Salim, M. R. (2011). The influenced of PAC, zeolite, and Moringa oleifera as biofouling reducer (BFR) on hybrid membrane bioreactor of palm oil mill effluent (POME). *Bioresource Technology*, *102*(6), 4341–6. doi:10.1016/j.biortech.2010.12.061

[10] Wibraham,C., Antony,Mata. dan S.Michael., 1982, *Introduction to Organic and Biological Chemistry*. Translated by Suminar Achmadi, ITB, 1992.

[11] Amagloh, Francis Kweku dan Amos Benang. 2009. Effectiveness of *Moringa oleifera* Seed as Coagulant for Water Purification. Full Length Research Paper. *African Journal of Agricultural Research Vol. 4 (1), pp. 119-123.* http://www.academicjournals.org/AJAR .ISSN 1991-637X © 2009 Academic Journals.

[12] WHO. 2006. *Guideline for drinking-water quality (electronic resources): incorporating first addendum. Vol. 1, Recommendations, 3rd edition.*

[13] Oluduro, A. O and B.I. Aderiye. 2007. *Efficacy of Moringa oleifera Seed Extract on the Microflora of Surface and Underground Water*. Department of Microbiology, University of Ado-Ekiti, Ado-Ekiti, Nigeria.

[14] Todar, Melnick. 2008. *Medical Microbiology*. 22th Ed. Lange Medical Books: New York

[15] Fayos, B. G., J. M. Arnal, G. Verdú, I. Rodrigo. 2010. *Purification of a Natural Coagulant Extracted from Moringa oleifera Seeds: Isolation and Characterization of The Active Compound.* Journal on Food Innovation. Institute for Industrial Universidad Politécnica de Valencia. Camino de Vera s/n 46022 Valencia (Spain)

[16] Kebreab, A., Ghebremichaela, K.R. Gunaratnab, H. Henrikssonc, H. Brumerc. 2005. *A Simple Purification and Activity Assay of The Coagulant Protein from Moringa oleifera Seed.* Journal of Water Research. Department of Biotechnology, Royal Institute of Technology (KTH), Albanova University Centre,106 91 Stockholm, Sweden
[17] Sahni, Pushpa dan Shalini Srivastava. 2008. A Systems Approach to Isolation and Characterization of Protein Content of Shelled *Moringa oleifera* Seeds Used for Decontamination of Arsenic From Water Bodies. *XXXII National Systems Conference, Nsc.*
[18] Broin M., C. Santaella, S. Cuine, K. Kokou, G. Peltier, T. Joët. 2002. *Flocculent Activity of Recombinant Protein from Moringa oleifera Lam. Seeds.* Journal of Appl Microbiol Biotechnol. Received: 19 April 2002 / Revised: 16 July 2002 / Accepted: 22 July 2002 / Published online: 23 August 2002 © Springer-Verlag 2002
[19] Schwarz D. 2000. *Water Clarification Using Moringa oleifera*. Technical Information W1e, Gate Information Service, Eschborn, Germany. http://www.gtz.de/gate/gateid.afp
[20] Madsen, M., Schlundt J, Omer EF. 2005. *Effect of Water Coagulation by Seeds of Moringa oleifera on Bacterial Concentrations*. http://www.ncbi.nlm.nih.gov/portal/utils/pageresolver.fcgi?log$=activity&recordid=1255923630823234
[21] Lurling, Miquel and Wendy Beekman. 2009. *Anti-cyanobacterial Activity of Moringa oleifera Seeds*. Aquatic Ecology & Water Quality Management Group, Department of Environmental Sciences, Wageningen University, the Netherlands

IOP Conf. Series: Earth and Environmental Science **31** (2016) 012034 doi:10.1088/1755-1315/31/1/012034

Techno-economical evaluation of protein extraction for microalgae biorefinery

Y W Sari[1,2], J P M Sanders[3], and M Bruins[3]
[1]Biobased Chemistry and Technology, Wageningen University, P.O. Box 17 6700 AA Wageningen, the Netherlands
[2]Biophysics Division, Department of Physics, Bogor Agricultural University, Jl. Meranti, Gedung Wing S, Kampus IPB Darmaga, Bogor 16680, Jawa Barat, Indonesia
[3]Food and Biobased Research, P.O. Box 17, 6700 AA Wageningen, the Netherlands

E-mail : yessie.sari@apps.ipb.ac.id

Abstract. Due to scarcity of fossil feedstocks, there is an increasing demand for biobased fuels. Microalgae are considered as promising biobased feedstocks. However, microalgae based fuels are not yet produced at large scale at present. Applying biorefinery, not only for oil, but also for other components, such as carbohydrates and protein, may lead to the sustainable and economical microalgae-based fuels. This paper discusses two relatively mild conditions for microalgal protein extraction, based on alkali and enzymes. Green microalgae (*Chlorella fusca*) with and without prior lipid removal were used as feedstocks. Under mild conditions, more protein could be extracted using proteases, with the highest yields for microalgae meal (without lipids). The data on protein extraction yields were used to calculate the costs for producing 1 ton of microalgal protein. The processing cost for the alkaline method was € 2448 /ton protein. Enzymatic method performed better from an economic point of view with € 1367 /ton protein on processing costs. However, this is still far from industrially feasible. For both extraction methods, biomass cost per ton of produced product were high. A higher protein extraction yield can partially solve this problem, lowering processing cost to €620 and 1180 /ton protein product, using alkali and enzyme, respectively. Although alkaline method has lower processing cost, optimization appears to be better achievable using enzymes. If the enzymatic method can be optimized by lowering the amount of alkali added, leading to processing cost of € 633/ton protein product. Higher revenue can be generated when the residue after protein extraction can be sold as fuel, or better as a highly digestible feed for cattle.

1. Introduction

Owing to the scarcity of fossil feedstock, there is an increasing demand for biobased fuels. Microalgae are considered one of the most promising biobased feedstocks, as their productivity in converting carbon dioxide into lipids exceeds that of oilseed crops [1-3]. However, currently, microalgae based fuels have not yet been produced at large scale. In depth works are being conducted to enable microalgae-based fuel production in a sustainable and economical way. Applying a biorefinery concept may aid here. Microalgae biorefinery uses the overall composition of the algae, and next to lipids, including carbohydrates and proteins [3].

Methods on microalgal protein extraction have been developed [4]. However, the yield is still considered low, limiting commercial protein production from microalgae. This low protein yield is due to the complexity of microalgae cell wall and microalgal protein properties. Extreme extraction conditions can be used to obtain substantial cell wall degradation, but this may reduce functional protein properties.

This paper discusses two relatively mild conditions for microalgae protein extraction. Green microalgae (*Chlorella fusca*) with and without lipid are used as the feedstock. Two protein extraction techniques were used; alkaline and enzyme assisted extraction. Data on the yield of extracted protein were used to calculate cost for producing 1 ton of microalgae protein.

2. Materials and Methods

2.1 Materials

Microalgae meal (*Chlorella fusca*) and non–deoiled microalgae (microalgae) (*Chlorella fusca*) were obtained from Ingrepro B.V., the Netherlands. Protex 40XL (activity: 52 MPU/ml) was obtained from Genencor (Danisco) International Oy, Denmark.

2.2 Protein content analysis

Protein content in starting material and hydrolysate was determined using DUMAS analysis (FlashEA 1112 series, Thermo Scientific, Interscience). Methionine was used as a standard for the calibration and a constant of 6.25 was used as a nitrogen-protein conversion factor to calculate the protein content in the samples.

2.3 Protein extraction

Alkaline protein extraction was conducted as described in [5] while enzymatic assisted protein extraction was conducted as described in [6] with some minor modifications.

2.3.1 Alkaline method

Protein extraction was conducted, according to [5], at three ascending temperatures: starting with a one day incubation at 25°C, followed by one hour at 60°C, and ending with a one hour incubation at 120 °C. Total protein yield obtained after three- step protein extraction was defined as mass of extracted protein relative to mass of protein in the starting material.

2.3.2 Protease assisted method

Protein extraction was conducted using Protex 40XL, a mixture of proteases, according to [6]. The enzyme dosage in this study was 1 and 5% (volume enzyme per weight of protein). Total protein yield obtained was defined as mass of extracted protein relative to mass of protein in the starting material.

2.4 Cost calculation

Cost analysis in this study focused on the main processing cost: biomass, NaOH, and heating. In the case of the enzymatic method, enzyme costs were added. Biomass costs were calculated based on the prices of microalgae meal and microalgae, which were € 230 [7] and € 1650 [3] per ton dry mass. NaOH cost were determined based the amount of NaOH required to extract protein per ton dry biomass. The required NaOH mass was determined by lab scale experiments. Mass biomass used on lab scale was 10 g. NaOH cost was determined as:

$$Cost_{NaOH} = m_{NaOH} \times p_{NaOH}$$

and

$$m_{NaOH} = \frac{m_{NaOH\ lab\ scale}}{m_{biomass\ lab\ scale}} \times m_{biomass}$$

with

$Cost_{NaOH}$ = cost NaOH (€)
m_{NaOH} = mass NaOH required (ton)
p_{NaOH} = price NaOH $\left(\frac{€}{ton}\right)$ = € 288/ton [8]
$m_{NaOH\ labscale}$ = mass NaOH required during lab scale experiment (g)
$m_{biomass\ labscale}$ = mass dry biomass during lab scale experiment (g)
$m_{biomass}$ = mass dry biomass (ton)

The energy costs were calculated in relation to the thermal energy required to heat up the mixture of biomass and solvent. Specific heat capacity for water is 4.2 $J.g^{-1}.K^{-1}$ and for biomass is assumed to be 1.3 $J.g^{-1}.K^{-1}$ (at 25 °C)[9]. The ratio of biomass to water used in the calculations is 1:9.

The thermal energy required was calculated as:

$$Q = \left((m_{water} \times c_{water}) + (m_{biomass} \times c_{biomass})\right) \times \Delta T$$

The thermal energy in Joules is then converted to kWh values (1 J = 2.7 $\times 10^{-7}$ kWh). The price per kWh used in the calculations; € 0.041/kWh, is the price of natural gas for a medium-sized industrial consumer [10].

Assuming that 50% of heat involved in the extraction can be recovered then the final energy cost is defined as 50% x calculated energy cost.

The enzyme costs used in this calculation were based on the amount of enzyme that was used

$$Cost_{enzyme} = m_{enzyme} \times p_{enzyme}$$

and

$$m_{enzyme} = \frac{m_{enzyme\ lab\ scale}}{m_{biomass\ lab\ scale}} \times m_{biomass}$$

with

$Cost_{enzyme}$ = cost enzyme (€)
m_{enzyme} = mass enzyme required (ton)
p_{enzyme} = price enzyme $\left(\frac{€}{ton}\right)$ = € 1000/ton enzyme solution.[1]
$m_{enzyme\ labscale}$ = mass enzyme required during lab scale experiment (g)
$m_{biomass\ labscale}$ = mass dry biomass during lab scale experiment (g)
$m_{biomass}$ = mass dry biomass (ton)

---

[1] Personal communication with Sanders, J.P.M (2014). Commercial price of enzyme is €100/kg active enzyme. Assuming that Protex 40XL has the concentration of active enzyme for 10 g/ L solution, than the solution costs are € 1/ L or € 1000/ton.

3. Results

3.1 Protein extraction

Microalgal protein has been extracted using alkali [5] and enzymes [6]. For both methods, microalgae meal (with prior lipid removal) protein was more easily extracted as compared to protein from microalgae (Fig 1.). This finding confirms the benefit of oil extraction prior to protein extraction [5].

The three-step alkaline extraction of microalgae meal resulted in 32.7% (w/w) protein extraction yield. A higher protein yield was obtained when enzymes were used to aid in the extraction. As much as 73.2% (w/w) of microalgal protein can be extracted from microalgae meal using 5% proteases (Fig. 1). Less microalgal protein was extracted when less enzyme was used. The protein extraction yield decreased to 62.7% w/w at 1% enzyme. Despite of this decrease, the 1% enzyme aided extraction still showed higher yield than the alkaline method, indicating the technical superiority of the enzyme assisted method in extracting microalgal protein.

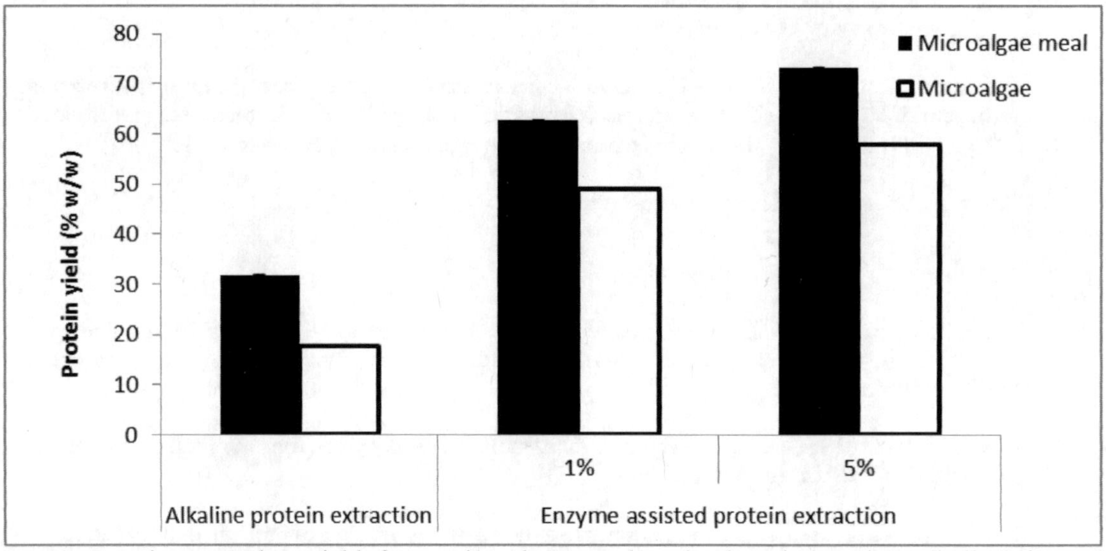

Figure 1. Protein extraction yield from microalgae meal and microalgae using alkali (left) and proteases at 1 and 5% (enzyme volume/protein weight).

3.2 Cost calculation

To gain insight in the cost effectiveness of alkaline and enzyme assisted protein extraction, economic calculations were performed on part of the processing cost including cost of biomass, chemicals (NaOH for alkaline protein extraction, NaOH and enzyme for enzymatic protein extraction), and energy for heating. Biomass cost are the main contributor to the processing cost (Table 1). Figure 2 indicates that lipid removal has substantial effect on costs for production. It is therefore advisable to de-oil microalgae prior to protein extraction.

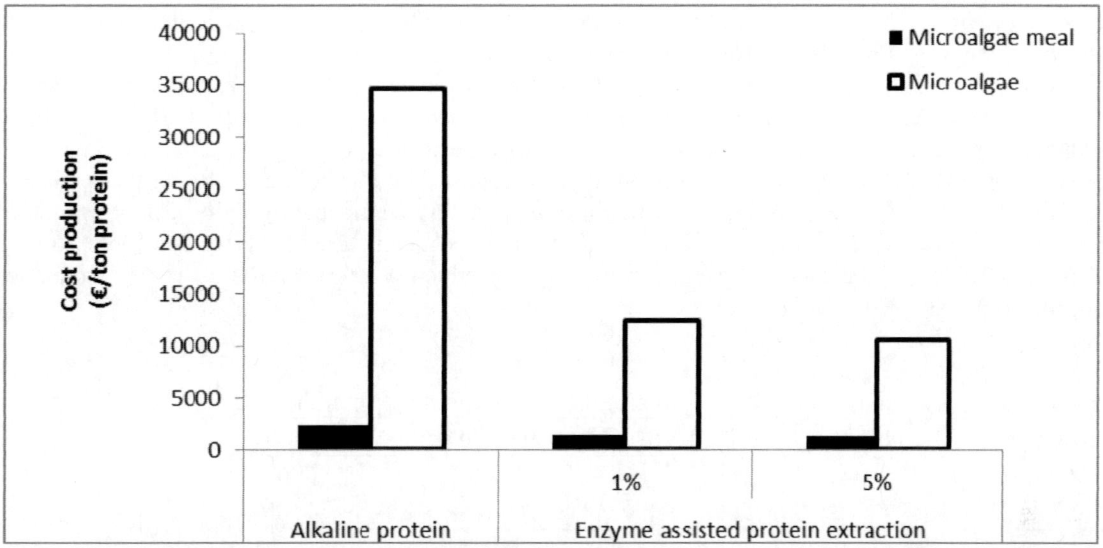

Figure 2. Cost required to produce 1 ton protein from microalgae meal and microalgae using alkali (left) and proteases at 1 and 5% (enzyme volume per weight of protein).

Table 1. Detailed processing cost calculation for producing 1 ton microalgal protein

		Cost required (€/ton protein)				
		Biomass	Chemicals	Heating	Enzyme	Total
Alkaline	Microalgae meal	1948	10	490	0	2448
	Microalgae	33422	77	1173	0	34672
Enzymatic (1%)	Microalgae meal	984	333	35	16	1367
	Microalgae	11970	426	45	27	12468
Enzymatic (5%)	Microalgae meal	858	308	30	70	1266
	Microalgae	10141	338	38	115	10631

Due to the unavailability of a reliable industrial microalgae market price, the processing cost for microalgae protein depicted in Fig. 2 will probably not reflect the actual possible industrial protein production cost. The deviation on the predictive market price for microalgae is very high. In the calculation we cite € 1650/ton microalgae [3], which is already very high, while some also mentioned values of microalgae as high as € 25000/ton microalgae [7, 11].

4. Discussion

With the current low 17.6% protein yield from the alkaline extraction from microalgae, industrial microalgae protein production is far from being feasible. De-oiling increases protein yield to 32% protein from microalgae meal (with alkaline method at elevated temperatures). Looking at the still low yield, it seems that further optimization is required to extract microalgae protein. Protease-assisted-protein extraction has been developed for extracting more protein from microalgae and microalgae meal. With protease addition, the protein yield for microalgae and microalgae meals increased from 17.6 to 49.1% and from 31.7 to 73.2%, respectively (using 5% enzyme dosage).

The enzymatic assisted method in general required lower cost as compared to alkaline method. Without protease, protein production from microalgae requires 26 k€/ton protein (using alkali, see Fig. 2). Addition of 1% and 5% protease reduces the cost to 13 k€ and 11 k€/ton protein, respectively. Yet, this value is still too high for industrial production. The benefit of de-oiling microalgae brings protein production closer to industry. Here the addition of 1% and 5% protease to microalgae meal solution reduces the cost to 1367 € and 1266 €/ton protein, respectively.

This makes the enzymatic method on microalgae meal economically preferred for microalgal protein production. However, 1266 €/ton protein is still much too high. In comparison, only 350 €/ton protein[1] is required to produce 1 ton soybean protein concentrate [12]. The details on the microalgal protein cost calculation (Table 1) points out that biomass cost contribute most to the processing cost per ton of product A further optimized extraction method may contribute to higher extraction yield. If protein extraction yield can be increased to 90%, biomass cost can be reduced to € 555/ton protein (see Table 2). Table 2 indicates total processing cost for producing 1 ton microalgae meal protein with optimized method (90% yield).

Table 2. Processing cost for producing microalgae meal protein with optimized method (€/ton protein)

Method	Biomass	Chemicals	Heating	Enzyme	Total
Alkaline	555	15	50	0	620
Enzymatic	555	562	18	45	1180

For a theoretical 90% yield, the alkaline method is less costly than the enzymatic one (see Table 2). However, in practice, it is rather difficult to increase the yield up to 90%. In this study, alkaline protein extraction was conducted overnight with increased temperature up to 120 ℃. Still, most protein remained in the biomass.

Optimization of the enzymatic method appears more feasible. Cost minimization of the enzymatic method should first focus on the chemical cost. The enzymatic method has higher chemical cost compare to that of alkaline method due to use of a pH-stat during the extraction process. If the extraction can be performed in the absence of pH-stat and optimized for a fixed, but lower amount of alkali, then less chemicals will be required. Assuming as low a cost as € 15/ton protein (similar to the alkaline method without enzyme, see Table 2) the production cost may be reduced to € 633/ton protein, using 5% enzyme on microalgae meal.

In addition to an improved extraction yield or lower chemicals usage, biorefinery of other components from the microalgae is required to meet economic feasibility. In this case, the residual microalgal fraction after protein and oil extraction can be used for fuel or for feed with increased digestibility, due to the alkaline treatment. By refining microalgae into more than one or two products a feasible process may be achieved.

5. Conclusion

We examined the technical and economic feasibility of protein extraction from microalgae meal and microalgae, using alkali or proteases. For both methods, de-oiling microalgae prior to protein extraction increased the protein extraction yield. Enzyme assisted extraction resulted in higher protein yield as compared to alkaline method. Therefore, enzymatic extraction is preferred from an economical point of view. A reduction in chemicals cost due to excessive buffering is however needed to further reduce production cost. When combined with additional revenue from the residue after protein extraction, biorefinery of microalgae may become a feasible process.

References

[1] Konur O, 2011 *Applied Energy* **88**(10) 3532-3540.
[2] Wijffels R H and M J Barbosa, 2010 *Science* **329**(5993) 796-799
[3] Wijffels R H, M J Barbosa, and M H M Eppink, 2010 *Biofuels, Bioproducts and Biorefining* **4**(3) 287-295

[1] Biomass, chemicals, and power cost only.

[4] Sari Y W, W J Mulders, J P M Sanders, and M E Bruins, 2015 *Biotechnology Journal* **10**(8): p. 1138-1157
[5] Sari Y W, U Syafitri, J P M Sanders, and M E Bruins, 2015 *Industrial Crops and Products* **70** 125-133
[6] Sari Y W, M E Bruins, and J P M Sanders, 2013 *Industrial Crops and Products* **43** 78-83
[7] FAO, 2010, Rome, Italy.
[8] Alibaba, *NaOH price*. [cited 2014 July 12] Available from: www.alibaba.com
[9] Dupont C. R Chiriac, G Gauthier, and F Toche, 2014 *Fuel* **115** 644-651
[10] Eurostat 2014 *Energy price statistics* [cited 2014 October 27] Available from: http://epp.eurostat.ec.europa.eu/statistics_explained/index.php/Energy_price_statistics
[11] Pulz O and W Gross, 2004 *Applied Microbiology and Biotechnology* **65**(6) 635-648
[12] Mustakas G C and V E Sohns, *Soy process, equipment, capital, and processing cost*. [cited 2014 August 27]; Available from: http://naldc.nal.usda.gov/

IOP Conf. Series: Earth and Environmental Science **31** (2016) 012035 doi:10.1088/1755-1315/31/1/012035

Outlier Detection on Hotspots Data in Riau Province using DBSCAN Algorithm

Pristi Sukmasetya[1]*, Imas S. Sitanggang[2]
Computer Science Department Bogor Agricultural University, Jalan Meranti Wing 20 level 5 Kampus IPB, Bogor, West Java, Indonesia

E-mail: pristisukmasetya92@gmail.com

Abstract. Indonesia has serious problems in forest fires. One of the potential factors which indicates forest fires is hotspot. Hotspot is a forest fires indicator that detects a location with relatively higher temperature in comparison with nearby positions. One possible prevention efforts for forest fires is by detecting outliers on hotspots data. This study detects outlier on hotspots data in Riau Province in between year 2001 to 2012 using the DBSCAN algorithm and determines the distribution of outlier hotspots by region and time. The experiment results show that the highest occurrence of outliers is in 2005. The number of outliers on hotspots data reaches 1241 hotspots with the sum of square error (SSE) is 0.084. Outlier hotspots in Riau Province in 2005 spread across 11 districts/cities and 136 districts. In 2005 the highest outlier are found in Rokan Hulu with the number of outliers is 186 points. The highest frequency of hotspot that is considered as outliers is found in August 2005, with a total of 355 outliers in which as many 97 of these outliers are occurred in Rokan Hulu District.

1. Introduction

Indonesia is the largest archipelago country which has serious forest fire problems. Historical data between year 2001-2012 shows about average of 20.000 hotspots warning in Sumatra region every year, with confidence level of detection is more than 30 percent [1]. This indicates the changeover of hotspots occurs dynamically in a region over time.

Ref. [2] shows one of indicator of forest fires that will likely occurs is the hotspot. Monitoring of hotspots is done using remote sensing by satellites. The hotspot data takes form time-series , which the hotspot data are periodically observed every day by satellites such as NOAA series with AVHRR sensor onboard or Terra/Aqua satellite with MODIS sensor onboard. It is possible that the distribution of hotspots gather in a space naturally, so that data can be analyzed using clustering techniques. However it is also possible that there are hotspots which scattered far from the hotspot distribution, so the probability of hotspot as outliers also can be analyzed. Some clustering techniques are K-means, hierarchical cluster, DBSCAN and ST-DBSCAN. Among all the clustering techniques mentioned before, DBSCAN algorithm can be used to find cluster from large spatial database [3].

*corresponding author

Outlier is a set of data which is considered to have different properties or deviate when compare with other majority data. Outlier analysis is also known as anomaly analysis, anomaly detection or detection of deviations [4]. Outliers often contain useful information about the characteristics of the abnormal data. An object is said to be an outlier if (1) the object is not belongs in any cluster, (2) there is a great distance between the object to the closest formed cluster, (3) the object is included in a small cluster or separate clusters. The object is not included in any cluster can be detected using the algorithm DBSCAN [4].

Therefore, outlier detection from hotspot data can give information about the distribution of hotspots. If it deviates than it can be concluded there are anomalous dispersion of hotspots occurrence. The frequency of occurrence of the hotspot outliers is analyzed as an indicator of forest fires.

In this present paper, the authors implemented outlier detection on hotspot data in Riau Province in between year 2001 to 2012 using the DBSCAN algorithm to locate outliers based on its location and time of occurrence.

2. Data and Method

2.1. Data

The attributes of the data consist of latitude, longitude and acq_date. Latitude and longitude attributes describe the location of hotspots in Riau Province, while the acq_date attribute describes the date of hotspot occurrence in a certain time. Fig. 1. represents the number of hotspots in Riau province.

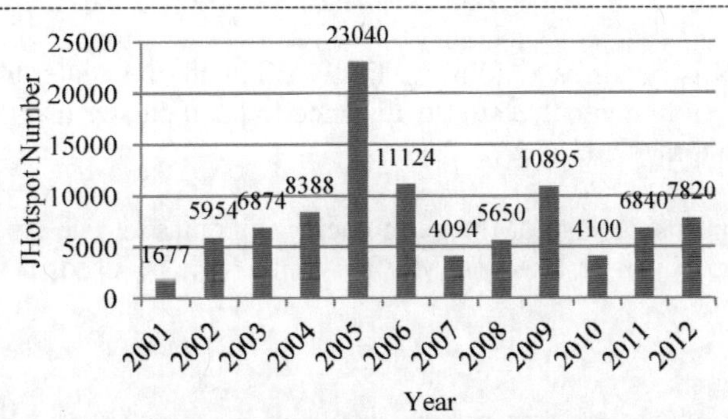

Figure 1. Hotspot data in Riau Province.

2.2. Spatial hotspot clustering using DBSCAN algorithm

DBSCAN (Density-Based Spatial Clustering of Applications with Noise) is a clustering algorithm which is developed based on the density level of data. DBSCAN find a central object which has a dense neighborhood. DBSCAN connects the center of the object with its neighborhood to create an area as a cluster [4].

DBSCAN algorithm was first introduced which is designed to find clusters and its outliers on spatial data. This algorithm has two main parameters, namely Eps and MinPts. Eps is the distance between center object of cluster and it's neighborhood, while MinPts is the minimum required number of neighbor objects from center object of cluster [6]. DBSCAN concept is as follows:

- The neighborhood which located within a radius Eps is called *Eps-neighborhood* of data objects,
- If the *Eps-neighborhood* of an object containing at least a minimum number *MinPts*, then the object is called a core object,

- A point *p* is said *directly density-reachable* from a point *q*, if *p* is adjacent with point *q* with a certain distance (ε) and point *q* is the center point,
- An object *p* is *density reachable* from object *q* with Eps and MinPts in a set of objects *D* if there is a chain of objects (p_1, p_2, ..., p_n), where p_1 = q and p_n = p, which $p_i + 1$ *density reachable* from p_i with Eps and MinPts, for $1 \leq i \leq n$, p_i member of D,
- An object p is *density-connected* to object q with Eps and MinPts in a set of objects D if there is an object *o* which member of D where p and q are density reachable from o with Eps and MinPts.

2.3. Outlier detection based on clustering

Outliers are objects or data that deviate significantly from the majority objects or data. In periodic data or time-series data, an outlier is defined as a data point that differs significantly from all the data [4].

One way to detect outliers is clustering. Clustering-based method assumes that the normal objects data belong to large and dense groups, while outliers belong to small and sparse group, or not belonging to any group. DBSCAN define outliers as data that do not include in any cluster, or noise [4].The result of clustering had been evaluated between Sum of Square Error (SSE) and silhouette index. This work uses SSE to evaluate clustering results. SSE formula is used to obtain the total number of clusters that minimizes the square error, $p \in C_i$ is each data point in cluster i, m_i is the centroid of the cluster i, and d is the distance to each cluster i [4]. The SSE formula is given as in (1).

$$SSE= \sum_{i=1}^{K} \sum_{x \in c_i} d(p,m_i)^2 \quad (1)$$

SSE formula on DBSCAN is obtained as follows, firstly calculate the center object (centroid) of each cluster which has been formed then calculate the distance to each cluster using the Euclidean distance formula and lastly find the total sum.

Calculation of SSE is done using RapidMiner. on each year dataset. The smallest SSE is used as a reference to determine the value of Eps and MinPts which will be used in DBSCAN algorithm to detect outliers.

2.4. Outliers Analysis

At this stage, an analysis of outliers was carried out from hotspots data in Riau Province based on location and time of occurrence of hotspots using the DBSCAN algorithm. The terms outliers in this paper is defined as a hotspot that stray away from the other hotspots location in general.

3. Results and Discussions

3.1. Spatial Hotspot Clustering

Various formation of experiments was carried out to determine parameter values for the DBSCAN algorithm. The values of Eps are 0.1, 0.06, 0.05, 0.04, 0.03, 0.02, and 0.01. The values of MinPts which were used in experiment are 1, 2, 3, 4, 5, 6.

The combination of Eps and MinPts used to cluster data hotspots in Riau Province from 2001 to 2012. Comparing of the these various Eps and MinPts produce the best Eps and MinPts value, and both will be used to form clusters of hotspot distribution data from 2001 to 2012 using the DBSCAN algorithm. For example, to find a cluster with variations in the value of Eps of 0.05 and MinPts of 2 we execute the R code as follows:

```
library(fpc)
tabel<-read.table ("HOTSPOT_2005.csv",header= TRUE,sep=",")
```

```
tabel$acq_date <- NULL
sapply(tabel, class)

#DBSCAN
x <- as.matrix(tabel)
ds <- dbscan(x, eps=0.05, MinPts=2)
plot(ds, x)
ds
```

Function *library* is used to call available packages in R software. In this case, function *library* is used to call the fpc packages that contains a module of DBSCAN algorithm and its toolbox. DBSCAN is executed using two parameters namely Eps and MinPts in accordance with the variation mentioned before to cluster hotspots data each year. After the clustering results are obtained, sum of square error (SSE) is calculated.

3.2. Outlier detection based on DBSCAN algorithm

Outlier detection was done after clustering using the DBSCAN algorithm. Outliers in clustering approach are defined as data that do not belong to any cluster. Table 1 shows the clustering result using the DBSCAN algorithm and outliers in the hot spot data from 2001 to 2012.

Table 1. Clustering result using DBSCAN algorithm.

Year	Eps	MinPts	Number of Cluster	Number of outlier	Percentage	SSE
2001	0.02	2	175	188	11.20%	0.048
2002	0.01	2	457	769	12.90%	0.036
2003	0.02	2	497	375	5.45%	0.022
2004	0.10	2	25	14	0.16%	0.026
2005	**0.01**	**2**	**864**	**1241**	**5.38%**	**0.084**
2006	0.02	2	862	1229	11.00%	0.032
2007	0.02	2	446	409	9.99%	0.021
2008	0.02	2	489	366	6.48%	0.018
2009	0.02	2	768	990	9.09%	0.022
2010	0.02	2	389	311	7.58%	0.021
2011	0.02	2	500	349	5.10%	0.021
2012	0.02	2	742	946	12.09%	0.024

3.3. Outlier Analysis

The highest frequency outlier in hotspots data is in the year 2005. In that year large and severe fires were occurred, so year 2005 was practically suitable and then chosen for analysis. Fig. 2. shows a map of the pattern of hotspots spread in 2005 in Riau Province.

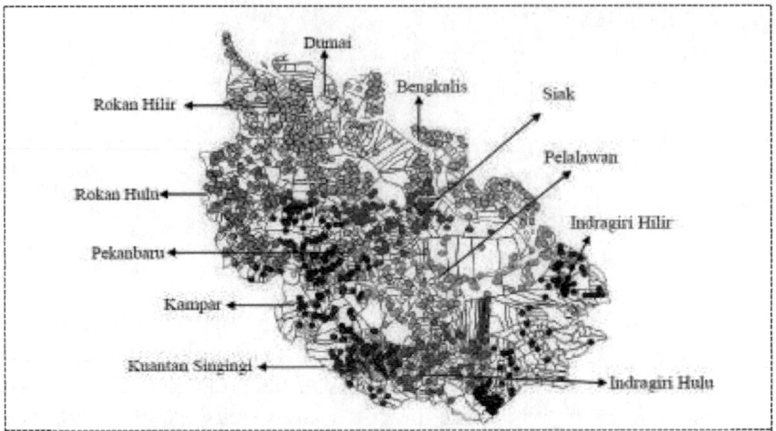

Figure 2. Hotspot spread in Riau province in 2005

Outlier detection on hotspots data could be analyzed by location or time of hotspots occurrence. Clustering using the DBSCAN algorithm results the highest occurrence of outliers in year 2005 with 1241 outlier with 864 clusters and SSE about 0.084. Outlier on hotspots data in Riau Province in year 2005 spread across 136 sub-districts and 11 districts/cities. The number of outliers based on districts/cities in Riau Province are as follows, Bengkalis-165 hotspots , Indragiri Hilir-97 hotspots, Indragiri Hulu-93 hotspots, Kampar-134 hotspots, Dumai City-37 hotspots, The capital city of Pekanbaru-20 hotspots, Kuantan Singingi-71 hotspots, Pelalawan-132 hotspots, Rokan Hilir-175 hotspots, Rokan Hulu-186 hotspots, and Siak-131 hotspots.

Fig. 3. shows a Riau map with outlier patterns in the year 2005 based on location. Each district / city is distinguished by distinct color.

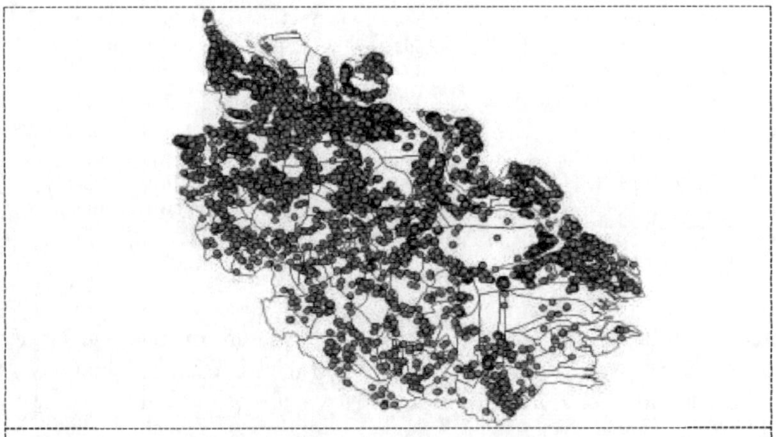

Figure 3. Riau map with outlier pattern in year 2005

In year 2005, the highest outlier of hotspots is occurred in Rokan Hulu district which has 186 outliers. The hotspots are scattered in 13 sub-districts. From all Rokan Hulu's sub-district, the as many 50 hotspot outliers are found in Tambusai sub-district. The outliers spread over 9 villages in which in West Tambusai village and East Tambusai village 8 outliers are detected. Fig. 4. shows the spread of outliers on hotspots data in Rokan Hulu.

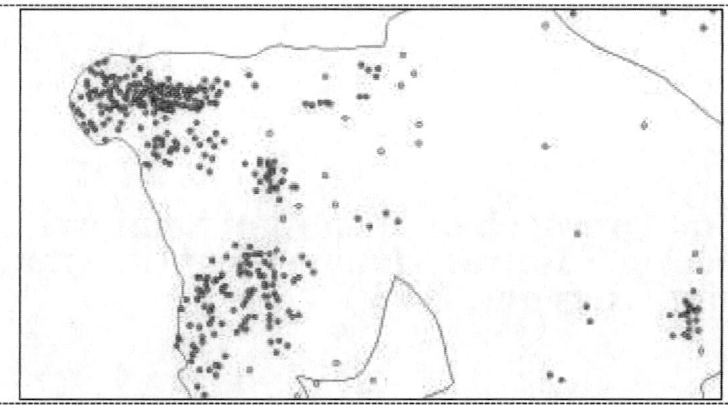

Figure 4. Rokan hulu outlier

In term of time of hotspot occurrence, as many 355 objects that are considered as outliers are occurred in August 2005. In this period Rokan Hulu district has the highest number of outliers i.e. 97 outliers compared to other districts.

Outliers in hotspots data, can be used as an alternative for consideration in keeping attention to the location of the hotspot which stray away that may become trigger to forest fires, judging from the number occurrence of hotspots outliers.

4. Summary

This work is able to obtain information of outliers occurrence on hotspots data in Riau Province from the year 2001 to 2012. The information gained including information about the name of villages, sub-districts, districts/cities and time of outliers occurrences. The highest number of outliers is occurred in 2005 with the number outliers of 1241, SSE of 0.084, Eps of 0,01 and minpts of 2. Outlier points spread in 11 districts/cities and 136 sub-districts in Riau province, with the highest is located in Rokan Hulu for 186 outliers. Outliers on hotspots data are scattered in 13 districts and are mostly found in Tambusai sub-district, as 50 points, with the largest is in West Tambusai and East Tambusai as 8 points respectively. Throughout year 2005, the occurrence of outliers reached 355 hotspots and most prevalent in Rokan Hulu district as many 97 points. This outlier refers to hotspot which its location faraway from majority hotspot. The information obtained from this research can be used as an alternative for consideration in keeping attention to the location of the hotspot outliers that may become trigger to forest fires.

References

[1] K. Austine, A. Alisjahbana, N. Sizer. (2013). *Data terbaru menunjukkan kebakaran hutan di indonesia adalah krisis yang telah berlangsung sejak lama.* [Online]. Available: http://insights.wri.org/news/2013/06/ data-terbaru-menunjukkan-kebakaran-hutan-di-indonesia-adalah-krisis-yang-telahberlangs#

[2] W.C. Adinugroho, *et al*. *Panduan Pengendalian Kebakaran Hutan dan Lahan Gambut*. Bogor, Indonesia: Wetlands International, 2005.

[3] M. N. Gaonkar and K. Sawant, "AutoEpsDBSCAN: DBSCAN with Eps automatic for large dataset," in *International Journal on Advanced Computer Theory and Engineering*. India, 2013, pp. 2319-2526.

[4] M. Ester, et al, "A density-based algorithm for discovering clusters in large spatial databases with noise," in *Proceedings of the 2nd International Conference on Knowledge Discovery and Data Mining.* Oregon, United States, 1996, pp. 226-231

IOP Conf. Series: Earth and Environmental Science **31** (2016) 012036 doi:10.1088/1755-1315/31/1/012036

The Ecosystem Approach of Wanatani Semi Arid Equatorial to Sustainable Agriculutral Management (in Amarasi District, Kupang Regency, NTT)

Rinda Yanti[1], Adi Basukriadi[2], Moh. Hasroel Thayib[3], Tri Edhie Budhi Soesilo[3]
[1]Agricultural Polytechnic of Payakumbuh, West Sumatera, Indonesia
[2]Faculty of Mathematics and Natural Sciences,(MIPA) Indonesia University, Depok
[3] Environmental Science Post Graduate Program, Indonesia University, Salemba

E-mail: rinda_yanti@yahoo.co.id

Abstract. The existing condition of the wanatani management in Amarasi District, Kupang Regency, NTT, has not optimized the welfare of the farmers yet, and the land degradation keeps happening. The objectives of this research was to analyze and obtain information on the ecological, social, and economic benefits of sustainable wanatani in dry land management. The research result shows that based on the observation from the ecological function including vegetation, land fertility, micro climate, erotion, and land suitability, wanatani is at present not optimal and not sustainable in supporting productivity and land conservation. From the economic function, the productivity in wanatani should be optimal, but the lack of institutional support and social function causes the agricultural management to be not optimal and not sustainable.

1. Introdution

Sustainable development is the conscious and planned effort, combining life environment including bio-natural resources into the development process to guarantee the capability, welfare, and quality of life of the present and future generations with the ecosystem approach and its environmental supporting power. Article 2 of the Bio-Natural Resources Conservation Law Number 5 of the Year 1990 states that bio-natural resources conservation and its ecosystem are having the principles based on preservation of the ability and the establishment of bio-natural resources and the ecosystem harmoniously and in balance. Furthermore, the conservation principle as contained in Article 2 is in accordance with the life environmental principle as put forward in Article 3 of the Life Environmental Protection and Management Law No.32 of the Year 2009 stating that the life environmental management is conducted with the principle of the state's responsibility and the principle of environmentally-friendly sustainability for the development Indonesian people and society entirely.

Moreover, Notohadiprawiro [1] explains that the land resources conservation is not meant to maintain the form or its appearance, but it is to maintain the use.The land form can be changed, the soil can be treated with numerous ways which alter its characteristics, and/or the cropping can be replaced as long as all those treatments aim to maintain or increase the land resources sustainably.

Most of the land in the semi-arid ecosystem like in NTT has a limited use for agriculture. The forms of the area are mountainous, having slopes and shallow soil, and still developing.The low precipitation and the long dry season limit the land productivity.Traditionally, most agricultural systems in NTT combine the commodity of forestry, agriculture, cattle food with or without animal husbandry/fishery on a piece of land known with the term agroforestry or wanatani.Theoretically, the activities of wanatani are conserving the agricultural resources, increasing the welfare of the farmers, and preserving the environment.However, in reality the wanatani management in Amarasi District of NTT has not reached an optimal and sustainable result.

Land degradation and critical land keep happening, and the poverty rates have not been reduced either.Land degradation tends to increase marked with more critical land found.Based on the Agreement on Forest Use Procedure (the Forestry Service of the East Nusa Tenggara Province 1999), the forest area width in NTT is 1,689,423 ha or 35.5% out of the area width, and out of that land width around 667,601 ha (40.62%) is the protected area.Until today, the total of the critical land width recorded has reached 1,313,897 ha or 28% out of the area total width (4,735,000 ha). The spreading of the critical land is in 14 Regencies and Towns, consisting of 297,322 ha in the forest area and 1,016,575 ha outside the forest area.Based on the landsat image information, in 2004 around 2.1 million ha out of the area width was already in a critical condition.

The use of firewood as the source of energy is still utilized by most people in NTT, especially in the rural areas or among the weak economic community, who generally use the wood for household need (cooking).This happens because the cost spent for kerosene is bigger than the cost for obtaining firewood.Based on the research result Widhana Susila et. al [2] ,the firewood use in Amarasi District is 1.13 kg per capita/day and in Kupang District is 1.11 kg/capita/day.The firewood used by the community is obtained from various places surrounding them, such as house yards, gardens, and the forest area.

The fact shows that the amount of poor population in NTT is relatively high.The number of poor people (BPS measurement) increased in 2004 to 20.86%.In 2007, the number of poor people was 27.58% out of the total amount of the population which was 4,448,873 people (BPS of NTT Province). Amarasi District had 15,302 residents.The number of the Family Heads was 3,677or 3,677 households. From that household number, the percentage of the poor family was 65.62% of the total number of Households (Annual Report of Amarasi District, 2007). The low income of the residents was connected to the low productivity, stability, and sustainability of the wanatani in Kupang Regency (NTT Forestry Balitbang, 2005). Therefore, the concept to manage the land resources with planning is required by paying attention to the aspects of production, quality, and preservation, so that at the same time the community can fulfill their economic needs, without ignoring its environmental and social roles.

The activity of wanatani which produces the commodity of forestry, agriculture, cattle food with or without animal husbandry/fishery formed artificially should not only provide environmental benefits but also provide economic and social benefits. Failure to combine those aspects will cause some obstacles to appear for the sustainable wanatani pattern development/application.Therefore, the formulation of the wanatani management concept should be more site-specific. The cultural, economic, and social reality empirically influences the land productivity.The wanatani management concept should also be sensitive with the segments of culture, economy, political characteristics, society, and scientific technical changes in the applied social context. The introduction of the wanatani management concept referring to the wet land ecosystem is admitted that it has not influenced the land management and use in NTT.This can make the planning and the application of the wanatani management concept not refer to the settlement of the roots of the problems, both biophysical and economic-cultural-local problems.The implication of the condition is it will be normal if then there is a process of a wrong adaptation, such as the force of applying the wet land wanatani management concept, which is generally not suitable with the dry land condition.This condition will lead to a situation where it is not success which appears, but instead clashes of values will often occur in a form of conflicts of interest.

Based on the explanation above, the study of wanatani ecosystem with the integration of all its ecological, social, and economic functions in managing dry land is necessary to be done so that a sustainable concept can be realized.This condition is supported with the non-existence of the research which formulates a concept of sustainable wanatani management in the semi-arid ecosystem

Ecosystem Approach Concept

The management activity with the ecosystem approach requires the understanding of the ecological system entirely, parts constituting it, and the interaction among the components inside it. Bruce Mitchell *et al.* [3] explains that the ecosystem concept is marked with the study of kinds of living creatures and their physical environment as an integrated unity.In managing environment, the importance of the ecosystem approach lies in the comprehensive, overall, and integrated approach.This definition also covers a system concept, including parts constituting it and the relations among those parts.

Furthermore, Jatna Suprijatana [4] states that the management with the ecosystem approach is knowledge about ecosystem and how every organism functions in it, limited by its physical environmental limitation.The objectives are to maintain the ecological process, bio diversity, and its complexity, to reduce external input, and to maintain the existing output.

Ideally, in the ecosystem management, the ecological process is expected to be able to keep continuing, although the area is exploited.In order that the ideal expectation can be realized, plenty of scientific and accurate information is required. Grumble in Jatna Suprijatna [4] explains that the management with the ecosystem approach is the management activity integrated with the scientific knowledge on ecological processes in the socio-political framework and complex values with the aim to protect the integrity of local ecosystem and the continuity in the long term. Grumble in Jatna, Suprijatna [4] also identifies 10 dominant tasks in the management with the ecosystem approach, which are (a) diversity system stages, (b) ecological limit knowledge, (c) ecological integrity, (d) research systematics and data collection, (e) monitoring, (f) adaptive management, (g) intersector cooperation, (h) organization change, (i) humans as the ecosystem component, and (j) humans' values in accomplishing goals.

The Wanatani Management Concept

The wanatani management concept refers to the environmental management concept.The environmental management is none other than the integrated resources management.The characteristics of resources based on the composition and based on the behavior in the use need to be observed in determining the sufficient environmental management ways.Because the kinds, availability, and capability of resources existing in an area can differ from the ones existing in another area, and because the importance of resources for humans can change from time to time, the environmental management system requires to have the special features of time and space.

Wanatani is a new name from the use system which has long been done practically.This system, according to King [5], is to avoid mistakes from the definition of agriculture and forestry, and at the same place and time to produce food ingredients and firewood.The modern concept from wanatani now is being developed, but the universal wanatani definition is not yet approved.In the beginning 1977 the Royal Tropical Institute in Amsterdam, the Netherlands, gave a definition of wanatani, and this definition was suggested by Bene *et al.* [6] which was then expanded by King and Chandler [7] as follows:

Wanatani is a land management system which can preserve and increase land productivity entirely, and is the combination of activities of agriculture (cultivation plants), forestry (tree plants), and animal husbandry, simultaneously or sequentially on the same land, by using practical management adjusted with the local cultural pattern. This definition was then developed again by Vergara in APY. Djogo [8] by adding a social factor, the economic condition of the local people, and the environmental ecology.It is also to accommodate all types of land use combining annual cultivation plants and parenial wood plants with cattle.

Dry Land Concept

Based on the formulation result of*Seminar Nasional Pengembangan Wilayah Lahan Kering*(the National Seminar of the Dry Land Area Development) in May 2002, the dry land area covers:rice field dependent on rainwater for irrigation, dry field, not irrigated rice field, mixed farming, plantation, forest, bushes, meadow, and ranch.Dry land is a spread-out area of land which is not welled up or inundated with water in most of the time in a year (Hidayat *et al.* 2000). One of the characteristics of dry land according to Rukmana [9], among others, is being sensitive to erotion, especially if the slopes are not coveredwith vegetation.

Dry land is differentiated based on precipitation as the following:

a) Wet climate dry land can be found in an area which has precipitation more than 200mm/month for 6 to 7 months, and its dry months have precipitation less than 100mm/month for 3 to 4 months, or the precipitation is minimally less than 2000mm/year.

b) Dry climate dry land can be found in an area having dry months for 7 to 9 months and wet months for 3 to 4 months.

According to Soil Survey Staff (1960) dry land is the land which more than half the time (for seven months) suffers from drought. Dry land is the land which can be used for agriculture business by using limited water, and usually the water comes from the rain.

2. Materials and Method

The population of this research is all farmers applying wanatani in Amarasi District, Kupang Regency, NTT.The research samples were determined with purposive sampling, which is 9 groups of wanatani farmers (8 villages and 1 sub-district) along with their land and vegetation owned. The technique of taking samples with purposive sampling is by taking the board and the members of the group of farmers joining the group of farmers in each village.The research samples are determined with the Slovin equation, which is:

$n = N/ (1 + Ne^2)$

Explanation: n = sample size; N= farmer's household size; e = percentage of thoroughlessness (10%). Based on the above equation, out of 3,461 farmer's households joining the group of farmers in Amarasi District, around 100 respondents were selected to become the targets of interview.

Sample compartment for soil analysis and vegetation was decided through 2 phases with purposive sampling.The first phase was to determine 3 sample areas, which were Iay, IIay, IIIay based on AEZ map.Every area is not similar with one another, but inside each area it is similar.The second phase was to determine 8 points of sample compartments for each area by using GPRS so that the number of all sample compartment points would be 24 sample compartments. Data analysis is an important part in this research because through this process the raw data collected could be given meanings and definitions in order to be useful to solve formulated problems (Table 1).

Table 1. Data Analysis Method

Objective	Data	Data Collection	Data Analysis
1) To analyze the ecological	a. Precipitation	Observation Station Climatology	Climate data tabulation Statistical Test
	b. Micro climate	Observation Station Climatology	Climate data tabulation Statistical Test
	c. Land texture	Measurement of Land Composite	Pipette Method
	d. KTK	Laboratory Analysis	Extractof NH4O AC pH7
	e. pH	Laboratory Analysis	Electrode Glass
	f. Base saturation	Laboratory Analysis	Extractof NH4O AC pH7
	g. C-organic	Laboratory Analysis	Walkey and Black
	h. N-tot	Laboratory Analysis	Kjeldahl
	i. Available P2O3	Laboratory Analysis	Bray I
	j. Available K2O	Laboratory Analysis	Extractof NH4O AC pH7
	k. Ca	Laboratory Analysis	Extractof NH4O AC pH7
	l. Mg	Laboratory Analysis	Extractof NH4O AC pH7
	m. Microorganism (Total fungi SPK/g)	Laboratory Analysis	Total plate count (TPC)
	n. Relative Density (RD)	Map observation of land use Sampling	
	o. Relative Frequency (RF)	Sampling observation &	
	p. Significant Value Index (SVI)	Sampling observation & measurement	SVI = RD+RF
	q. Type Diversity Index ofMargalef (R)	Sampling observation & measurement	$R = \frac{(S-1)}{NL.N}$
	r. The number of types observed (S)	Sampling observation &	
	s. The number of individuals (all types) observed = N	Sampling observation & measurement	

IOP Conf. Series: Earth and Environmental Science **31** (2016) 012036 doi:10.1088/1755-1315/31/1/012036

	t. Natural Logarithm (NL)	Sampling observation &	
	u. pi= proportion of the number of species	Sampling observation &	
	v. Ratio Distribution Index of Hill (E)	Sampling observation & measurement	$E = \frac{(1/\lambda - 1)}{e^{H'} - 1}$
	w. The number species individual number i (ni)	Observation Measurement Parameter Calculation of	
	x. Abundance Index of Simpson (λ)	Sampling observation & measurement	$\lambda = \sum_{i=1}^{S} \frac{[ni(ni-1)]}{N(N-1)}$
	y. Diversity Index of Shannon (H')	Sampling observation & measurement	$H' = \sum_{i=1}^{S} piNL\ pi$
2)To analyze the socio-economic of the wanatani ecosystem	X2 = Knowledge of farmers X 3= Farmers' motivation to accomplish success in farming effort X1 = Counselling frequency	Survey	*Regresy Analyze* *Y=a+b1X1+b2X2+b3 X3*
	Π = Benefits TR=Total income TC = Total cost	Survey	Π = TR - TC
3)To compare ecological, social, and	Ecological Function: Land fertility Vegetation, erotion, &climate Socio-economic function	Table of land fertility ratingof IPB laboratory.	Categorize: Optimal and sustainable if in accordance with the indicator
The indicator.		AEZ map of Kupang Regency and requirements of growing plants in the semi-	sustainable if not in accordance with the indicator.

IOP Conf. Series: Earth and Environmental Science **31** (2016) 012036 doi:10.1088/1755-1315/31/1/012036

3. Results and Discussions

The Sustainability of the Ecological, Social, and Economic Functions of Wanatani

The wanatani ecosystem approach has empirically been analyzed including the ecological, social, and economic functions.The sustainability indicator of the wanatani ecosystem function refers to the land supporting power and the semi-arid natural forest ecosystem function. The sustainability of those functions holistically is elaborated in Table 2.

Table 2 . The Sustainability of the Ecological, Social, and Economic Functions of Wanatani

Ecosystem Function	Measurement		Sustainability Indicator		
			Research Result	*Reference (Semi-Arid Climax Forest*	*Explanation*
I. Ecological	1.	Vegetation Diversity (Distributio n Index/E)			
	a.	Tree	0.55	>0.5	*Sustainable*
	b.	Sapling	0.37	>0.5	*Not Yet Sustainable*
	c.	Seedling	0.28	>0.5	*Not Yet Sustainable*
	2.	Micro climate			*Sustainable*
	a.	Air temperature	25.36°C	<27°C	*Sustainable*
	b.	Soil temperature	25.5°C	<27°C	*Sustainable*
	c.	Humidity	88%	>80%	*Sustainable*
	3.	Soil Fertility			
	a.	Macro nutrients: N, P,	Medium	Medium	*Sustainable*
	b.	Micro nutrient:	Very low	Medium	*Not Yet*
		Mg			*Sustainable*
	c.	Na	High	Medium	*Sustainable*
	c.	C/N, KTK, Base	Medium-high	Medium	*Sustainable*
	d.	Land microorganis m (The Number	3	>5	*Not Yet Sustainable*
	e.	Erotion level (tonnes/ha/year	60-480		*Not Yet Sustainable*

II. Social	1.	Conservation Behavior (R^2) Y = 8,54 + 0,38 X1 + 0,01 X2 + 0,25 X3	0.75		*Sustainable*
III. Economic	1.	Productivity Potential (R/C)	7.68		*Sustainable*
	2.	The economic institutional role	Lack of	Adequate	*Not Yet Sustainable*

Source: Managed Data (2011)

Descriptively, the ecological, social, and economic functions presented in Table 2 are relatively the same between semi-arid natural forest and wanatani. The finding shows that wanatani application in Amarasi is not yet sustainable. Observed from the ecological aspect, variety of vegetation cultivated by farmers fall into the category in accordance with the land suitability, but it is not yet optimal. This happens due to the fact that the food plants still dominate the wanatani land.It will cause the micro nutirent elements contained in the wanatani land to be completely gone so that its content is relatively low.On the other hand, wanatani orientation for food supply from seasonal plants still becomes the main goal.The fact is very far different from the semi-arid natural forest ecosystem, whose primary food can be obtained from breadfruit trees and other seasonal plants which are relatively rare, even non-existent.Besides the breadfruit plant and other seasonal plants which have been extinct, NTT area is grown with endemic plants, such as the sandalwood plant.

This sandalwood plant is a plant which can only be discovered in NTT area, and it is not found in other areas.This is caused by the influence of some environmental factors, such as 1) temperature, 2) precipitation, 3) types of land, and 4) topography in NTT area.In a macro scale, the logical consequences from the influence of those four factors result in a lot of vegetation having different natural biota. In an environment generally we discover that a plant only grows in one area and cannot be found in other areas; this is called endemic, which means the plant only exists in one area and its spreading is limited.The endemic plant for NTT area is, among others, the sandalwood tree.There are a few things influencing the spreading of living creatures.The environmental factors, such as temperature, precipitation, types of land, and topography, really influence the distribution pattern of living creatures.The presence of these factors causes many plants to have different natural biota.

The sandalwood plant is a local specific plant which is more adaptive.This plant has a high economic value, so that it becomes the NTT export mainstay commodity.There are at least 4 obstacles and 1 solution offered by previous researcherson the development of the sandalwood plant in West Timor of NTT area, and they are:

a. The sandalwood resources in the West Timor Island have been completely gone all being cut down.
b. Although it is endemic and has a long history with Timor community, the regeneration, the seed quality, and the techniques of sandalwood plantation are still problematic.
c. The policies and regulations governing the classification and ownership of sandalwood trees become a big obstacle based on the effort of a more effective replantation.
d. The cultivation of sandalwood trees in a big scale does not succeed.

e. The research in the future is expected to develop a small scale farmer-based system which integrates sandalwood trees in synchronized wanatani with the provisions of agro-ecology;the determination of good quality seeds and minimalism of in-breeding; and on-farm research with the supportive government regulations which assist poor farmers.

The farmers' tendency to harvest trees (for the wood) whose age is still young to be used for energy (as firewood) and building material makes those trees lose the potential as the erotion control.This condition causes the erotion level to become high exceeding the total erotion level which is tolerable.The high erotion level also causes the big runoff of the land surface to occur, which dissolves the land nutrients and leads to the decrease of land fertility.Moreover, erotion happens on the farmers' land which is managed for food plants and vegetables, such as corns, tomatoes, beans, chillies, and garlic & shallot. The monoculture practice on the agricultural land provides an opportunity for erotion to occur.

To obtain harvesting results in a year, the farmers will do numerous ways, including cutting down certain production plants and burning them.The practice is done by the farmers so that the food plants can grow well.The land clearing by cutting down trees and burning them on the wanatani land leads to the potential of erotion to occur.The field burning causes the loss of vegetation as land covering, and the smoke released is a harmful pollutant for environment and farmers' health.

The solution offered to deal with the problems above (on the sandalwood plant and other kinds of wood) is to have the technology which is suitable with the local wisdom, cheap, and easily done by farmers, and it is wanatani.Wanatani empirically can improve the micro climate and the land nutrient cycle because of the interaction of trees and other plants.Consequently, the land productivity will be better.

The reality shows that due to relatively wide land ownership it becomes an obstacle to manage the land (because of the limited work force).That obstable can be overcome by applying wanatani.With wanatani, the spirit of helping one another and working together which has been inherited traditionally from one generation to another and is something positive can be applied in the land management so that the load of work can be lighter.The behavior of the farmers in applying wanatani land conservation is the inherited habit which is continuously applied.The farmers really understand the custom related to land conservation. However, their low level of education and skills make the conservation behavior not yet optimal.

The economic aspect is the wanatani economic function which provides direct and indirect advantages.The economic analysis from the research result shows that wanatani productivity potential can afford to cover the high costs of food plant cultivation, but it cannot guarantee the sustainable condition.The high production cost for agricultural production facilities spent by the farmers requires improvement with the environmentally friendly and easily applied agricultural production facilities by farmers, such as the composting technology of cattle waste and harvest remains as the source of fertilizer.According to the researcher, the natural resources existing surrounding the land have not been used optimally by the farmers.

Food plants (rice and corn) are the basic necessity which must exist.The ecological obstacle is less considered by the farmers in cultivating these two kinds of plants so that the high cost which they have to spend is not a problem for them.If this condition keeps happening, the land productivity will be lower and lower, and this will lead to low income (life quality keeps decreasing) so that the poverty rates will keep increasing.This condition also does not become the concern of the farmers because they rely on other wanatani productions.Every year farmers also gain some fruits, plantation productions such as betel nuts and coconuts, and once in a while they will get some wood and some cattle.

The sustainable agricultural management by wanatani application empirically can be accomplished by the effort to optimize the positive interaction among the various constituting components (trees, agricultural plant production, cattle/animals) or the interaction among those components with the environment referring to the semi-arid natural forest ecosystem.Observed from the principles of environmental science, several wanatani advantages in Amarasi compared with the other land use systems are:

i) Productivity

Based on the research result, it is proven that the total products of the combined systems in wanatani are much higher than the food plants (monoculture).This happens because the output is from one land field which is in diversity and is evenly all year long. The combination of plants provides advantages because the failure of one component/kind of plants can be covered up by the success of other components/kinds of plants.

ii) Diversity

The combination of more than two components in the wanatani system produces high diversity, covering both products and service.Observed from the economic aspect, it can reduce the risk of losses due to market price fluctuation.From the ecological aspect, it can prevent the failure from happening in the monocultural cultivation. More than one species can live in a hollow, and every species creates a condition to maintain the life of other species.The agroecosystem with many different hollows and inhabited by various kinds of species tends to be more stable than the one inhabited by one species in the monocultural environment.

Reducing the diversity of species and increasing the variety through genetic methods make the agricultural land unstable.The functional diversity in agroecosystem is accomplished by combining species of plants and animals having complementary characteristics in the positive-synergetic interaction resulting in ecosystem stability, supporting the hydrological function, andmaking the community of farmers survive.

iii) Self-regulation

High diversification in wanatani is expected to be able to fulfill the farmers' basic necessity and to release them from their dependence on outside products.The system independence to function will be better; in other words, it will not require much input from outside (fertilizer, pesticide) with high diversification compared with the monocultural system.

iv) Stability

Wanatani practice having optimal diversity and productivity will be able to provide balanced results as long as the land is exploited so that it can guarantee the income stability and sustainability of the farmers.Even though the wanatani diversity in Amarasi has fulfilled the criteria, in reality the productivity is not yet optimal.

The principles of sustainability start by having two basic foundations, ecoliteracy and ecodesign.Ecoliteracy is the understanding of organizational principles developed by ecosystems through evolution to support life networks, such as moving the ecological cycles, as diversity guarantees resistance.Ecodesign is to introduce an era which is not based on what we can obtain from the nature but on what we can learn from it.

For sustainable development, what needs to be preserved is not the suitability and the balance but the environmental supporting power which can support sustainably the growth and the development.The increase and the decrease of the supporting power depend on the factors influencing it.Those factors, among others, are: (1) the geographical factors, which are climate, land fertility, erotion, and others; (2) the sociocultural factors, which are knowledge, technology, and behavior influencing environment.

The ecological supporting power is to optimize the food productivity by minimizing the use of agricultural production facilities, such as chemical fertilizer and pesticide observed from the volume, time, and frequency aspects.The implication is the production cost and land degradation are expected to decrease.Moreover, optimizing wanatani productivity is by determining the proportion of various kinds of trees,plantations, and fruits with the balanced combination.Cutting down trees will not be the priority, and the advantages of the positive interaction between trees and other vegetation will be prioritized more so that the direct and indirect advantages of wanatani will be also optimal.The ecological supporting power is the support system of the semi-arid ecosystem with the application of agriculture with forest ecosystem.

The socio-economic supporting power is to optimize the income of the farmers by increasing the capacity of the social and financial institutions.The social institution includes the customary institution roles in increasing the conservation behavior so that land degradation decreases and the land becomes productive and increases income. Besides that, the capacity improvement of human resources is by enhancing knowledge and introducing innovation and local specific technology in utilizing resources surrounding the land. The financial institution is to provide funding by soft loan so that it can optimize the farming capital.Increasing the capital in a form of funds and human resources quality is expected to raise income and reduce the poverty in the population.

Based on the research result, it can be proven that what causes land degradation and the high rates of proverty in Amarasi is the non-optimal socio-economic supporting power.The ecological supporting power is actually quite optimal to support the wanatani productivity.However, the farmers' behavior is not yet optimal to apply conservation making the ecological function disturbed, as it can be shown especially with the high total erotion value in Amarasi.

Observed from the ecological supporting power, the wanatani development is very suitable and potential in increasing the income and the welfare of the farmers compared with the food productivity including paddy and nonstaple food crops.The fact is also supported by BPS data (2003) that the agricultural productivity in Indonesia provides a clear picture that there is a pretty wide economic disparity between the farmers of food plants and the horticultural farmers.

Time seriesdata of Susenasfor the periodof 1996-2002 show that the demand of the consumers based on high value food grows fast, while the demand of food ingredients sourced from various seeds suffers from stagnation.The 2003 Agricultural Census Data also show the pretty fast growth of farming in Indonesia is from the horticultural subsector.The number of the horticultural farmers also increased sharply, from 23% in 1993 to 36% in 2003, out of the total number of farmers.On the other hand, the number of the farmers of paddy and nonstaple food crops decreased from 84% in 1993 to 71% in 2003.

The productivity development of several agricultural commodities, especially in the last 20 years, indicates that the productivity growth trend has kept decreasing, although the level of productivity itself still increases (Table 3).

Table 3. Agricultural Productivity Development in Indonesia

Productivity (tonnes/ha)	Average 1980-1985	Trend (%)	Average 1986-1995	Trend (%)	Average 1996-2000	Trend (%)
Paddy	3.70	3.55	4.22	0.91	4.34	0.45
Corn	1.62	3.98	2.10	1.89	2.63	2.30

Source: Data managed from BPS (2011)

The agricultural sector requires media or a quite wide area in a form of plain land, and the requirements of growing plants, especially the main food ingredients, are pretty complicated.With a quite wide area, the complicated requirements lead to also a pretty high opportunity cost for agriculture.The requirements required, for example, are physical structure and land chemistry, topography, nutrient elements, water content, climate, and others, so that not all land is proper and appropriate or suitable to be developed for certain plants.The ecology of NTT is empirically very suitable for wanatani, which is agriculture with the forestry system compared with food plant development, especially paddy and nonstaple food crops.

The literature study result explains that main food plants (paddy and corn) cannot be entirely called as the indigenous variety of Indonesia since all of those come from and have been mixed with the products from outside.Based on the history, the place of birth or the Motherland of paddy was in the Southwest border of China, Yung Nan – Sour Hill, India, in 7000 B.C.The first batch of rice diffusion happened along Mekong river and other rivers connecting Indochina peninsula followed with human migration.The second batch occurred from the Sour Mountain entering Brahmaputra, Gangga, Bengal to Chennai, Bengal Bay,Melayu peninsula,and Indonesia.

The third batch went through Himalaya Route.The fourth batch went through Yangtze river flowing from Yung Nan as long as 6,400 km heading to Shanghai and then to Korea and Japan.Corn comes from South America, and after the arrival of Europeans, it was spread to all over the world.It means what we consume now is the result of ethnic, race, and cultural cross scientific breeding.

Based on the prioritized food plant kinds, especially paddy in NTT, it causes the farmers to give all their funds and energy to manage very wide land.The single crop approach forces the farmers to use pesticide and chemical fertilizer which, on one hand, have been proven to be able to increase production but, on the other hand, can cause environmental problems.In line with the research result ofNurmalina R. (2008), she states that the sustainable index value of the rice availability system (ecological, economic, socio-cultural, institutional, and technological) between areas in Indonesia really varies in the range of 33.37-67.23. NTT area is included in the category of 'less sustainable'. The daily solar (sun) radiation energy in Indonesia is around 20 mega joule (MJ) every square meter of land.The energy will be lower and lower if the sun is covered with cloud.The dillema faced in the agriculture is the high solar energy happening in the dry season when the precipitation is relatively low, unless there is irrigation.

On the contrary, the planting season is done in the rainy season as the solar radiation energy is less because the sun is covered with cloud so there is tradeoff between water necessity and the solar energy availability for photosynthesis.Generally, the fruit quality and plantation commodities are higher in the dry area becausethe growth of fruit or sugar content happens during the dry season when the solar energy is high. In contrast, severalplants are very vulnerable to drought, whereas some can adapt like teak which has its leaves fall during the dry season to reduce the transpiration speed.

The long-day variation for a year in Indonesia is not too big (11 to 12 hours) if compared with temperate areas like in Europe which can reach 18 hours. For certain long-day plants, although farmers plant them three times in a year and in Europe they can only be planted once in a year, by using the plant seeds with the same quality, the result obtained will be far higher if they are planted in Europe which has a long day longer and lower temperature compared if planted in Indonesia.The long day controls the change of plant development phases, while the interphase time will then determine the period of plant biomass accumulation which eventually determines not only the productivity but also the quality of plant result.Therefore, the farmers need to consider the agricultural commodities which will be developed or produced in certain climate environment.

In Indonesia, the solar radiation energy received on the surface can support agricultural production for the whole year as long as water is not an obstacle. NTT has low precipitation so that water availability is an obstacle, but this area has bigger solar radiation energy than other areas.The energy is the main energy source in the photosynthesis process to support plant growth and also is the factor which limits and filters which kinds of plants are really suitable for certain areas (local specific).

In the ecological principle, the use of agricultural technology, such as chemical fertilizer, pesticide, excellent seeds, is basically the process of energy subsidy increase towards the human-made ecosystem which has some side effects that must be considered. To maintain the preservation of the life environmental function and to prevent pollution and environmental damage, it is necessary to propose an economical agricultural system in using energy subsidy whose basic

IOP Conf. Series: Earth and Environmental Science **31** (2016) 012036 doi:10.1088/1755-1315/31/1/012036

foundation can be searched from the local wisdom. Wanatani is one of the local wisdom which has been applied from one generation to the next by the local farmers, and the research result proves that it is ecologically, socially, economically suitable and requires management improvement to achieve its sustainability.

With the tendency of fuel price increase, the energy problem in agriculture becomes a more serious and interesting discussion because modern agricultural technology cannot be separated from the use of high energy production inputs, such as excellent variety, artificial fertilizer, irrigation, pesticide, and mechanism.Those production inputs are categorized as high energy since the process of finding, setting, making, distribution, and use of those requires high energy, especially from oil. It is common if those agricultural production energy inputs are used economically, efficiently, and optimally by remaining to do the effort for the optimal productivity goal to be accomplished.This can be done by using the energy economical agricultural system, which is the agricultural system using efficient energy, such as wanatani in Amarasi.

Food availability is not identical with rice availability, and food resistance is not identical with self-supporting in rice.The food balance inequality means challenges in food supply in the future will be harder and more complex.Therefore, it is necessary to have technological innovation, conventionally and biotechnologically.NTT has variety of ecosystems and biological resources.This is the basic capital in providing food supply in the future.It is necessary to have a dynamic agricultural system, depending on commodity diversification having a more advantageous market prospect by integrating variety of plants and cattle.

4. Conclusions and Recomendations

Conclusion

Wanatani management has not completely referred to the semi-arid natural forest ecosystem concept. Observed from the ecological function including vegetation, land fertility, micro climate,dan land suitability,the wanatani is not yet optimal and not yet sustainable to support land productivity and conservation.From the economic function, the wanatani productivity is already optimal, but it is lack of institutional support, and the social function makes the management not yet optimal and not yet sustainable. Based on the empirical testing in the field and the analysis done, it can be formulated that the sustainable agricultural management is the application of wanatani referring to forest ecosystem by integrating the ecological, social, and economic functions.

Recommendations

This needs to be implemented, especially for the relevant institutions, such as the Ministry of Agriculture and the Ministry of Forestry, done cross-sectorally involving other institutions (the Ministry of Life Environment, the Ministry of Research and Technology, Higher Education, Regional Governments, and Society Self-Funding Institutions) to formulate various technologies and agricultural productivity increase programs with the agricultural pattern applying the local specific forest ecosystem.

References

[1] Notohadiprawiro, T. Exposure As Shape Wanatani Land Use Criteria According to the Land and Water Preservation. Seminar Agroforestri dan Pengendalian Peladangan.IlmuTanah.UGM.1998.http://www.gadjahmada.edu/index.php?page=rilis&Artikel=1773 PDF: 23 September 2009. pk.11.20 WIB

[2] Widhana *et al*. Firewood Consumption in Kupang district. Case Study Oenoni Village, Oeipuah, and Kotabes. Forestry Research and Development Center for Kupang. Balai Penelitian dan Pengembangan Kehutanan Kupang. Santalum No. 17.1994 [3]Mitchell, B.

(2007). Resources Management and Environment. Gadjah Mada University Press. 2007
[4] Supriatna, Jatna. Preserving Nature Indonesia. Yayasan Obor Indonesia. Jakarta. 2008
[5] King, K.F. (1979). *Wanatani.* Proc. 50th TropicheLandbouwdag. Bull. 303 Kon. Inst. Trop. Amsterdam. h.1-10
[6] Bennet, E.L,, J.G. Robinson, and E.Wilkie. "Why is eating bushmenat biodiversity crisis?. Conservation in Practice 3(2): 28-29 pp.1999
[7] King, K.F. & Chandler. *The Waste Lands*. ICRAF. Nairobi. Kenya. 1978 [8]Djogo, A.P.Y. Models Wanatani Potential For Dryland Agriculture. Wanatani development in eastern Indonesia. Prosiding Lokakarya Nasional Wanatani II. APAN Report N0.18.1995
[9] Rukmana, R. Hilly Land Management Techniques and Critical. Kanisius. Jakarta.1994
[10] Bisseleua. Biodiversity Conservation, Ecosistem Functioning, and Economic Incentives Under Cocoa Wanatani Intensification. Conservation Biology Vol.23, no. 5, pp. 1176-1184, Oct 2009
[11] Carroll, C.R. Agroecology. Mc-Graw Hill, New York. 1990
[12] Conway, G.R., and E.B. Barbier. After The Green Revolution; Sustainable for Agriculture. Earhscan Publications. 1990
[13] Chistinesen, N. The report of the Ecological Society of America Committee on the Scientific Basis for Ecosistem Management. Ecological Applications: 6: 665-691. 1996
[14] Celine, L. Virtual trees and light capture: a Method for Optimizing Wanatani Stand Design.Wanatani sistems.Vol.77. No. 1. Pp. 37-47, Sept 2009.
[15] De ForestaH, and Micho G. The Agroforest Alternative To Imperata Grasslands: When Smallholder Agriculture and Forestry Reach Sustainability. Wanatani Sistems 36: 105-120. 1997
[16] Hidayat, A. Potential and Management of Dryland Lowland. Prosiding Seminar Nasional Sumber Daya Lahan Indonesia dan Pengelolaannya. Cisarua, 9-13 Februari. Pusat Penelitian Tanah dan Agroklimat. Badan Penelitian dan Pengembangan Pertanian. Departemen Pertanian. Bogor. 2000 pp.192-199
[17] FAO. Planning for Sustainable Use of Land Resources. Toward s New Approach.FAO Land and Water Bulletin.Food and Agriculture Organization. Rome. 1995
[18] Odum, E.P.Basic Ecology.SaundersCollege Publishing.Holth-Saunders, Tokyo, Japan.1983

Morphology and isozyme band-profile as sexual determinant of nutmeg plant *(Myristica fragrants* Houtt*).*

Karmanah [1] , M Maslahat [2] dan L Nurhayati [2]
[1] Faculty of Agriculture Nusa Bangsa University Bogor
[2] Faculty of MIPA Nusa Bangsa University Bogor

E-mail : karmanahs@yahoo.com

Abstract. Sex information in nutmeg plant is important to distinguish between male and female species. The recommendation of optimum sex-ratio for a nutmeg plantation is 1:10. This research is aimed to discover the morphological characteristic of nutmeg plant and the difference of leaf pattern as sexual determinant between male and female plant with the comparison of isozyme analysis. The result of morphological identification provided from 3 research farm location indicated that female nutmeg was higher, with longer and wider leaf, more vein, bigger diameter of stem and branching angle as compared to male nutmeg. Two parameters which were widht of leaf and branching angle showed significant differences, where female nutmeg plants have wider leaves and branching angle. Isozyme analysis suggested that the analysis of peroxidase enzyme (PER) was better and able to provide more information than aspartarte amino transferase (AAT) and acid phospatase (ACP) Enzymes.

1. Introduction

Based on the inflorescence, nutmeg plants can be characterized as dioecious, where male and female flowers are located on different trees. In the field, male and female flowers are sometimes also found in the same tree, and therefore there are three types of nutmeg: male, female, and hermaphrodite nutmeg. For the plantation, the information of nutmeg sexuality is important, because the recommended sex-ratio minimum for a nutmeg plantation is 1:10 (male : female). Therefore, it's required to conduct test to distinguish nutmeg plant sexuality through leaf morphological identification and isozyme analysis.

The aims of this research was to discover nutmeg plant morphological characteristic and isozyme analysis as determinant feature to distinguish male and female nutmeg. The desired target was obtaining nutmeg plant morphological characteristic and that isozyme becomes the sexual determinant of nutmeg plant in the field.

2. Methods

Nutmeg plant morphological identification was conducted in 3 (three) nutmeg farm locations which were Kebun Percobaan Balitro located in Cimanggu, Bogor and Cicurug, Sukabumi, and also nutmeg farmer field in Sukamantri area Tamansari sub-district Bogor. This research was conducted by survey method. Sample selection was conducted by purposive sampling. The number of observed sample was

20 female nutmeg plants and 10 male nutmeg plants from each research farm location. The observation was conducted by choosing 10 old and normal leaves that randomly chosen from each male and female nutmeg plant sample in 3 (three) research farm locations. The observed parameters including: length of leaf, wide of leaf, number of veins, length of branch, height of tree measured from the ground to the canopy, diameter of stem on 50 cm height from the ground, and branching angle measured 0° from tree stands. Observation result data were analyzed by Least Significance Difference test.

Isozyme analysis was conducted in PAU Laboratory of Bogor Agricultural University. Samples were provide from young and fresh leaves (buds) of male and female nutmeg plants from the 3 research farm locations. The enzymes analyzed were *Peroxidase* enzyme (PER), *Aspartate Amino Transferase* (AAT) and *Acid Phospatase* (ACP).

3. Results and Discussions

A. *Nutmeg Plant Morphological Identification*

Morphological analysis result showed that there were some morphological characteristics that can be used to distinguish the feature of male, female and hermaphrodite nutmeg plants (Figure 1). The Figure 1 shows that female nutmeg was taller, with bigger diameter of stem, longer and wider leaves, had more veins, and had wider branching angle compared to the male ones. This distinction of morphological character was consistent among 3 research farm locations, even though the size of the plants were different (Figure 1). Based on logistic regression analysis by Stepwise method, prediction accuracy level was 85.4% for female and 83.9% for male respectively. It means that the data confidently can be used as a parameter to determine nutmeg plant sex. This result is in accordance to the research carried out by *Marzuki et al.* (2006), that male nutmeg was featured by smaller habitus, shorter branches, and smaller leaves.

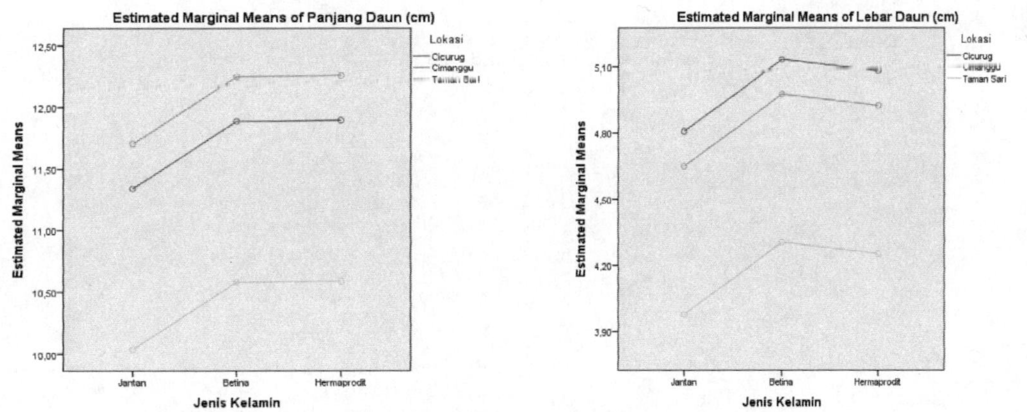

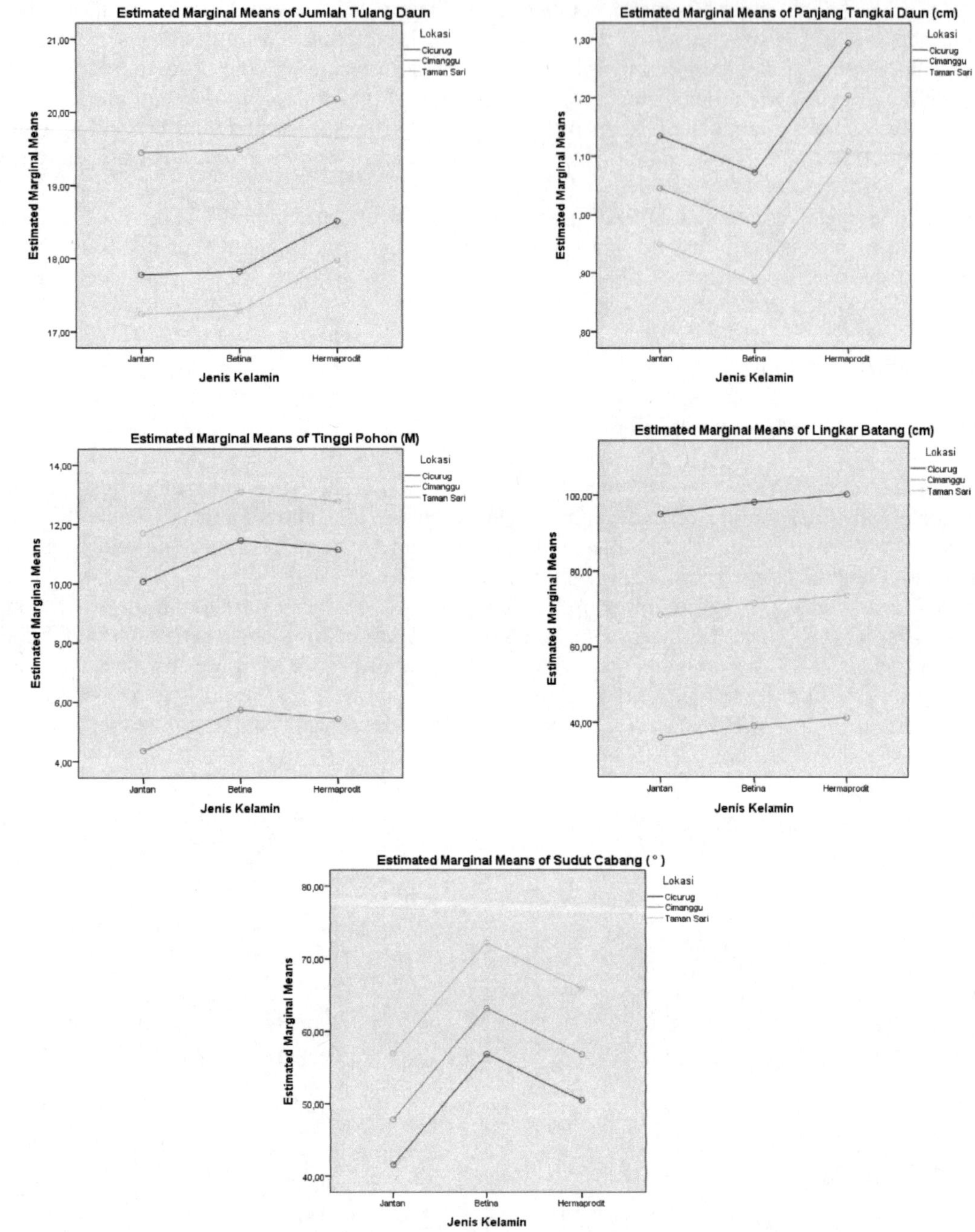

Figure 1. Male, Female, and Hermaphrodite Nutmeg Plant Morphological Characteristics.

Based on the observation, there was significant differences (α, 5%) on width of leaf and branching angle parameters, where the female nutmeg plants had wider leaves and wider branching angle as compared to the male plants (Table 1).

Table 1. Nutmeg Plant Morphological Identification Result

NUTMEG SEXUALITY	PARAMETERS						
	Length of leaf (cm)	Leaf wide (cm)	Number of veins (piece)	Length of branch (cm)	Tree Height (m)	Diameter of Stem (cm)	Branching Angle (°)
Male	11.00 A	4.05 A	17.98 A	1.05 A	9.31 A	73.08 A	47.96 A
Female	11.66 A	**4.89 B**	18.14 A	1.00 A	10.32 A	75.85 A	**62.30 B**
Hermaprodhite	11.52 A	4.74 AB	18.77 A	1.20 B	10.18 A	74.45 A	57.83 B

Note: t-student analysis (α 5%),

According to Randriani *et al.* (2007), the characteristic of leaf wide can be used as a simple selection method, because leaf wide positively correlated with nut and mace weight. The results also showed wider female nutmeg branching angle, this is associated to the result provided by Rostiana *et al.* (2013) that female nutmeg plant branching angle tended to be wider and can be used as sexual determinant between male and female. Similar opinion was also suggested by Wahid (2012) and Plantation General Directorate (1995), that male nutmeg plants have pointing branching angle (acute) while female nutmeg plants have more horizontal/obtuse branching angle.

B. Isozyme Analysis

Isozyme analysis was carried out for 3 different enzymes which were *Peroxidase* enzyme (PER), *Aspartarte Amino Transferase* (AAT) and *Acid Phospatase* (ACP). The result unfortunately did not provide a clear and distinctive feature of sex differentiation of nutmeg plant. Meanwhile, isozyme analysis using PER enzyme relatively provided more information as compared to AAT enzyme and ACP enzyme (data not shown).

Isozyme analysis using PER enzyme showed the number of patterns appeared varied from one to another. Based on dendogram from cluster analysis that was used to discover the similarities of tested varieties, we found three groups obtained with different varies patterns (Figure 2). Sample with the code of 1,2,3,4,5,6,7,10,11,12, and 18 were in the similar level (we call Group 1). Sample with code 8,9, and 13 were in the similar level (Group 2) while sample code 14,15,16,17,19, and 20 were in the similar level (Group 3).

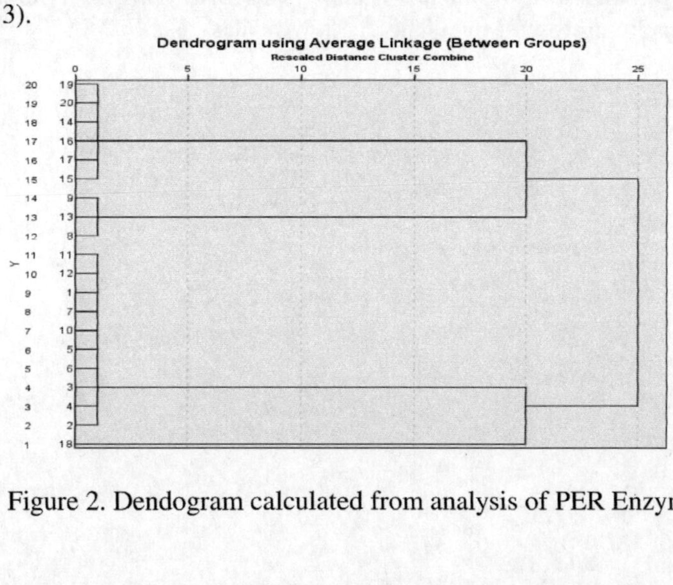

Figure 2. Dendogram calculated from analysis of PER Enzyme

4. Conclusion

The analysis of nutmeg morphology consistently provide sex determination. Female nutmeg typically tended to have higher canopy, bigger diameter of stem, with longer and wider leaf, have more vein and bigger branching angle as compared to male nutmeg. Two parameters which were wide of leaf and branching angle showed significant different. Isozyme analysis suggested that the analysis of peroxidase enzyme (PER) was better and able to provide more information than aspartarte amino transferase (AAT) and acid phospatase (ACP) Enzymes

5. Acknowledgement

This research was funded by competitive grant research through DIPA DIKTI, Research and Community Service Directorate DIKTI.

6. References

[1] Hadad M and Wahid P 1995 *Budidaya Rempah-Rempah dan Tanaman Penyegar Pala* (*Myristica fragrants* Houtt) (Universitas Terbuka Fakultas MIPA Jakarta).

[2} Nayar BK, Rajendar R and Vathsala P 1977 A Simple Morphological Technique to Distinguish the Sex of Nutmeg Seedling *J Current-Science* **46 (5)** pp 156-157.

[3] Pasteur N, Pasteur G, Bonhomme F, Catalan J and Britton-Davidian J 1988. *Practical Isozyme Genetics* Ellis Horwood Limited England 213 p.

[4] Phadnis NA and Choudhari KG 1971 Sex Determination in the Seedlings Stage of Nutmeg (*Myristica fragrans* Houtt) *Tropical Science* **13 (40)** : 265-274.

[5] Randriani E, Tresniawati C, Yuniyati N, Izzah NK, Wardiana E and Luntungan HT 2007 *Keragaan dan Korelasi Fenotipe Beberapa Nomor Aksesi Pala Hasil Eksplorasi di Sulawesi Utara* Prosiding Seminar Nasional Rempah Badan Penelitian dan Pengembangan Pertanian Pusat Penelitian dan Pengembangan Perkebunan Hal 279-2836.

[6] Shibu MP, Ravishankar KV, Anand L, Ganeshaiah KN and Shaanker U 2000 Identification of Sex-Specific DNA Markers in the Diecious Tree, Nutmeg (*Myristica Fragrans* Houtt) Publicado En El **121** pp 59-61.

[7] Syukriani L and Kusumawati A 2007 *Analisis Isoenzim Sebagai Penanda Biokimia Untuk Mengkarakterisasi Tanaman Gambir* (*Uncaria gambir* (Hunter Roxb)) Laporan Ringkasan Penelitian Dosen Muda Universitas Andalas.

Utilization of Anting-Anting (*Acalypha indica*) Leaves as Antibacterial

Irmanida Batubara, Wulan Tri Wahyuni, Imam Firdaus

Department of Chemistry, Faculty of Mathematics and Natural Science, Bogor Agricultural University, Darmaga Campus, Bogor, Indonesia
Biopharmaca Research Center, Bogor Agricultural University, Jl Taman Kencana No 3 Bogor 16128, Indonesia

ime@apps.ipb.ac.id

Abstract. Anting-anting (*Acalypha indica*) plants is a species of plant having catkin type of inflorescence. This research aims to utilize anting-anting as antibacterial toward *Streptococcus mutans* and degradation of biofilm on teeth. Anting-anting leaves were extracted by maceration technique using methanol, chloroform, and *n*¬-hexane. Antibacterial and biofilm degradation assays were performed using microdilution technique with 96 well. *n*-Hexane extracts of anting-anting leaves gave the best antibacterial potency with minimum inhibitory concentration and minimum bactericidal concentration value of 500 µg/mL and exhibited good biofilm degradation activity. Fraction of F3 obtained from fractionation of n-hexane's extract with column chromatography was a potential for degradation of biofilm with IC_{50} value of 56.82 µg/mL. Alkaloid was suggested as antibacterial and degradation of biofilm in the active fraction.

1. Introduction

Anting-anting (*Acalypha indica*) plants is a species of plant having catkin type of inflorescence. This plants traditionally used to treat dysentery, diarrhea, malnutrition, and malaria[1]. The activities of anting-anting is related to the chemical constituents such as saponins, tannis, flavonoid, and essential oil[2]. The ethyl acetate extract of Anting-anting was reported as an anti-bacterial againts *Staphylococcus aureus, Staphylococcus epidermidis, Bacillus cereus, Streptococcus faecalis,* and *Pseudomonas aeruginosa*[3]. However, this extract is not active as antibacterial against *Klebsiella pneumoniae, Escherichia coli,* and *Proteus vulgaris*.

Oral and dental disease is one of health problem which highly correlated to bacteria. On the tooth surface, bacteria makes solid biofilm called as plaque. Dental plaque adheres to the teeth and consists of many species microbe, salivary polymers and microbial extracellular products. The most common bacteria play a role in the formation of plaque is *Streptococcus mutans*[4]. This research aims to utilize anting-anting as antibacterial toward *S. mutans* and biofilm degradator on teeth. Antibacterial could decrease the number of *S. mutans* in mouth to less than normal levels. Eventually, the oral and dental would be healthy. On the otherhand, biofilms acts to protect and improve the nutrition of bacteria that live in it, the matrix of the biofilm protects the bacteria from antibacterial effect. That is why the product which has antibacterial and biofilm degradator activities is needed.

2. Materials and Methods

Materials used on this research are anting-anting leaves, solvents such as water, *n-hexane*, chloroform, methanol, dichloromethane, acetone, media for antibacterial activity test such as *typtic soy broth* (TSB), *typtic soy agar* (TSA), reagent for phytochemical analysis and *Streptococcus mutans*. The anting-anting leaves was collected from Biopharmaca Research Center Conservation and Cultivation unit, Darmaga Campus on December 2013 and determined by Biology Research Center of Indonesian Institute of Research, Cibinong, Jakarta. The *Streptococcus mutans* was obtained from Faculty of Medicine, University of Indonesia No. 63301.

The research steps including extraction, phytochemical test, fractionation, antibacterial test, degradation of biofilm test, as well as the identification of active component. The anting-anting leaves were extracted by maceration method using 3 different solvents. The first solvent was *n*-hexane, continue with chloroform and finally with ethanol to result 3 different types of extract. All extracts were determined the phytochemical content qualitatively and the antibacterial activities as well as the biofilm degradation activity. The most active extract then separated by column chromatography followed by preparative thin layer chromatography. The identification of compounds using the Fourier transform infrared spectrometry (FTIR) was performed to the most active fraction.

2.1. Antibacterial test

The antibacterial test against *S. mutans* was performed using micro-dilution method[5]. Extracts or fractions diluted in DMSO to obtain a concentration stock of 10000mg/mL. The stock samples were made into several concentrations (15.63-2000mg/mL). In each well of sterile 96 well plates, samples, TSB medium and bacterial inoculant were added. The mixture were incubated at 37 °C for 24 hours. The minimum inhibitory concentration (MIC) then determined. The Minimum bactericidal concentration (MBC) was determined after 24 hour incubation of MIC clear zone on new media.

2.2. Biofilm degradation activity test

The method used is micro-dilution method[6]. Biofilms are formed by synthetic saliva (Mc Dougall solution) put in a 96 well plate together with TSB medium, 3% glucose and bacterial inoculant. The mixture was incubated for 24 h at 37 °C. Once a biofilm is formed, the remaining medium is discarded. Extracts or fractions are added at a concentration of 16-2000 mg/mL and then incubated 24 hours at temperature of 37 °C. Biofilms attached to the wall of the wells is washed using phosphate buffer. Crystal violet 1% was added to the wells and left for 15 minutes. Well rinsed with sterile water three times and 95% ethanol was added. The suspension was incubated for 45 minutes and the solution was transferred to a new micro-plate. Suspension absorbance of each well was measured using a micro-plate reader at a wavelength of 595 nm to determine the % degradation. Chlorhexidine was used as positive control and 20% DMSO as a negative control.

3. Results and Discussion

3.1. Samples preparation

Anting-anting leaves were extracted by maseration technique using methanol, chloroform, and *n*-hexane. The yields of extraction on different solvents are varied. The highest yield found in methanol extract, while the lowest in *n*-hexane extract (Table 1). Based on qualitative phytochemical analysis, all extracts consisted of alkaloid. Flavonoid was only found at chloroform and methanol extract, while tannin was only found in methanol extract. The total phenolic contents of chloroform and methanol extracts were determined by spectrometric method. The results showed that the content of the two extracts were almost the same, about 1.3%. The tannin content on the methanol extract is about 2.21%. The tannin content was not found in the *n*-hexane and chloroform extracts. All extracts were used to determine its activity as antibacterial and biofilm degradation. The activities of each extract will be varied based on the phytochemical content in each extract.

Table 1 The yield and phytochemical content of anting-anting leaves extracts

Solvent of extraction	Phytochemical content				Yield (%w/w)	Phenolic content (%w/w)	Tannin content (%w/w)
	Tannin	Saponin	Flavonoid	Alkaloid			
n-hexane	-	-	-	+	1.54	-	-
Chloroform	-	-	+	+	2.44	1.34	-
Methanol	+	-	+	+	4.67	1.35	2.21

note: (+) present (-) absent

3.2. Screening and Fractionation the active extract

Antibacterial and degradation biofilm assay were cariied out in order to screen the most active extract between the 3 extracts. The results showed in Table 2. *n*-Hexane extracts of anting-anting leaves gave the best antibacterial potency with minimum inhibitory concentration and minimum bactericidal concentration value of 500 μg/mL. This extract also exhibited a good biofilm degradation activity. The chloroform extract could not inhibit the bacterial growth till the highest concentration of 2000μg/mL. The methanol extract inhibit the bacteria growth at concentration of 500μg/mL, otherwise this extract could not kill the bacteria. The most active as biofilm degradation was found in chloroform extract.

Table 2 Antibacteria and biofilm degradation activities of Anting-anting extracts and fractions

Sample name	Antibacteria		Biofilm degradation
	MIC(μg/mL)	MBC(μg/mL)	IC_{50} (μg/mL)
Methanol extract	500.0	-	214.6
Chloroform extract	-	-	149.8
n-hexane extract	500.0	500.0	196.9
F1	250.0	-	202.5
F2	250.0	2000.0	156.3
F3	250.0	2000.0	56.8
F4	-	-	76.7
F5	-	-	225.5
F6	-	-	141.5
F7	-	-	-
F8	-	-	-
F9	-	-	-
F10	-	-	-
F11	-	-	-
F12	-	-	-
F13	-	-	-
F3.1	500.0	2000.0	139.2
F3.2	500.0	1000.0	180.7
F3.3	500.0	1000.0	138.7
Tetracycline	15.6	15.6	
Chlorhexidin			2.58

note: (-) concentration >2000.0 μg/mL

Based on the phytochemical content in the extracts, alkaloid would be the active component because only alkaloid is found in the *n*-hexane extract. Tannin in methanol extract would be active to inhibit the bacteria growth[7], but on this extract could not kill the bacteria. The active component for biofilm degradation would be from flavonoid on chloroform extract. To get the two activity from one prospective extract, *n*-hexane extract was selected.

n-hexane extracts was separated with silica gel column chromatography resulted 13 fractions. The fractionation was performed using column chromatography with silica gel as stationary phase and elution by dichloromethane:chloroform using step gradient method. All of fractions resulted from this fractionation was used to analyze the activity as antibacterial and biofilm degradation. Fraction 7 – 13 were not active as antibacterial and biofilm degradation, meanwhile fraction F3 was good biofilm degradator with IC_{50} value of 56.82 μg/mL (Table 2).

Fraction F3 was futher separated by preparative thin layer chromatography eluated with dichloromethane:choloroform (3:7). The fractionation of F3 resulted 3 subfractions, F3.1 – F3.3. All of the subfractions showed antibacterial activity but not as good as tetracycline as positive control. All subfractions also showed biofilm degradation activity but not as good as chlorhexidine as positive control. To determine the active ingredient as antibacterial and biofilm degradation, F3 was used.

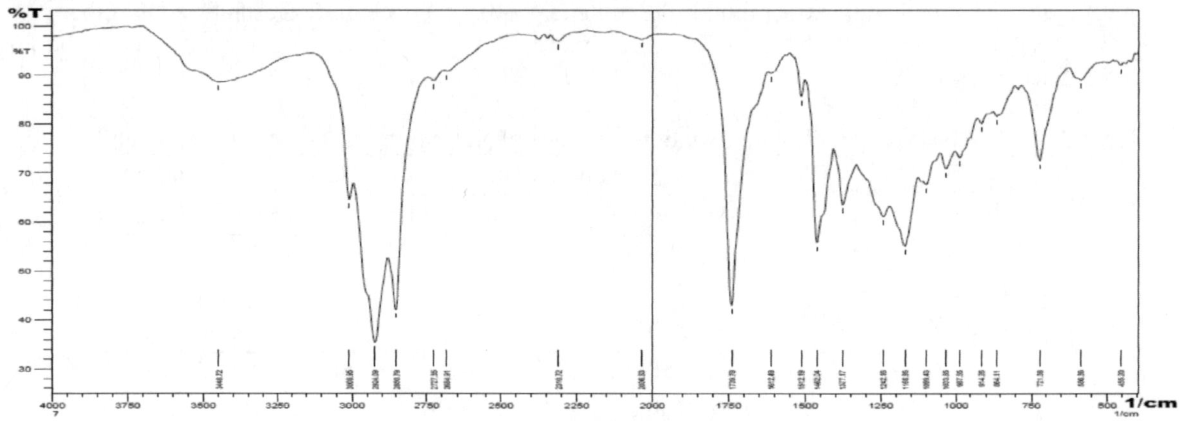

Figure 1. Spectrum FTIR of F3

FT-IR analysis was performed to F3 (Fig 1) in order to determine the active component in this fraction. The spectrum at wavenumber 3448 cm^{-1} stalling indicated amine NH_2, 1512 cm^{-1} bend for NH_2 (amine secondary), 1377 and 1242cm-1 for CN (amine) and 1169cm^{-1} for ester. This information indicates the presence of alkaloid in F3. Based on phytochemical qualitative analysis and FT-IR analysis, F3 contain an alkaloid and was suggested as antibacterial and biofilm degradator in the active fraction.

4. Conclusion

n-Hexane extracts of anting-anting leaves is the most potent extract as antibacterial and good biofilm degradation. Fraction F3 obtained from *n*-hexane's extract was good antibacterial and biofilm degradator. Alkaloid was suggested as antibacterial and biofilm degradator in the active fraction.

References

[1] Arisandi Y, Andriani. 2008. *Khasiat Tanaman Obat*. Jakarta (ID): Pustaka Buku Murah

[2] Azmahani A, Somchti MN, Rosyilah AR. 2002. *In Vitro* Anti Bakterial and Anti Fungal Properties of Acalypha Indica (Kucing Galak). *Proceedings of The Regiona Symposium on Environment and Natural Resources*. Department of Biomedical Sciences, Faculty Medicine and Health Sciences, University Putra Malaysia, 43400 UPM Serdang, Selangor Darul Ehsan.

IOP Conf. Series: Earth and Environmental Science **31** (2016) 012038 doi:10.1088/1755-1315/31/1/012038

Malaysia
[3] Govindarajan M, Jabanesan A, Reetha D, Amsath R, Pushpanathan T, dan Samidurai K. 2008. Antibacterial Activity of *Acalypha indica* L. *Eur Rev Med Pharmacol Sci.* 12:299-302.
[4] Tarigan R. 1990. *Karies Gigi*. Jakarta (ID): Hipokrates.p.17, 41-46.
[5] Batubara I, Mitsunaga T, Ohasi H. 2009. Screening antiacne potency of Indonesian medicinal plants; antibacterial, lipase inhibition, and antioxidant activities. *J Wood Sci* 55: 230-235.
[6] O’ Toole G, Kolter R. 1998. Initiation of biofilm formation in pseudomonas fluorescens wcs365 proceeds via multiple, convergent signaling pathways: a genetic analysis. *Mol Microbiol.* 28(3): 449-461.
[7] Rodrigues, C. G., Ferreira, P. R. B., Mendes, C. S. O., Reis-Jr, R., Valério, H. M., Brandi, I. V., & de Oliveira, D. A. (2014). Antibacterial activity of tannins from Psidium guineense Sw. (Myrtaceae). Journal of Medicinal Plants Research, 8(35), 1095-1100.

Studies on Electrical behavior of Glucose using Impedance Spectroscopy

Jajang Juansah†[1] and Wina Yulianti*[2],

[1,] Biophysics Division, Physics Department, Bogor Agricultural University, Jalan Meranti Wing S Kampus IPB Dramaga, Indonesia.
[2] Analytical chemistry, Diploma program Department, Bogor Agricultural University, Indonesia.

E-mail: j_juansah@ipb.ac.id

Abstract. In this work we report the electrical characteristics of glucose at different frequencies. We show the correlation between electrical properties (impedance, reactance, resistance and conductance) of glucose and glucose concentration. Electrical property measurements on glucose solution were performed in order to formulate the correlation. The measurements were conducted for frequencies between 50 Hz and 1 MHz. From the measurements, we developed a single-pole Cole-Cole graph as a function of glucose concentration.

1. Introduction

Glucose is a very important material in various aspects of life. In food technology, glucose is often used as a food sweetener. Glucose content in fruits is one of the determining factor in taste and quality. In Yeast cells, glucose also trigger unexpected, hormone-like effects, including the activation of cellular growth, the mobilization of storage compounds and the diminution of cellular stress resistance [1]. In medical science, regulation of glucose content in the blood is an important role for diabetic patients [2]. Glucose is used as an energy source in most organisms, from bacteria to humans. Use of glucose may be by either aerobic respiration, anaerobic respiration, or fermentation. Glucose is the human body's key source of energy, through aerobic respiration, providing approximately 3.75 kilocalories (16 kilojoules) of food energy per gram.[3]

The electrical impedance spectroscopy provide important information on ion-ion and ion-solvent interactions [4] , such as limiting molar conductivity and the ion pair formation constant of Ammonium Bromide [5], specific conductivity of solution of sodium chloride in pure water and ethanol water mixed solvent media [6]. Impedance spectroscopy has been proposed as possible approach for non-invasive glycaemia monitoring. Other applications require the measurement of total dissolved solids (TDS), which is related to conductivity by a factor dependent upon the level and type of ions present [7]. However, few quantitative data are reported about impedance variations related to glucose concentration variations, especially below the MHz band [8]. This research will be assessed

†The author recently deceased, please conduct further correspondence to wjuans@gmail.com

on the electrical characteristics of glucose for various frequencies in the low electrical signals. It aims to examine the electrical behavior of glucose as the data base information for the benefit of other advanced.

2. Results and Discussions

Soluble Solid Content of Glucose Solution

Glucose is a carbohydrate, and is the most important simple sugar in human metabolism. Glucose is called a simple sugar or a monosaccharide because it is one of the smallest units which has the characteristics of this class of carbohydrates. Glucose is one of the primary molecules which serve as energy sources for plants and animals. It is found in the sap of plants, and is found in the human bloodstream where it is referred to as "blood sugar".
Total soluble solids of fruit such as sweet orange, mandarin, grapefruit, and pummelo juices constitute mainly sugars. TSS of citrus juices indicates all the soluble solids. It is not a measure of sugars only [9] Other applications require the measurement of total dissolved solids (TDS), which is related to conductivity by a factor dependent upon the level and type of ions present [7]. So, it is necessary to test the correlation between TSS and glucose concentration in the sample used in this study. Figure 1 shows the results of the linear correlation for the glucose concentration in the sample solution and TSS parameters. The figure shows that the increase in glucose concentration will increase the value of TSS. This occurs in a linear relationship with a high coefficient determinant (R^2=0.9991).

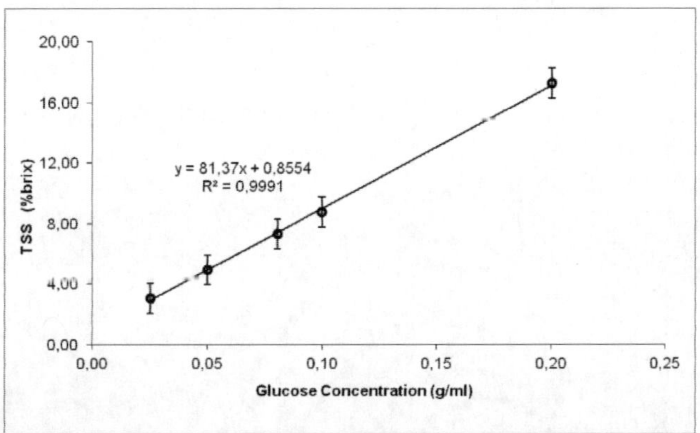

Figure 1. Correlation between glucose concentration and TSS parameters in the sample solution.

Electrical Impedance of Glucose Solution

Any electrode–material system in a measuring cell has a geometrical capacitance and a bulk resistance in parallel with it. Electrical impedance (Z) is defined by a complex quantity in terms of resistive (R) and capacitive (C) components as $|Z| = \sqrt{(Z')^2 + (Z'')^2}$, where $Z'' = [\omega C(\omega)]^{-1}$ and $Z' = R$. It is only real when $\omega = 0$ and thus $Z = Z'$, that is, for purely resistive behavior. In this case the impedance is completely frequency-independent. When Z' is found to be a variable function of frequency ($Z'(\omega) = R(\omega)$) which holistically connect real and imaginary parts with each other, ensure that Z'' cannot be zero over all frequencies but must vary with frequency as well.

In essence, dielectric studies are based on quantifying the response of a material to an electric field applied to it. The response is typically described by the material's conductivity and permittivity. Conductivity (σ), measured in S/m, quantifies the ability of the material to conduct the electrical charge. Permittivity (ε), measured in F/m, is the amount of charge that is stored by the

IOP Conf. Series: Earth and Environmental Science **31** (2016) 012039 doi:10.1088/1755-1315/31/1/012039

material due to the polarization of its components. Permittivity of the material is often expressed as relative to the permittivity of vacuum (ε0= 8.854 × 10–12 F/m), giving the dimensionless relative permittivity (also called dielectric constant), εT= ε/ε0. By dividing conductivity and permittivity by the probe constant (d/Ain m–1, the ratio of the distance between the electrodes and the electrodes' area), one obtains the corresponding conductance (G in S) and capacitance (C in F) of the material, respectively. The permittivity of a material tends to fall (and its conductivity to rise) in a series of step-like shifts as the frequency of the electrical field rises. These step changes, called dispersions, are due to losses of certain characteristic polarization abilities of the substance.[10]

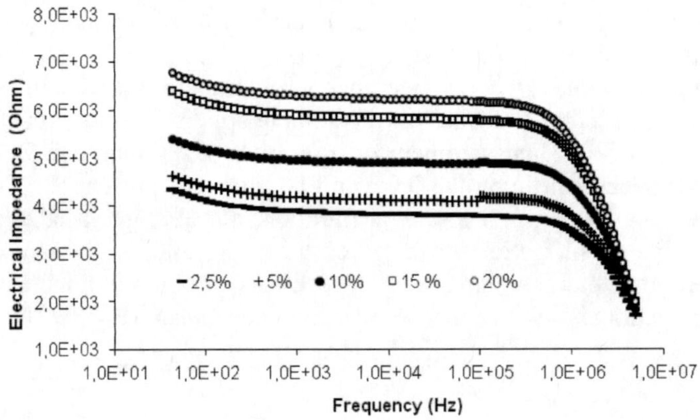

Figure 2. Spectrum of electrical impedance in the sample solution.

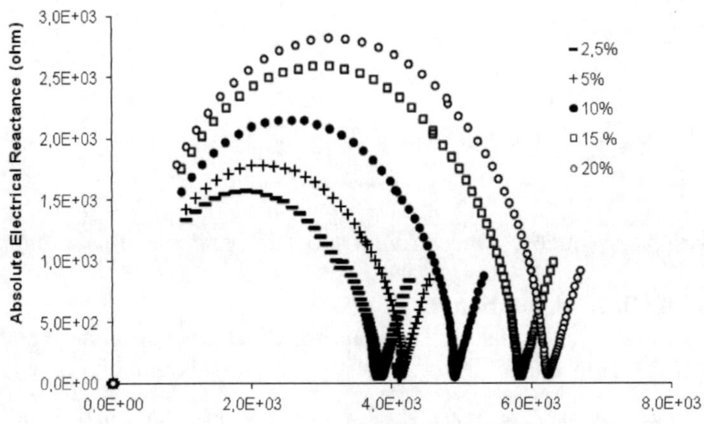

Figure 3. Cole-cole mode . in the sample solution.

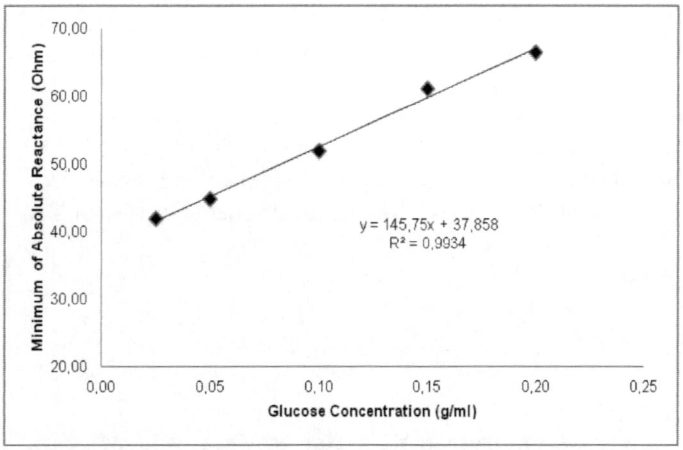

Figure 4. Correlation between glucose concentration and minimum of reactance in the sample solution.

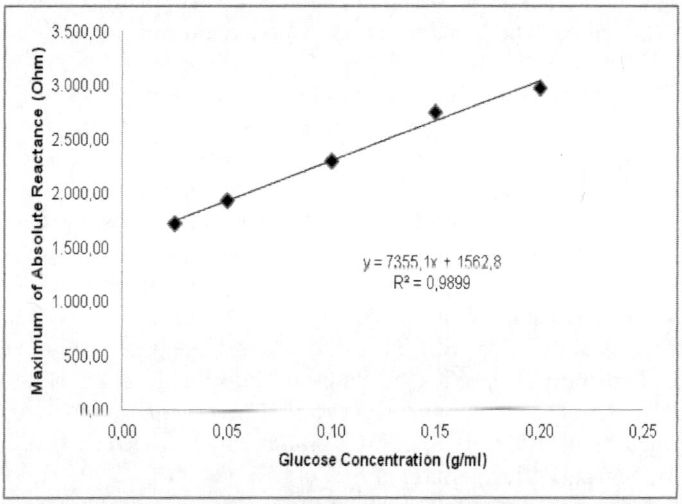

Figure 5. Correlation between glucose concentration and maximum of absolute reactance in the sample solution.

Studies on electrolytic impedance or conductance in mixed solvents can provide useful information and sensitive indications of ion-solvent and ion-ion interactions and solvent structure. The effect of ion association on the conductance behavior of electrolytes has been a subject of extensive investigation. Although numerous conductance measurements have been reported in the literature, such studies in mixed solvents are relatively rare [11]. In recent, several studies have reported that the nature of the spherical ions, having a large variation in size in aqueous mixtures of the alcohols, has received considerable attention

On the other hand, it was recognized, that the sulfamate function is essential for cyclamate sweetness although the cation seems to have some effect on the sweet taste [10] The behaviour of an ion in a solvent depends on the ion-ion and ion-solvent interactions. It is to be expected that the taste of a sweet substance could be interpreted by an understanding of these interactions in the medium.

3. Summary

We have performed electrical property measurements on glucose solution in order to formulate the correlation. We showed that the permittivity of a material tends to fall (and its conductivity to rise) in a series of step-like shifts as the frequency of the electrical field rises. These step changes, called dispersions, are due to losses of certain characteristic polarization abilities of the substance.

References

[1] Kevin J, Verstrepen, Iserentant D, Malcorps P, Derdelinckx G, Dijck PN, Winderickx J, Isak, Pretorius, Johan M. Thevelein and Freddy RD. 2004. Glucose and sucrose: hazardous fast-food for industrial yeast, TRENDS . *Biotechnology 22*,10

[2] Park JH, Kim CS, Choi BC. 2003. The correlation of the complex dielectric constant and blood glucose at low frequency, Ham. *Biosensors and Bioelectronics* (19) 321-324

[3] [ISBN 92-5-105014-7]. 2003. Chapter 3: Calculation of the Energy Content of Foods – Energy Conversion Factors", *Food energy-methods of analysis and conversion factors*, FAO Food and Nutrition Paper 77, Rome: Food and Agriculture Organization

[4] Bhattarai A, Neupane S. 2012. Effects of concentration and solvent composition on the electrical conductivity of sodium bromide in pure water and ethanolwater mixed solvent media. *Journal of Science, Engineering, and Technology* 8 (II) 1-6.

[5] Sokol V, Tomas R and Tominic I. 2009. Conductometric Study of Ammonium Bromide in 2-Butanol Water Mixtures. *Acta Chim. Slov* 56.(773-779)

[6] Bhattarai A and Kumar SS, 2011. Effects of Concentration and Relative Permittivity on the Transport Properties of Sodium Chloride in Pure water and Ethanol-Water Mixed Solvent Media. *Research Journal of Chemical Sciences*, V 1(6), 48-52,

[7] Radiometer Analytical SAS. 2004. *Conductivity theory and practice*, France

[8] Tura A, Sbrignadello S, Barison S, Conti S, Pacini G. 2007. Impedance spectroscopy of solutions at physiological glucose concentrations. *Biophysical Chemistry* 129. 235–241

[9] Ladaniya MS. 2008. Citrus fruit, Biology, *Technology and Evaluation*, Elsevier Inc. USA.2008; 574 pp

[10] Desouky OS. 2009. Rheological and electrical behaviour of erythrocytes in patients with diabetes mellitus. Romani. *Biophys 19* (4) 239–250

[11] Janz G T and Tomkins RPT.1972. Non Aqueous Electrolytes Handbook. cademic Press: London 1

Comparison of physiological and anatomical changes of C3 (*Oryza sativa* [L.]) and C4 (*Echinochloa crusgalli* [L.]) leaves in response to drought stress

Hamim Hamim[1], **Sri Banon and Dorly Dorly**
Department of Biology, Faculty of Mathematics and Natural Sciences, Bogor Agricultural University, Kampus IPB Darmaga, Bogor, Indonesia 16680.

[1]Corresponding author: hamim@ipb.ac.id

Abstract. The experiment aimed to analyse the different response of C3 (*Oryza sativa* L.) and C4 (*Echinochloa crusgalli* L.) species to drought stress based on physiological and anatomical properties. Seeds of rice (*Oryza sativa*) and Echinochloa (*Echinochloa crusgalli*) were grown in 15 cm (D) pot for 6 weeks under well-watered conditions. After 6 weeks the plants were divided into two groups, (1) well-watered which were watered daily, and (2) drought stress which were withheld from watering for 6 days. After 6 days of drought, the plants were then re-watered to analyse plant recovery. During drought period, the plants were analysed for growth, leaf relative water content (RWC), photosynthesis, and leaf anatomy. Drought stress significantly reduced leaf RWC of both species, but the reduction was bigger in rice than in Echinochloa. The maximum efficiency of photosynthesis (Fv/Fm) was decrease significantly in response to drought stress by about 48.04% in rice, while it was only 34.40% in Echinochloa. Anatomical analysis showed drought treatment tended to reduce leaf thickness in the area of bulliform cell, major- as well as intervein and xylem diameter, more in Echinochloa than in rice, suggesting that the decrease of vein and xylem diameter is among the anatomical parameters that is important to overcome from drought stress in Echinochloa. The number of chloroplast in the mesophyll cell and bundle sheath cell (BSC) was different between these two species, where in Echinochloa chloroplast was found in both mesophyll as well as BSC, while in rice it was only found in mesophyll cell, confirmed that Echinochloa is a C4 and rice is a C3 species. Interestingly, in Echinochloa, the number of chloroplast was significantly increased due to drought stress in BSC, but not in mesophyll cell. The number of starch granules also dramatically increased in response to drought in the mesophyll cells of rice and Echinochloa, and in the bundle sheath cell of Echinochloa which indicate that C3 cycle may be occurred in C4 species, at least in Echinochloa, especially during drought stress.

Keywords: C3 and C4 plant, drought stress, *Echinichloa crusgallii*, rice, photosynthesis, leaf anatomy, starch, chloroplast.

1. Introduction

Under natural conditions, plants usually undergo environmental stress due to lower precipitation and soil water content that is known as “drought stress”. Drought stress or water stress may arise in several ways including high rate of evaporation, the existence of osmotically active compounds, and inadequate water uptake due to shallow soil[1]. Drought is a major abiotic stress which causes

IOP Conf. Series: Earth and Environmental Science **31** (2016) 012040 doi:10.1088/1755-1315/31/1/012040

detrimental effect on most plant growth stages and decreases crop yield, and therefore being a severe threat to sustainable agriculture[2]. Water stress, due to limitation of water supply from root or because of higher transpiration rate, severely impair growth and development of plants. Even at moderate water deficit, the rate of CO_2 assimilation in the leaves is depressed, mostly due to stomatal closure[3,4].

The effect of water stress to plant depends on the plant genetic and the degree of the stress. The difference of plant morphology, anatomy and metabolism may result in variations of response to water deficit. Due to differences in metabolism, there are some differences between C3 and C4 plants in response to water stress. In C3 species, drought stress is normally followed by photosynthetic rate reduction due to lowering stomatal conductance which consequently increases photorespiration[5]. On the other hand, C4 species is able to maintain photosynthetic rate under mild drought due to its mechanism known as CO_2 concentrating mechanism (CCM) which enables the plant to suppress photorespiration[6,7]. These two species in many cases have differences in leaf anatomy especially the existence and the role of bundle sheath cell (BSC) in the plant photosynthesis[8]. We believe that this differences may cause variations in response to drought stress indicated by physiological and metabolic processes. Therefore, we compared *Oryza sativa* L. (C3) and *Echinochloa crusgalli* L. (C4) to analyse the different response of these two species to drought treatment in a pot series of experiment.

The objective of this experiment was to analyse the different response of C3 (*O. sativa*) and C4 (*E. crusgalli*) species to drought stress based on physiological and anatomical characteristics.

2. Methods

Seeds of rice (*O. sativa*) and Echinochloa (*E. crusgalli*) were germinated in plastic container for 2 weeks. The seedlings then were transplanted and grown in 15 cm (D) pots. The plants were watered daily, and fertilizer was applied at the planting time and three weeks after planting using NPK (12-12-12) fertilizer of 12.5 gram per pot. After 6 weeks the plants were divided into two groups, (1) well-watered which were supplied with water daily, and (2) drought stress which were withheld from watering for 6 days. After 6 days drought treatment, the plants were then re-watered to analyse plant recovery.

During the period, the plants were analysed for growth, leaf relative water content (RWC), photosynthesis, and several leaf anatomy parameters. Leaf RWC was analysed using 10 circular leaf that was cut by choke-borer from the sample leaves. The leaves were measured for their fresh weight (FW), soaked in the distilled water for 24 hours, and then they were weighted to find water-saturated weight (WSW). Later on, the leaves were dried in a 80 °C oven for 3 days to find dry weight (DW). RWC was calculated as follows[9]:

$$RWC = ((WSW - FW)/(WSW - DW)) \times 100\% .$$

Photosynthesis was measured using Chlorophyll Fluorescence Analyser (Qubit systems) to analyse the maximum efficiency of photosynthesis *Fv/Fm*[10]. Leaf anatomy was analysed using light microscope with paraffin method[11] as well as using Transmission Electron Microscope (TEM)[12]. Paraffin method was carried out to determine leaf thickness in major vein, intervein, and leaf blade, and the area of bulliform cells; and to determine xylem diameter and the number of bulliform cells. The measurement using TEM was carried out to analyse the number of chloroplast and starch in the mesophyll and bundle sheath cells.

The experiment was arranged using completely randomized design with two factors and 3 replications. The first factor was environmental factor that contained two levels: well-watered (control) and drought stress, while the second factor was plant species which comprised two levels: *O. sativa* (rice) and *E. crusgalli* (Echinochloa). The collected data were then analyzed by t-test analysis using SPSS 15.

3. Results

3.1. Media water status

Drought treatment for 6 days caused reduction of media water content of all plants from nearly 30% before drought to an average of 13.40% for rice and 15.55% for Echinochloa. During this period, droughted plants exhibited stress and loss their turgor specified by wilting in all part of the leaves, while control plants were entirely turgid. Three days after re-watering, media water content increased back to the level similar to that of control plants (Figure 1). Water content recovery in the droughted treatments caused the plants refreshed again and recovered from wilted.

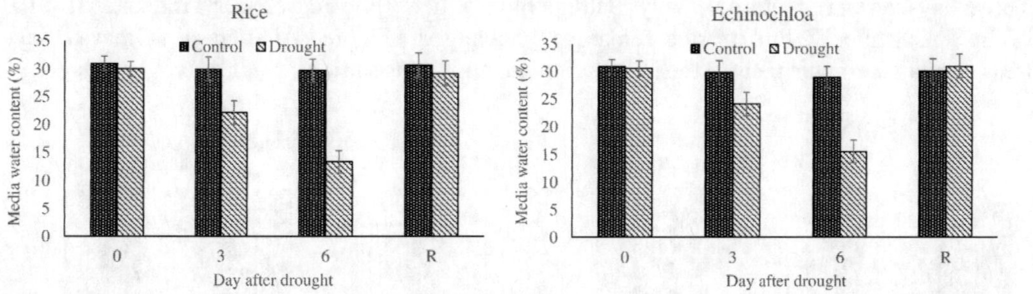

Figure 1. Media water content of rice and Echinochloa during 3 and 6 days of drought stress and 3 days after re-watering (recovery, R) between well-watered (control) and drought stress. (Bar indicative of standard error calculated from 5% of t-test).

3.2. Plant Growth

To compare plant growth of both species, several parameters including plant height, leaf number and tiller number were measured. Drought stress during 6 days caused the decrease of all growth component in both species including plant height, the number of leaves and tillers. However, based on statistical analysis, only plant height and the number of leaves of rice but not of Echinichloa that significantly decreased in response to drought treatment, while the others were not significantly different (Table 1).

Table 1. Plant height, leaf number and tiller number of rice and Echinochloa grown in well-water (control) and drought stress during 6 days.

Plants	Plant height (cm)		Leaf number		Tiller number	
	Control	Drought	Control	Drought	Control	Drought
Rice	81.30 a	75.00 b	12.40 a	10.00 b	6.10 a	4.40 a
Echinochloa	113.70 a	111.00 a	11.00 a	10.30 a	3.60 a	2.60 a

Note: The numbers followed by similar letter is not significantly different based on 5% of t-test analysis.

3.3. Relative water content

3.4. Drought stress exposed during 6 days caused the decrease of relative water content (RWC) of both species, even though each plant had different response. Leaf RWC of rice was declined significantly since 3 days after drought, and the decrease continued until 6 days of drought when it reached 54.73%, while it was still more than 92% for well-watered (control) plants (Figure 2). On the other hand, in Echinochloa, 3 days of drought slightly decreased RWC, even though it continued to decrease until 65.55% after 6 days of drought. During this time, both species were severely wilted suggesting that the plants were under severe stress. Three days after re-watering, both plants were able

IOP Conf. Series: Earth and Environmental Science **31** (2016) 012040 doi:10.1088/1755-1315/31/1/012040

to recover from stress specified by the increase of RWC back to the level of almost similar to that of control plants (Figure 2).*Photosynthetic parameter*

Drought stress for 6 days which caused all plant to wilt surely affected plant photosynthesis. To analyze the impact of the stress to the photosynthetic apparatus, we observed the maximum efficiency of photosynthesis (Fv/Fm) using fluorescence analyzer. In general, drought stress caused the decrease of Fv/Fm of both species. After 6 days drought exposure the value of Fv/Fm declined significantly in both species. The decrease of Fv/Fm was high in droughted rice which reached 48.04% as compared to the control well-watered plants, even though it was almost unchanged when the drought was exposed for 3 days (Figure 3). In Echinochloa, 6 days drought caused the reduction of Fv/Fm by only 34.5%. Three days after re-watering, only Echinochloa which showed recovery indicated by the increase of Fv/Fm, while Fv/Fm of rice remained unchanged (Figure 3), suggesting that 6 days drought stress caused an impairment to the photosynthetic apparatus in rice.

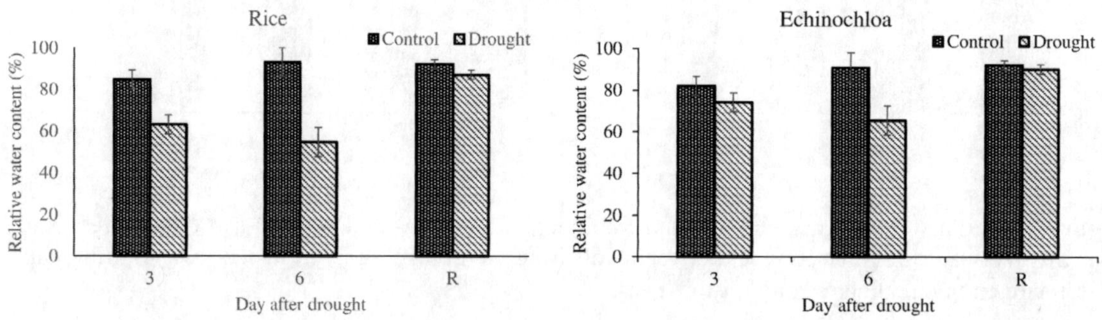

Figure 2. Relative water content (RWC) of rice and Echinochloa during 3 and 6 days of drought stress and 3 days after re-watering (recovery, R) between well-watered (control) and drought stress. (Bar indicative of standard error calculated from 5% of t-test).

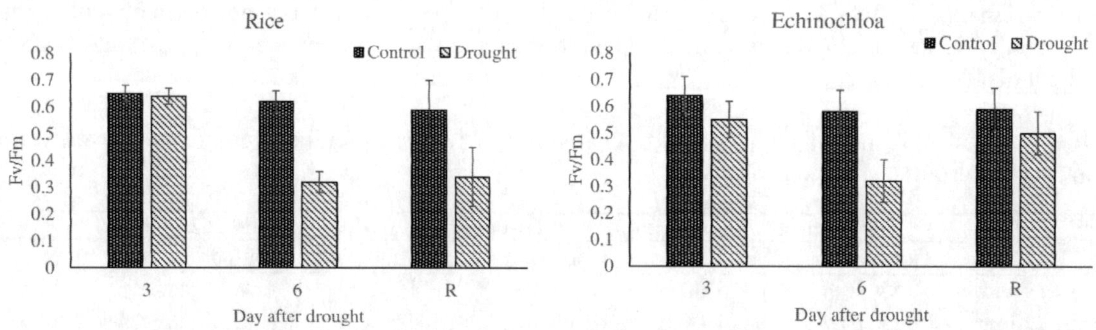

Figure 3. Maximum efficiency of photosynthesis (Fv/Fm) of rice and Echinochloa during 3 and 6 days of drought stress and 3 days after re-watering (recovery, R) between well-watered (control) and drought stress. (Bar indicative of standard error calculated from 5% of t-test).

3.5. Leaf Anatomy measurement

Leaf anatomy investigation was carried out by measuring leaf thickness, xylem diameter, and the bulliform cell number and height. Leaf thickness was measured in three area: major vein, bulliform cell and inter-vein. In general, drought treatment exposed during 6 days did not significantly influence leaf thickness of both species either at the area of major vein, bulliform cell as well as inter-vein (Table 2).

Table 2. Leaf thickness of rice and Echinochola in the are of major vein, bulliform cell and intervein in response to well-water (control) and drought treatments (stress).

Plants	Major vein (µm)		Change	Bulliform (µm)		Change	Intervein (µm)		Change
	Control	Stress	(%)	Control	Stress	(%)	Control	Stress	(%)
Rice	119.86	120.00	100.11	58.00	62.00	106.89	68/67	70.83	103.40
Echinoc	190.00	168.33	88.40	128.17	119.00	92.85	127.33	112.00	87.97

Note: The numbers followed by similar letter is not significantly different based on 5% of t-test analysis.

The analysis of xylem diameter of major vein and intervein was also not significantly affected by drought treatment for 6 days. Moreover, even though not significantly different, the decrease of xylem diameter due to 6 days drought stress tended to be higher in Echinochloa than in rice, and intervein xylem diameter tended to be affected more than major vein in both species (Table 3). The number and height of bulliform cell also remained unchanged due to drought stress. Despite not significantly different, however, the drought tended to slightly increase the number of bulliform cell in rice (table 4).

Table 3. The xylem diameter of rice and Echinochola in major vein and intervein in response to well-water (Control) and drought treatments (stress).

Plants	Major vein (µm)		Change (%)	Intervein (µm)		Change (%)
	Control	Stress		Control	Stress	
Rice	28.68 a	27.85 a	97.11	6.83 a	6.25 a	91.51
Echinochloa	32.64 a	29.44 a	90.20	8.42 a	6.50 a	77.20

Note: The numbers followed by similar letter is not significantly different based on 5% of t-test analysis.

Table 4. The number and height of bulliform cell of rice and Echinochola in response to well-water (Control) and drought treatments (stress).

Plants	Bulliform number		Change (%)	Bulliform height (µm)		Change (%)
	Control	Stress		Control	Stress	
Rice	10.34 a	12.95 a	125.24	21.05 a	20.97 a	99.70
Echinochloa	9.77 a	9.56 a	97.86	35.20 a	35.65 a	101.27

Note: The numbers followed by similar letter is not significantly different based on 5% of t-test analysis.

TEM investigation was also carried out to analyze chloroplast and starch granules in the mesophyll and bundle sheath cell of rice and Echinochloa. The data showed that both mesophyll and bundle sheath cells contained a lot of chloroplast in Echinochloa, while in rice chloroplast it was found only in mesophyll cell. Drought stress for 6 days caused the increase of chloroplast in bundle sheath cell of Echinochloa. The drought also tended to increase the number of chloroplast in mesophyll cell of both species, even thought it was not significantly different (Table 5).

The starch granules inside the chloroplast (chloroplastic starch) was also altered in response to 6 days drought treatment. Drought caused the increase of chloroplastic starch in rice up to 9 fold as compared to that of control plants. In Echinochloa the significant increase of chloroplastic starch also occurred not only in the mesophyll but also in the bundle sheath cell in response to drought treatment. The drought improved chloroplastic starch almost 5 times and 3 times in the mesophyll and in the bundle sheath cell of Echinochloa, respectively (Table 6 and Figures 4 and 5).

Table 5. The number of chloroplast in the mesophyll and bundle sheath cell of rice and Echinochola in response to well-water (Control) and drought treatments (stress).

Plants	Mesophyll cell		Change (%)	Bundle sheath cell		Change (%)
	Control	Stress		Control	Stress	
Rice	5.4 a	5.8 a	107.40	0	0	0
Echinochloa	5.2 a	5.8 a	111.53	5.4 a	7.8 b	144.44

Note: The numbers followed by similar letter is not significantly different based on 5% of t-test analysis.

Table 6. The number of chloroplastic starch in the mesophyll and bundle sheath cell of rice and Echinochola in response to well-water (Control) and drought treatments (stress).

Plants	Mesophyll cell		Change (%)	Bundle sheath cell		Change (%)
	Control	Stress		Control	Stress	
Rice	0.6 a	5.4 b	900.00	0	0	0
Echinochloa	1.2 a	5.8 b	483.33	6.0 a	16.2 b	270.00

Note: The numbers followed by similar letter is not significantly different based on 5% of t-test analysis.

Rice

Echinochloa

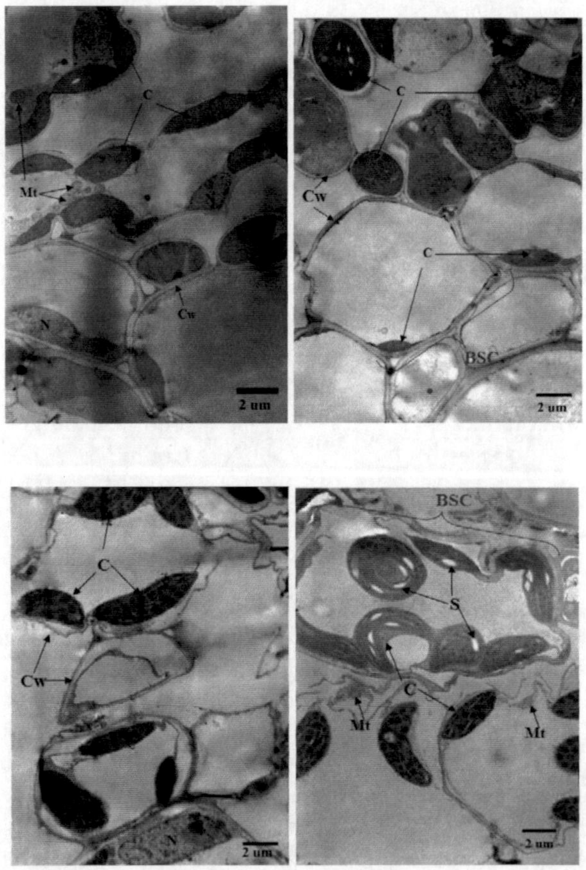

Figure 4. Electron micrograph of chloroplasts of rice and Echinochloa in the mesophyll cell (left) and bundle sheath cell (right) (3000x of magnification). Chloroplast (c), Nucleus (N), Bundle Sheath Cell (BSC), Mitochondria (Mt), Cell wall (Cw).

IOP Conf. Series: Earth and Environmental Science **31** (2016) 012040 doi:10.1088/1755-1315/31/1/012040

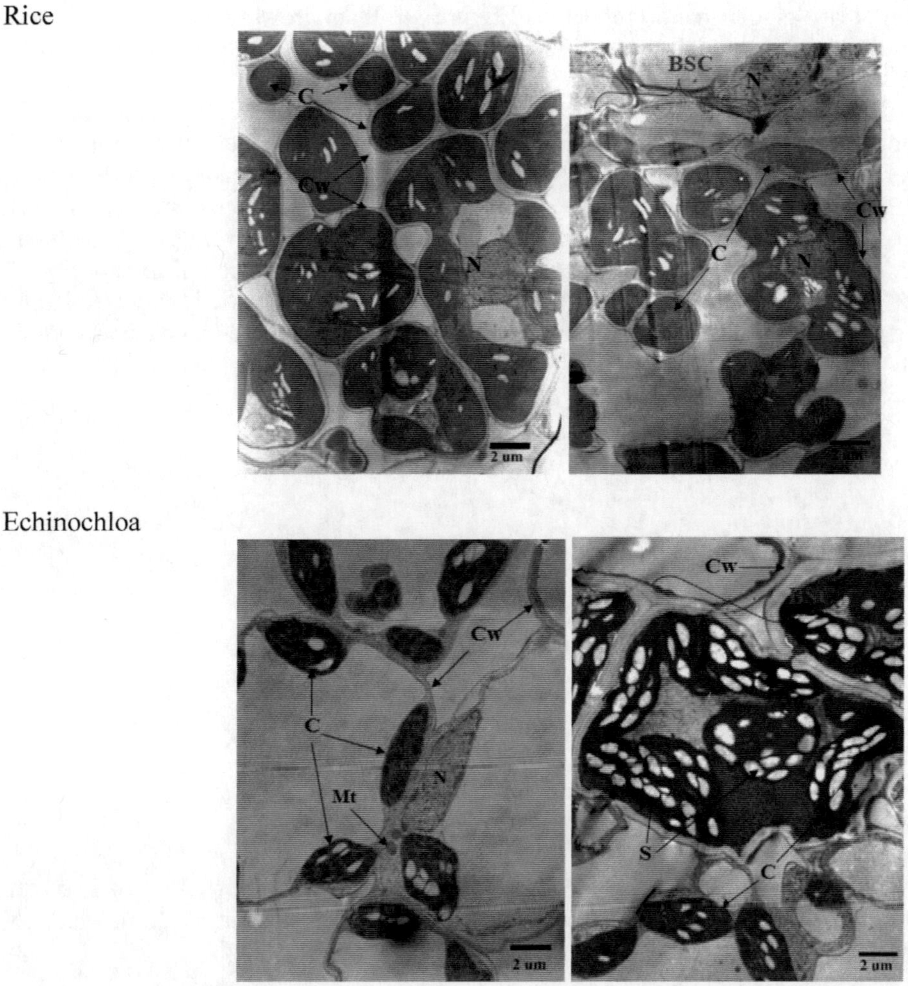

Figure 5. Electron micrograph of chloroplastic starch of rice and Echinochloa in the mesophyll cell (left) and bundle sheath cell (right) during well-watered and drought stress (3000x of magnification). Chloroplast (c), Nucleus (N), Bundle Sheath Cell (BSC), Mitochondria (Mt), Cell wall (Cw).

Discussion

3.6. Growth, physiological and anatomical responses of C3 and C4 plant to drought stress

Drought stress becomes major constraint of plant production, because drought influences many aspects inside the plant starting from cellular until organism level including physiology, metabolism, anatomy and even morphology. In physiological aspect, drought will decrease plant water potential which has negative impact on cell expansion and consequently plant growth[13]. The decrease of plant water potential due to drought also triggers the reduction of stomatal conductance which results in the decrease of photosynthetic CO_2 assimilation[4]. From these two aspects, it become understandable that drought or water stress reduces plant growth and production drastically, especially in rice[14].

In this experiment, drought stress provided by withholding water for 6 days significantly reduced media water content of rice and Echinochloa plants in average up to 50% of control (Figure 1). MWC of rice was slightly lower than that of Echinochloa in response to the last period of drought. This stress also decreased RWC of both rice and Echinochloa, but the decrease was bigger in rice than in

Echinochloa (Figure 2). Consequently, the drought caused the decline of maximum efficiency of photosynthesis (*Fv/Fm*) of rice bigger than Echinochloa (Figure 3). Plant growth data also confirmed that drought significantly inhibited growth of rice indicated by plant height and leaf number, but not Echinochloa (Table1).

To eliminate the effect of different water status that was happen during drought treatment, we analyze correlation between MWC and RWC of both species, so that we can evaluate the physical and physiological response of both species to the drought (Figure 6). From the graph shown in Figure 6, we conclude that the decrease of RWC in response to MWC declining was higher in rice than in Echinochloa. It means that drought caused rice plant underwent stress faster than Echinochloa, because RWC is an important physiological parameter of stress due to water deficit. The capacity of plant to maintain RWC during drought stress is an important tolerant-characteristic[15]. Some tolerant plants sometimes have capacity to avoid the stress by maintaining RWC with the mechanism known as osmotic adjustment[16].

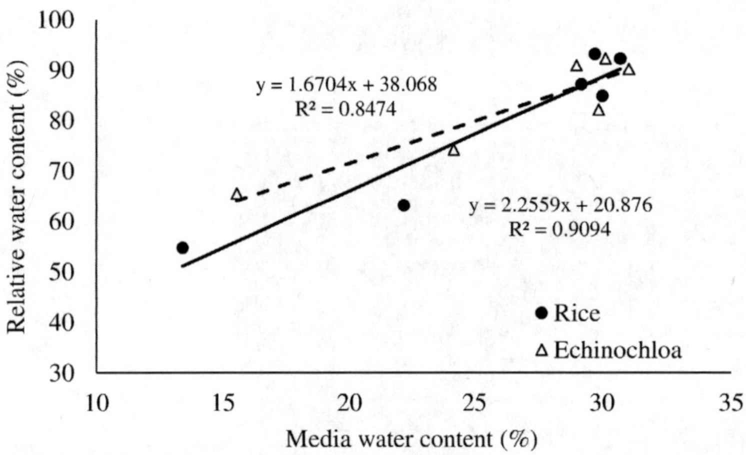

Figure 6. Correlation of media water content (MWC) and relative water content (RWC) of rice and Echinochloa during drought treatment. RWC of rice was decrease higher than that of Echinochloa in response to decrease of MWC.

The question was arise whether this difference has correlation with anatomical characteristics of both species. Xylem diameter, for example, is among the parameter that is suggested to have a role in drought tolerance[17]. This may associated to the development of cavitation and embolisms due to drought stress which cause the plant undergo stress[18]. An analysis on some tree species in Africa suggested that the trees with smaller vascular tissues were less vulnerable to cavitation and embolism and thus more tolerant of drought indicated by lowering dieback[19]. Anatomical analysis indicated that during drought treatment xylem diameter of Echinochloa either in major- as well as intervein was more elastic and decreased greater than that of rice (Table 3). This data also associated to the decrease of leaf thickness in the area of major vein as well as bulliform cell and intervein of Echinochloa that was greater than that of rice (Table 2). These were probably the reasons why Echinochloa underwent less stress than rice in response to 6 days drought.

3.7. Chloroplast and starch accumulation in response to drought stress

Plants usually response water stress by complex mechanisms including alteration of morphology and anatomy as well as physiology and metabolism. Chloroplast and starch accumulation inside the chloroplast is among several mechanisms observed in this experiment. The data clearly verified the existing chloroplast in the bundle sheath cell of Echinochloa, while it was absent in that of rice, suggesting that Echinochloa is a C4 species, and rice is C3 species. This is also in agreement with the classification of C4 species made by Sage and Manson[20] that Echinocloa is a C4 species. Interestingly, the number of chloroplast increased significantly in the bundle sheath cell of Echinochloa but not in mesophyll cell, during the drought stress (Table 5). This may associated with the increase of metabolic (carbohydrate) that was required to produce various substances to overcome stress. Several compounds such as trehalose, glycine betaine and proline are among the osmolyte substances abundantly accumulated during drought stress in many species that have function as osmoprotectant[16].

Starch produced during photosynthesis is a dynamic substance which is temporally accumulated in the chloroplast or exported from mesophyll cell to other cell of the plant through vascular tissues[21]. We compared chloroplastic starch accumulation in these two species to understand the different carbohydrate metabolism between C3 and C4 species in response to drought stress. Interestingly, in Echinochloa, starch was existed in mesophyll (even though only minor) as well as in bundle sheath cell (Table 6 and Figure 5). The existence of starch granules in mesophyll as well as bundle sheath cell of Echinochloa suggesting that C3 pathway (Calvin cycle) may be happened together with C4 pathway in Echinochloa. The C3-like characteristic of C4 species have been discussed by many authors as a specific phenomenon in C4 species such as *Panicum*[22]. The substantial increase of chloroplastic starch due to drought also occurred in mesophyll cell as well as in bundle sheath cell of Echinochloa as it was happened in mesophyll cell of rice (Table 6). This also even becomes strong evidence that C3 cycle might be occurred in mesophyll cell of Echinochloa (C4 species) especially during drought stress.

Conclusion

Rice exhibited more stress than Echinocloa due to drought stress indicated by the reduction of leaf relative water content, maximum efficiency of photosynthesis and plant growth. Drought treatment tended to reduce leaf thickness in the area of bulliform cell, major- as well as intervein and xylem diameter more in Echinochloa than in rice, suggesting that the decrease of vein and xylem diameter is among the anatomical parameters that is important to overcome from drought stress in Echinochloa. The drought also significantly increased the number of chloroplast in bundle sheath cell of Echinochloa, but not in mesophyll cell. The number of starch granules also increased in response to drought in mesophyll cells of rice and Echinochloa and in bundle sheath cell of Echinochloa, an indication that C3 cycle might occurred in C4 species at least in Echinochloa, especially during drought stress.

References

[1] Lambers H, Chapin III FS and Pons TL. 2008. Plant Physiological Ecology (2nded.). New York: Springer-Verlag. P.604.

[2] Pandey V and Shukla A. 2015. Acclimation and tolerance strategies of rice under drought stress. Rice Sci. 22:147-161.

[3] Hamim. 2004. Underlying drought stress effects on plant: Inhibition of photosynthesis. Hayati 11:164-169.

[4] Tezara W, Mitchell VJ, Driscoll SD and Lawlor DW. 1999. Water stress inhibits plant photosynthesis by decreasing coupling factor and ATP. Nature. 401:914-917.

[5] Wingler A, Lea PJ, Quick WP, Leegood RC. 2000. Photorespiration: Metabolic pathways and their role in stress protection. *Phil. Trans. R Soc. Lond.* 355:1517-1529.

[6] Dai Z, Ku MSB, Edwards GE. 1993. C4 photosynthesis, the CO2 concentrating mechanism and photorespiration. Plant physiol. 103:83-90.

[7] Raghavendra AS and Sage RF. 2011. Introduction pp. 17-29. In Raghavendra AS and Sage RF (Editors). Advance in Photosynthesis and Respiration: C4 Photosynthesis and Related CO2 Concentrating Mechanisms. Springer. The Netherland.

[8] Brown RH and Hattersley PW. 1989. Leaf anatomy od C3-C4 species as related to evolution of C4photosynthesis. Plant Physiol. 91:1543-1550.More references

[9] Prochazkoza D, Sairam RK, Srivastava GC, Singh DV. 2001. Oxidative stress and antioxidant activity as the basis of senescence in maize leaves. *Plant Sci.* 161:765-777.

[10] Genty B, Brantais JM, Baker NR. 1989. The relationship between the quantum yieldof photosynthesis electron transport and quenchingof chlorophyll fluorescence. *Biochem Biophys. Acta.* 990:87-92.

[11] Johansen DA. 1940. Plant microtechnique, McGraw-Hill Book Company, Inc. New York.

[12] Bozzola JJ and Russell LD. 1998. Electron microscopy principlesand techniques for biologist.2nd Edition. Massachusetts. Jones and Bartlett Publishers.

[13] Claeys H and D Inze. 2013. The agony of choice: How plants balance growth and survival under water-limiting conditions. Plant Physiol. 162: 1768-1779.

[14] Centritto M, Lauteri M, Monteverdi MC, and Serraj R. 2009. Leaf gas exchange, carbon isotope discrimination, and grain yirld in contrasting rice genotypes subjected to water deficits during the reproductive stage, J Exp Bot. 60:2325-2339.

[15] Virginia S. Pagan M, Cooper M, Kantartzi SK, Lightfoot DA, Meksem K and Kassem MA. 2012. Genetic analysis of relative water content (RWC) in two recombinant inbred line populations of soybean [Glycine max (L.) Merr]. J Plant Gen Sci 1: 46–53.

[16] Sanders GJ and Amdt SK. 2012. Osmotic adjustment under drought conditions. Pp. 199-229. In. Plant Response to Drought Stress. Aroca R (Editor). Springer-Verlag, Berlin.

[17] Comstock JP and Sperry. 2000/ Theoretical considerations of optimal conduit length for water transport in vascular plants. New Phytol. 148:195-218.

[18] Ennajeh M, Tounekti T, Vacel AM, Khemira H and Cochard H. 2008. Water relations and drought-induced embolism in olive (Olea europaea) varieties 'Meski' and 'Chemlali' during severe dought. *Tree Physiol.* 28:971-976.

[19] Kondoh S, Yahata H, Nakashizukas T, and Kondoh M. 2006. Interspecific variation in vessel size, growth and drought tolerance of broad-leaved trees in semi-arid regions of Kenya. Tree Physiology 26, 899–904.

[20] Sage RF, Li M and Monson RK. 1999. The taxonomic distribution of C4 photosynthesis. Pp.551-586. In: Sage RF, Monson RK. C4 Plant Biology. London: Academic Press.

[21] Taiz, L. and Zeiger E. (2010). Plant Physiology (5th Edition). Massachusetts: Sinauer Associates, Inc. Publishers. Sunderland, Massachusetts, USA.

[22] Ghannoum O, Siebke K, von Caemmerer S, Conroy JP. 1999. The photosynthesis of young *Panicum* C4 leaves is not C3-like. *Plant Cell Environ.* 21:1123-1131.

Mercury depuration effectiveness on green mussel (Perna viridis L.) and blood cockle (Anadara granosa L.) from Jakarta Bay using ozone, chitosan and hydrodynamic technique

Endar Widiah Ningrum[1], **Dedy Duryadi Solihin**[2], **Nurlisa A. Butet**[3]
[1,2] Biology Department, Bogor Agricultural University, Darmaga Campus 16680, Indonesia
[3] Department of Aquatic Resources Management, Bogor Agricultural University, Dermaga Campus 16680, Indonesia

E-mail: dduryadi@yahoo.com[2], endarwidiahningrum@gmail.com[1]

Abstract. Depuration has been used to eliminate microorganism and toxic chemical contaminants in bivalve. However, scientific research still needs to discover the effectiveness of depuration. This research aimed at assessing the best depuration effectiveness in decreasing mercury (Hg) concentration level in two species of bivalves, green mussel (*Perna viridis* L.) and blood cockle (*Anadara granosa* L.). The depuration treatments applied 1.5 ppm ozon, 0.5 ppm chitosan, hydrodynamic technique (1.3 m/s), combination between hydrodynamic-ozon, hydrodynamic-chitosan and ozon-chitosan. The experiment were conducted in mini aquaria for 60 minutes. Mercury concentration was measured in 10 g dry weight of green mussel (4.05±0.020 ppm) and blood cockle (3.27±0.666 ppm). The result showed that mercury depuration were highly effective by combination of ozone-chitosan in green mussel (96.51%) and 1.5 ppm ozone in blood cockle (87.06%).

1. Introduction

The abundance of Indonesia marine resources was nowadays putting fisheries sector as a major concern. Annual statistical reported by the General Directorate of Fisheries, Ministry of Marine Affairs and Fisheries (DJPB KKP RI) in the latest version that the mariculture is on the top position with 8,3 billion ton production while others are the 1-2 million ton. Bivalve molluscs (e.g. mussels, oysters and clams) commonly named as shellfish is the top commodity produced from the coastal Indonesia following fish and seaweed commodities. Shellfish commodity [1] as many as 48, 449 ton national product within which 25, 935 ton were produced throughout Jakarta Bay, the northern coast of Java.

Despite of its major contribution for shellfish production, Jakarta Bay was founded to be chemically polluted, Riyadi *et al.* reporting that. The concentration of mercury (Hg) is as high as 0.067 ppm in Jakarta Bay while [2] the standard concentration of mercury for security marine product should be less than 0.001 ppm [3]. Therefore, Jakarta Bay is not safe for marine production, especially for the shellfish commodity. This problem becomes a concern for the seafood consumers.

Shellfish are filter feeder organisms. They feed on filtered seawater and accumulate food in their mantle. Like other filter feeder organisms, green mussels (*Perna viridis* L.) inhabit substrate by attaching to wood or bamboo in the industrial cultivation, while the blood cockles (*Anadara granosa* L.) stay in the sediment. Chemical pollution can be found in the shellfish bodies as a contaminant. The shellfish that have been contaminated with mercury are not safe for consumption. This leads to a variety of health problems due to accumulation of heavy metals in the human body. Therefore, the Food and Drug Monitoring Agency (BPOM) [4] has set the value of 1.0 ppm as the maximum levels of mercury contaminant on shellfish to be safe for consumption.

In order to eliminate the contaminants, such as microorganism and their residual toxicity, depuration is commonly conducted on marine products. Depuration, on the lab scale, has been conducted to eliminate heavy metals. The conventional methods applied were soaking, washing and boiling. The latest result was reported by Anacleto *et al.* [5] by using filtered seawater and UV radiation in the three species, *Ruditapes philippinarum*, *Mytilus galloprovincialis* and *Scrobicularia plana*. This research resulted on decreasing level of mercury up to 32.2% in the shellfish body within 6 days of depuration. However, the research on depuration effectiveness are still need to be observed.

Human exposure to mercury contaminants can be dangered with comparative risk of fetal developmental neurotoxicity [6]. The mecanism of mercury exposure come from sea food consumed that contaminated. The aim of the research is to find the effectivity of depuration method by treatments ozone, chitosan, hydrodynamic tecnique and combintion both of treatments.

2. Materials and methods

2.1. Materials

Materials used in this research were green mussel (*Perna viridis* L.) and blood cockle (*Anadara granosa* L.), HNO_3 65%, $HClO_4$ (60%), aquades, 1.5 ppm of ozone, and 0.5 ppm of chitosan. Digital refractometer HACH sension5 and Atomic Absorption Spechtophotometer (AAS).

2.2. Methods

2.2.1. Sample collection. Green mussell (*Perna viridis* L.) and blood cockle (*Anadara granosa* L.) were freshly collected from Cilincing and Marunda (Jakarta Bay), DKI Jakarta (Picture 1). Samples were brought to laboratory using container and ice box. Dead or damaged specimens were eliminated prior to mussell size standarization. The green mussels with 2-3 cm in length, 1-1.5 cm in width, and 0.75-1.25 cm in thickness were used as samples. The blood cockle with 4-7 cm in length, 3-4 cm in width, and 1.5-2 cm in thickness were also used as samples.

2.2.2. Depuration experiment. The experiment was conducted in mini aquarium for 60 min with two replications. Weight of fresh sample were 500 g of green mussell and 400 g of blood cockle in used. The treatments were ozon, chitosan, hydrodynamics, combining between hydrodynamics-ozon, hydrodynamics-chitosan, ozon-chitosan, and control. Ozon liquid concentration was 1.5 ppm. Chitosan liquid concenteration was 0.5 ppm. The technique of hydrodynamic was approached by pumping the water to constant velocity of 1.3 m/s. Combination of two treatments was conducted together. Physical parameters was measured three times, at 9 am, 2 pm, and 9 pm. Flesh samples were separated from cockle and scaled for preparation. Preparation samples were dried at temperatures of 40 ^{0}C for 24 hours, and average water concentration was scaled.

2.2.3. *Mercury concentration analysis [7].* Depurated samples (10 g) was added 10 mL HNO_3 65% and $HClO_4$ 60% solutions. Samples were soaked for 48 h and warmed for 4 h. Afterwards samples were filtered and reduced into 100 mL liquid measure. Mercury concentration was measured by Atomic Absorption Sphectophotometer (AAS).

2.2.4. Statistical analysis. The results were presented as descriptive statistic. One-way analysis of variance (ANOVA) [8] was used to test the significance of depuration treatment. Post hoc tes was

IOP Conf. Series: Earth and Environmental Science **31** (2016) 012041 doi:10.1088/1755-1315/31/1/012041

used to analyze the multiple comparisons among control, treatment of ozone, chitosan, hydrodynamics, and combined between both. Stastical anaylses were performed using the SPSS 22.0 software.

3. Results

3.1. Mercury concentration

Mercury concentration was measured on green mussel and blood cockle under laboratory experiment. Depuration experiment was conducted in mini aquaria set for these conditions, i.e., temperature (28±1 ^{0}C), salinity (34.83 ± 2.369 ‰), and pH (7.63 ± 0.152). Mercury concentration was decreased after 60 minutes of depuration experiment and presented in the Table 1.

Table 1. Mercury concentration in green mussel and blood cockle

	Mercury concentration (mean ± SD)			
	Green mussel (ppm)		Blood cockle (ppm)	
Treatment	Dry weight	Wet weight	Dry weight	Wet weight
Control	4.05±0.020^a	0.83±0.004	3.27±0.666^a	0.53±0.109
1.5 ppm Ozone	0.90±0.024^c	0.13±0.003	0.33±0.005^d	0.07±0.001
0.5 ppm Chitosan	0.85±0.004^d	0.17±0.000	0.39±0.011^d	0.09±0.002
1.3 m/s Hydrodynamic	3.22±0.032^b	0.50±0.005	0.75±0.006^c	0.16±0.000
Hydrodynamic-Ozone	0.50±0.003^e	0.17±0.001	2.09±0.004^b	0.42±0.000
Hydrodynamic-Chitosan	0.20±0.002^f	0.03±0.000	0.43±0.022^d	0.06±0.003
Ozone-Chitosan	0.15±0.004^g	0.31±0.000	0.45±0.011^d	0.10±0.002

Note :Mercury concentration was measured in 10 g dry weight and converted in wet weight

The highest concentration of mercury was found in control sample (without any treatment, see Table 1). Concentration of mercury in dry weight was 4.05±0.020 ppm in green mussel and 3.27±0.384 ppm in blood cockle. Mercury concentration level decreased in the treatment of ozone-chitosan combination in green mussel and in dissolve ozone in blood cockle (1.5 ppm) .

The effectof depuration method on mercury level in green mussel is despited in Figure 1. Mercury concentration is significantly (P<0.000) decreased in the order of control > ozon > chitosan > hydrodynamic-ozon > hydrodynamic-chitosan > ozon-chitosan. Combination of ozone-chitosan treatment is the most effective depuration method, decreasing mercury concentration by 96.51%, and then followed by the treatment of hydrodynamic-chitosan (94.94%), hydrodynamic-ozone (87.79%), chitosan (79.11%), ozone (77.46%). Hydrodynamic has the least effective method, decreasing mercury concentration only by (21.53%).

The effect of depuration method on mercury concentration in blood cockle is presented in Figure 2. Concentration mercury are significantly (P<0.000) decreased in the order of control > hydrodynamic-ozone > hydrodynamic > ozone-chitosan > hydrodynamic-chitosan > chitosan > ozone. Combination of ozone treatment is the most effective depuration method, decreasing mercury concentration by 87.06%, followed by the treatment of chitosan (85.07%), hydrodynamic-chitosan (83.71%), ozone-chitosan (82.28%), hydrodynamic (70.03%) hydrodynamic-ozone has the least effective methode, decreasing mercury concentration only by (17.25%).

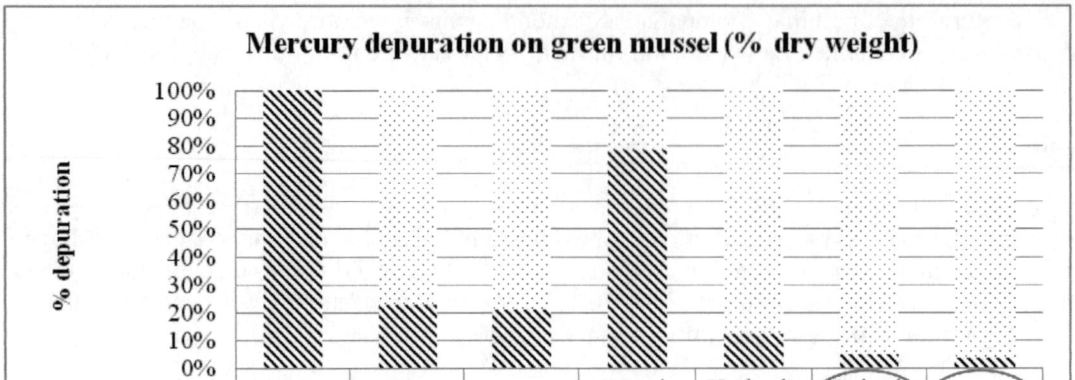

	Control	1.5 ppm Ozone	0.5 ppm Chitosan	1.3 m/s Hydrody namic	Hydrody namic-Ozone	Hydrody namic-Chitosan	Ozone-Chitosan
Depuration (%)	0	77.46	79.11	21.53	87.79	94.94	96.51
Concentration	100	22.54	20.89	78.47	12.21	5.06	3.49

Figure 1. Effect of depuration method on mercury concentration in green mussel

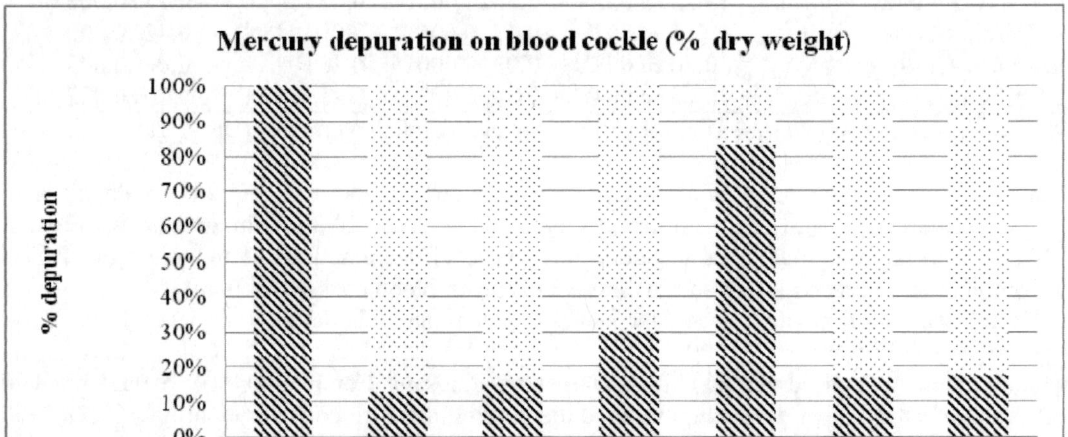

	Control	1.5 ppm Ozone	0.5 ppm Chitosan	1.3 m/s Hydrody namic	Hydrody namic-Ozone	Hydrody namic-Chitosan	Ozone-Chitosan
Depuration (%)	0	87.06	85.07	70.03	17.25	83.71	82.28
Concentration	100	12.94	14.93	29.97	82.75	16.29	17.72

Figure 2. Effect of depuration method on mercury concentration in blood cockle

4. Discussion

Depuration technique was conducted in the laboratory to eliminate microorganism contaminants and their chemical toxic levels. Depuration treatment was reported by Yokohama and Park [9] using the temperature treatment (15^0C and 25^0C), in five days might reduce Cyanobacterial toxicity. The latest result was reported by Anacleto et al [5] that mercury levels in three species were also reduced, *Ruditapes philippinarum*, *Mytilus galloprovincialis* and *Scrobicularia plana.* The treatment were using filtered seawater and UV radiation and could decrease the mercury levels up to 32.2% within six days of depuration. In this study, three depuration methods were tested. These were hydrodynamic tecnique, ozone and chitosan in this research. The combination among three types of depuration are used in the group of treatment to find out the depuration effectiveness.

Ozone has been used in the pesticides eliminating process of agricultural production [10] (e.g. on fruits and vegetables). The use of dissolved ozone (1.4-2.0 ppm) is reported that it can eliminate pesticide residues from the surface of vegetable. The low concentration of dissolve ozone are effective to reduce 60-99% residual pesticide in the first 5 minutes. The concentration of dissolved ozone used in this research treatment is low (1.5 ppm).

Chitosan has been used to eliminate mercury from contaminated water. Chitosan particles are reported can decrease the mercury concentration in the water up to 0.29±91.55%, [11], while Kyzaz and Kostoglou [12] reported that effective absorbent chitosan used to adbsorb heavy metals from polluted water was microsphore cross-linked chitosan. Its derivative of chitosan, i.e., chitosan beads, has a weak dilute acid in matrix of chitosan bead. This causes –NH_2 of chitosan beads has higher affinity than normal chitosan and can increased the heavy metal adbsorption. Chitosan beads can be made by dissolving chitosan powder into 5% acetic acid to form a gel. Thus, the use of low chitosan was determined by 0.5 ppm in this research treatment.

The technique of hydrodynamic was approached by pumping the water to constant velocity of 1.3 m/s. Hydrodynamic technique has been used [13] in combination with UV radiation to decrease contaminant in type of bivalves. However, this technique is only intended to remove microorganism contaminants while its use in heavy metals depuration has not been carried out.

The depuration of the experiment was fast and resulted satisfying outcomes. Statistical test conducted on the depuration rate showed statistical different (P<0.000) between control and the treatment in both samples. Mercury concentration in green mussel were decreased with percentage up to 96.51%, from 4.05 ppm to 0.15 ppm, by the treatment of ozon-kitosan. Meanwhile, the treatment of ozone in blood cockle decreased mercury concentration from 3.27 ppm to 0.33 ppm, with percentage up to 87.06% are decreased by the treatment of ozon-kitosan up to 96.51% (4.05 ppm to 0.15 ppm) in green mussel and the treatment of ozone in blood cockle can be decrease mercury level up to 87.06% (3.27 ppm to 0.33 ppm). The depuration treatment in green mussel were significantly different among control and groups of treatment. The result in blood cockle showed different response (P<0.000) between control and groups of treatment while variance of ozon and chitosan, also combination between hydrodynamic-chitosan and ozone-chitosan have not significantly different. However, the high depuration rate was showed in ozon-chitosan treatment (green mussel) and ozone treatment (blood cockle). The low depuration rate were on hydrodynamic (green mussel) and hydrodynamic-ozone (blood cockle).

5. Conclusion

The effective of mercury depuration in green mussel were by the treatment of ozone-chitosan, hydrodynamic-chitosan and hydrodynamic-ozone, in respect to 96.51%, 94.94% and 87.79%. The treatment of ozone, chitosan, hydrodynamic-chitosan and ozone-chitosan, in respect to 87.06%, 85.07%, 83.71% and 82.28% were effective to decrease mercury levels in blood cockle.

Acknowledgements

We are grateful to Dr Dedy Duryadi Solihin, DEA, Animal and Ecology Division (Department of Biology), and Dr Nurlisa A. Butet, MSc from Aquatic Resources Management, Bogor Agricultural University, Indonesia, for their supervision throughout this research.

References

[1] DJPB KKP RI 2011.
[2] Riyadi AS, Itai T, Hayase D, Isobe T, Horai S, Miller TW, Omori K, Sudaryanto A, Ilyas M, Setiawan IE. 2015. Comparison of trophic magnification slopes of mercury in temperate and tropical regions Case studies in Oregon coast, USA, Sanriku coast, Japan and Jakarta Bay, Indonesia. (Japan: *J-STAGE*).
[3] Ministry Degree No. **51** LH RI 2004.
[4] Regulated BPOM RI No. **06** 2009.

[5] Anacleto P, Luisa A, Leonor M. 2015. *J Food Control*. **47** 493.
[6] Sheehan MC, Thomas AB, Ana NA, Patric NB, John M, Mary AF. 2014. *Bull World Health Organ*. **92**: 254.
[7] [AOAC] Association Official Agriculture Chemists. 2002. *Official Methods of Analysist of Chemicals Contaminants, Drugs*. Ed ke-7. Meryland (USA): AOAC International.
[8] [SPSS] 2013. SPSS Base 22.0*0 User's Guide*. Chicago (USA): SPSS Inc.
[9] Yokohama A, Park H. 2002. *J Environ. Toxicol*. **18** 61.
[10] Wu J, Luan T, Lan C, Wai T, Lo H, Yuk G, Chan S. 2007. *J Food Control*. **18** 466.
[11] Minu M, Kumar N, Shilpa J. 2015. *J Chitin and Chitosan Science*. **9** 68.
[12] Kyzas GZ, Kostoglou M. 2015. *J Separation and Purification Technology*. **149** 92.
[13] Lee R, Lovatelli A, Ababouch L. 2008. *Bivalve Depuration: Fundamental and Practical Aspects.* Rome (ITA): Food and Agricultural Organization of the United Nation.

IOP Conf. Series: Earth and Environmental Science **31** (2016) 012042 doi:10.1088/1755-1315/31/1/012042

Development of an Integrated Agricultural Planning Model Considering Climate Change

I P Santikayasa
Department of Geophysical and Meteorology, Bogor Agricultural University
Kampus IPB Dramaga, Bogor, Indonesia

E-mail: ipsantika@apps.ipb.ac.id

Abstract. The goal of this study is to develop an agriculture planning model in order to sustain the future water use under the estimation of crop water requirement, water availability and future climate projection. For this purpose, the Citarum river basin which is located in West Java – Indonesia is selected as the study area. Two emission scenarios A2 and B2 were selected. For the crop water requirement estimation, the output of HadCM3 AOGCM is statistically downscale using SDSM and used as the input for WEAP model developed by SEI (Stockholm Environmental Institute). The reliability of water uses is assessed by comparing the irrigation water demand and the water allocation for the irrigation area. The water supply resources are assessed using the water planning tool. This study shows that temperature and precipitation over the study area are projected to increase in the future. The water availability was projected to increase under both A2 and B2 emission scenarios in the future. The irrigation water requirement is expected to decrease in the future under A2 and B2 scenarios. By comparing the irrigation water demand and water allocation for irrigation, the reliability of agriculture water use is expected to change in the period of 2050s and 2080s while the reliability will not change in 2020s. The reliability under A2 scenario is expected to be higher than B2 scenario. The combination of WEAP and SDSM is significance to use in assessing and allocating the water resources in the region.

1. Introduction

Climate change, population growth and economic development are three factors likely to affect the availability of water resources for agriculture in the future. The demand and supply of water for irrigation will be influenced not only by the changes in climate and subsequent changes in hydrology but also by the increase in the competition for water in the future between the agricultural and the non-agricultural users, given the increasing population and associated economic development would impose pressure on water resources.

Adaptations relating water resources to agriculture are very important. A number of existing options can be used to adapt water resources for agriculture under climate change conditions in the future. In most studies, the sustainability of water uses is not incorporated in the evaluation of adaptation strategies of water resources for agriculture under possible climate change scenarios.

The assessment of future climate change impact at regional scales is vital for climate change adaptation issues. However, many researches were done on how to bridge the gap between the Global Circulation Model (GCM) output and the need of catchment scale assessment. Accordingly, downscaling methods have been developed to link GCM outputs at coarse resolutions with surface weather variables at finer resolutions. One of the statistical downscaling, Statistical Downscaling Model (SDSM), was developed by Wilby [1].

The hydrological response to climate change has been studied through the application of watershed-scale hydrological models driven by Global Circulation Model (GCM) scenarios of future climate (e.g. [2], [3], [4], and [11]). Water Evaluation and Planning (WEAP) as the distributed hydrology model integrates a range of physical hydrology processes with the management of the demands and installed infrastructure in a seamless and coherent manner [5]. It was used to assess the multiple scenario analysis, including alternative climate scenarios and changing of human intervention. Thus, WEAP was successfully applied to assess future climate impact on hydrology ([6], [11]).

In this study, the agriculture planning model is developed to determine how reliable water use in agriculture considering the future climate change. The model was developed using SDSM and WEAP. SDSM is used to downscale the future climate and WEAP allows the simulation and analysis of various scenarios of water allocation and user behaviour.

2. Method

In this study, the impact of climate change on water resources and agriculture was assessed using an integrated assessment approach (Figure 1). Three steps were applied to project the future climate in the study area, assessing the impact of future climate to the water availability and evaluate the irrigation management scenarios to the reliability of future water use in agriculture. The first step is the future climate projection. In this step, the Global Circulation Model namely Hadley Centre Coupled Model version 3 (HadCM3) was assigned. Two emission scenarios A2 and B2 were applied to project the future climate in the study area. The statistical downscaling SDSM was applied to generate the watershed scale spatial resolution of future climate in the study area. The second step is assessment of the future water availability. In this step, the WEAP hydrological model was applied to evaluate to water availability based on the downscaled climate output resulted from the first step. The model also assessed the future irrigation water requirement based on the irrigation management scenarios. Furthermore, the model evaluates the water allocation for the irrigation based on the irrigation water demand and supply. The last step is applying the reliability index to evaluate which irrigation management option is reliable for the irrigation water use under climate change in the future. The detail of each step is explained more detail as follows:

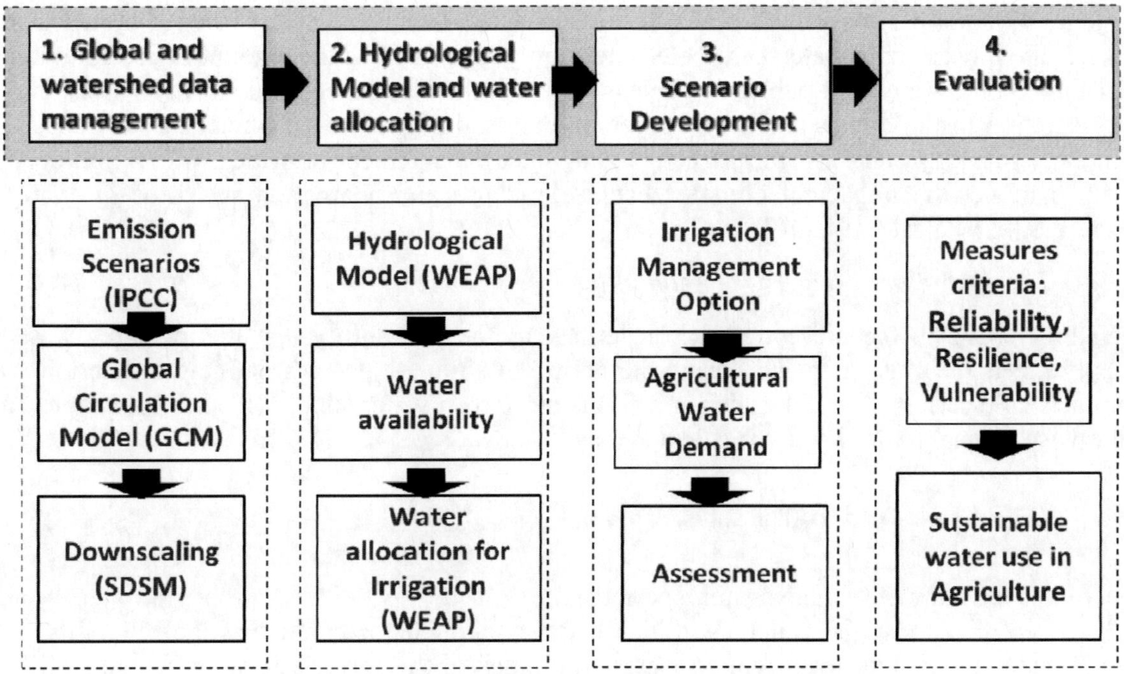

Figure 1. Agriculture Planning Model Framework

2. 1. Future Climate Projection

The future climate was projected in the period of 2020s (2011 – 2040), 2050s (2041 – 2070) and 2080s (2071 – 2099). To describe the future climate, the temperature and precipitation were projected based on the output of Hadley Centre Coupled Model version 3 (HadCM3) GCM developed by Hadley Center, UK. The climate was projected under two emission scenarios A2 and B2 [2]. To increase the resolution of global to local scale, we used SDSM to downscale the climate variables. The SDSM establish the empirical relationship function (F) between the predictors and the predictand. The function is a deterministic/stochastic function which is conditioned by predictors and predictand as the follow equation:

$$a = F\,(b) \tag{1}$$

where a is the predictand and b are the predictors

In this study, the GCM output employed as the predictors and the local climate variables as the predictand. To calibrate the SDSM, the atmospheric predictor variables from the National Center for Environmental Prediction (NCEP) reanalysis were used [1]. The temperature and precipitation of year 1961 – 2000 from four stations (Citarum, Cibeet, Cikarang and Bekasi) was used as the predictand.

2. 2. The hydrological model

The WEAP package tool is used to model the hydrological process in the study area. WEAP is the computer modelling package design, which is used for simulation of water resources

IOP Conf. Series: Earth and Environmental Science **31** (2016) 012042 doi:10.1088/1755-1315/31/1/012042

system and trade-off analysis. It operates on the basic principle of water balance model that defines the process on a watershed scale. The supply is defined as the amount of precipitation that falls on the watershed, which is depleted through the watershed process, water uses, and the accretion to the downstream. Moreover, it enables integrated assessment of watersheds' climate, hydrology, land use, infrastructure and water management priorities. WEAP has been used to model the impact of climate change, land use and adaptation scenarios on water resources (e.g. [7], [8], [9], [10]).

2. 3. The Agricultural Water Use Reliability

Reliability (Cr), in the present study, is defined as the probability that the available water supply meets the water demand during the period of simulation. Let's define the criterion x, for each time period (t), if water demand (D) is more than water supply (S) than x equal to 1, otherwise x equal to 0. Than Cr is calculated as:

$$Cr = \left(1 - \frac{\Sigma x}{n}\right) x\ 100\% \qquad (2)$$

where n is the number of analysis time periods.

In this study, the water use reliability was applied for evaluating the reliability of future water availability for irrigation water. The index of the agriculture water use reliability is applied to evaluate two irrigation management scenarios in the future period. The first irrigation scenario is by increasing the irrigation area, and the second scenario is by increasing the cropping intensity.

2. 4. Case study- Citarum River Basin, Indonesia

The West Tarum Canal (WTC) irrigation system in the Citarum river basin Indonesia has been selected as the study area. The canal extends from east to west and passes agricultural, industrial and urban areas. The area is categorized as a tropical area with the average temperature of 24.7° – 27.3° C, the average annual precipitation range of 1500 – 4000 mm and the humidity is about 80 – 95 %. The rainfall mostly falls in the wet season during November to April. The wet season is followed by the dry season during May to August.

The water in WTC canal is supplied by nearby sub-catchments: namely, Citarum, Cibeet, Cikarang and Bekasi (Figure 2). The WTC canal distributes water to the irrigation area, domestic and industry along the canal.

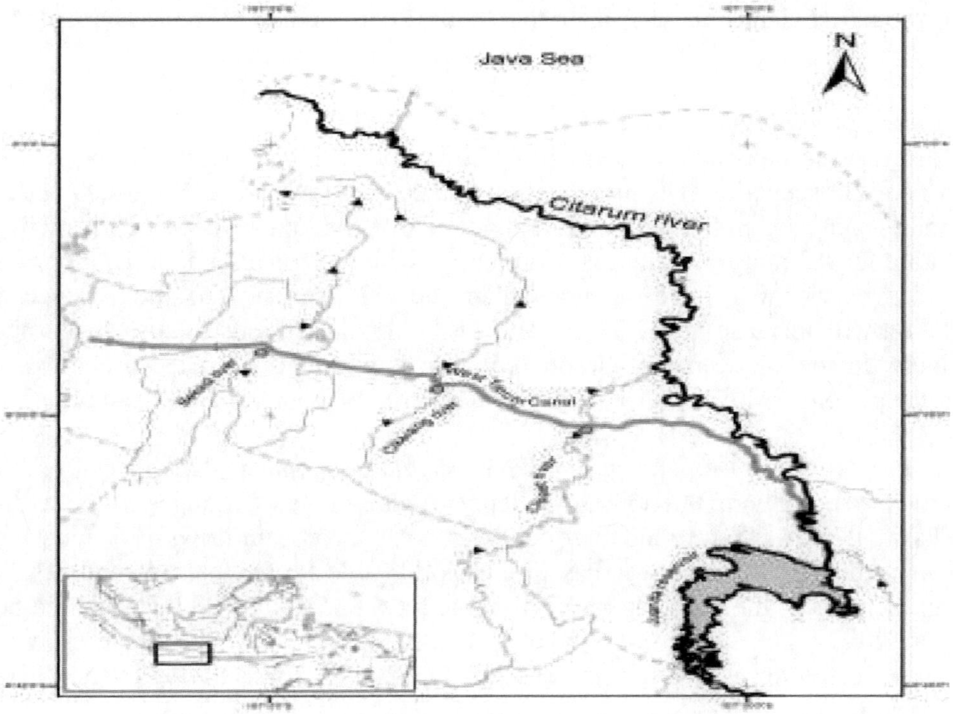

Figure 2 Map of Citarum River Basin, Indonesia

3. Result and discussion

3. 1. Hydrology Model and Future Climate Projection

WEAP model indicates the reasonable ability in simulating long term monthly time series of stream flow. At Citarum sub catchment, monthly stream flows from year 1994–2009 were simulated with a bias of -0.24 %, Nash-Sutclife of 0.58 and R=0.78 (n=180 months). At Cibeet sub catchment, the monthly flows from year 1987 – 2005 were simulated with a bias of -0.34%, Nash-Sutclife of 0.64 and R=0.81 (n=216 months). At Cikarang sub catchment, the monthly flows of year 1987 – 2005 were simulated with a bias of 1.5%, Nash-Sutclife of 0.52 and R=0.73 (n=216 months). At Bekasi sub catchment, the monthly flow of year 1987-2005 was simulated with a bias of 0.08%, Nash-Sutclife of 0.66 and R=0.79 (n=216 months).

Future climate was projected based on the output of HadCM3 GCM and downscaled using SDSM application under A2 and B2 emissions scenarios. Result shows, the temperature and precipitation are projected to increase in the future periods compared with the baseline period. The A2 and B2 climate projections are warmer and wetter than the baseline period.

3. 2. Future temperature

The temperature at the Cikarang and Bekasi sub basin under A2 scenario increased the most i.e. by 0.33, 0.99 and 1.63 ^{o}C, respectively, in the 2020s, 2050s and 2080s. On the other hands, the temperature in the Cibeet sub basin will decrease about 0.36 oC and investigated an increase about 0.24 in 2050s and 1.00 ^{o}C in 2080s under A2 scenario. As the average, the temperature will increase about 0.15, 0.74 and 1.47 oC in the 2020s, 2050s and 2080s respectively under A2 scenario. Under the B2 scenario, the temperature was investigated to

IOP Conf. Series: Earth and Environmental Science **31** (2016) 012042 doi:10.1088/1755-1315/31/1/012042

increase about 0.16, 0.50 and 0.97 oC in the 2020s, 2050s and 2080s periods, respectively.

3. 3. Future precipitation

The SDSM outputs show an increasing of precipitation in all sub basins for the future periods under A2 and B2 scenarios. The precipitation under DJF, JJA and SON periods under A2 and B2 scenarios increased in the future while the precipitation under MAM periods will decrease in the future for A2 and B2 scenarios. Comparing with the baseline periods, the precipitation under A2 scenario will increase more than the B2 scenario. In the A2 scenario, the precipitation will increase about 23, 55 and 88 % for the periods 2020s, 2050s and 2080s, respectively. In the B2 scenario, the precipitation is expected to increase about 27, 35 and 45% for the periods 2020s, 2050s and 2080s, respectively, compared with the baseline period.

3. 4. Water Availability and Irrigation Water Requirement under Climate Change

It is seen that the A2 and B2 scenarios produce a wide range of changes in the hydrology of the sub basin. It shows that for the three future periods, the stream flows are increasing in both scenarios when compared with the baseline period (Table I). The periods under A2 scenario shows an increase of the main stream flow of about 6.6, 53.1 and 111.7% for the periods of the 2020s, 2050s and 2080s, respectively.. Under B2 scenario, the mean flow of future periods is expected to respectively increase about 12.4, 37.3 and 67.1% in the 2020s, 2050s and 2080s periods..

The stream flow in the month of the whole year is projected to increase under A2 and B2 scenarios for the periods of the 2020s, 2050s and 2080s. However, the flows of the months of October, November and December are expected to decrease under A2 and B2 scenarios for the 2020s periods.

Table 1. Water availability under a2 and b2 scenario

Period	Emission Scenario	Water availability (x10^6 m^3)				
		Citarum	Cibeet	Cikarang	Bekasi	Total
Baseline		1253	1660	1757	2344	7013
2020s	A2	1625	1656	1971	2225	7477
	B2	1759	1735	2058	2331	7884
2050s	A2	2421	2409	2807	3106	10742
	B2	2183	2178	2485	2784	9630
2080s	A2	3523	3505	3786	4038	14852
	B2	2628	2741	3054	3292	11715

Source: [11]

Table 2 shows the comparison of annual irrigation water requirement between the historical period and future projection under two emission scenarios. It clearly shows that by considering the change in climate, the irrigation water requirements (IWR)will decrease in the future in both emission scenarios. Under A2 scenario, the total IWR will decrease about 16%, 37% and 63% compared with the historical in the period of 2020s, 2050s and 2080s, respectively. Similarly, under B2 scenario, the IWR will decrease about 20%, 24% and 44% on the period of 2020s, 2050s and 2080s, respectively. Thus, under A2 scenario the IWR is expected to decrease more than under B2 scenario.

Table 2. Irrigation water requirement under A2 and B2 scenario

Period	Emission scenario	Average Annual Irrigation Water Requirement ($x\ 10^6\ m^3$)				
		Citarum	Cibeet	Cikarang	Bekasi	Total
Baseline		58.2	390.4	344.8	119.9	913.3
2020s	A2	55.8	336.1	279.5	95.2	766.7
	B2	52.3	323.3	262.1	90.0	728.5
2050s	A2	47.5	257.2	202.2	69.8	574.7
	B2	51.2	304.5	251.1	84.2	691.0
2080s	A2	29.9	138.6	120.3	40.8	329.6
	B2	43.2	207.8	207.9	57.6	515.6

Source: [11]

3. 5. Irrigation Water Use Reliability under Future Climate Change

The water use reliability was assessed based on the impact of future climate only. The future water allocation for the irrigation and the change of future water requirement for the irrigation was compared and the reliability index was calculated. In future period, the water use reliability is expected to decrease compared with the historical period. As shown in the Figure 2, the reliability will not change in the period of 2020s under both A2 and B2 climate projection because the water allocation for the irrigation is able to fulfil the irrigation water requirement. It means that there will be no water shortages for the irrigation in the period of 2020s in the future. However, the decreasing of the water allocation for the irrigation in the period of 2050s and 2080s impact to the decreasing of the reliability index for the water use for irrigation in the future period. As shown in the Figure 3, under A2 emission scenario, the water use reliability is expected to decrease about 15% and 26% in the period of 2050s and 2080s, respectively. Similarly, under B2 scenario, the water use reliability is expected to decrease about 20% and 43% in the period of 2050s and 2080s, respectively, because of the decreasing water allocate to the irrigation.

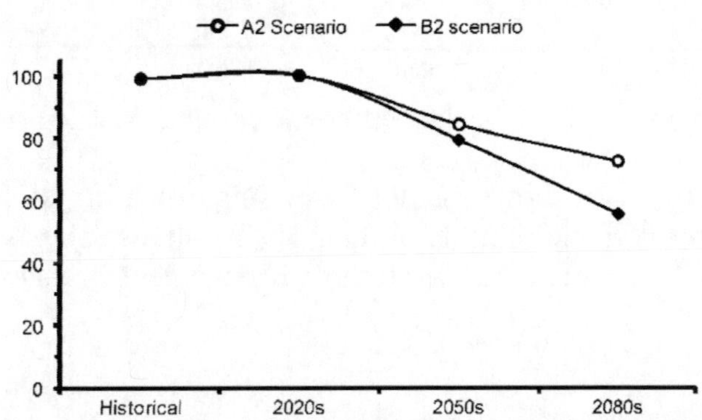

Figure 3 Agriculture water use reliability under historical and future climate projection

3. 6. Impact of irrigation management scenario on the reliability of irrigation water use

This study focuses on the development of the integrated agriculture planning model considering the future climate change. The framework is developed to incorporate the climate change on the water management especially in agriculture. The result of the implementation of the integrated agriculture planning model to evaluate the future management scenario is shown in the Figure 4.

Considering the climate change itself and in combination with the various irrigation management scenarios, it is found that the reliability is expected to decrease in the future. Even without any changes in the irrigation area and crop, the shortage in irrigation water supply is likely to increase in the future due to less water that will be allocated for irrigation as a consequence ofthe simultaneously increasing demands from domestic and industrial users. Therefore, in combining climate change with different scenarios, the irrigation water use reliability is expected to decrease in the future.

However, the decreasing of the reliability index is expected to be higher when the irrigation area is increasing rather than the increasing of the crop intensity. The increase of irrigation area will increase the annual IWR without change the monthly IWR. On the other hand, the increase of cropping pattern will increase not only the annual IWR but also the monthly pattern of the IWR. Therefore, the water shortage where the irrigation water supply is not able to fulfil the irrigation water requirement, is expected to increase in the future if the increasing of cropping pattern is applied.

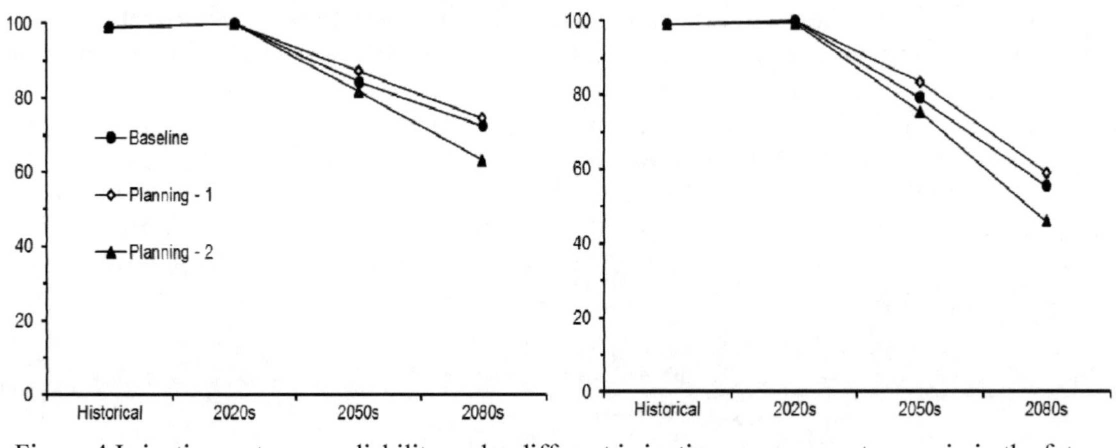

Figure 4 Irrigation water use reliability under different irrigation management scenario in the future

The result of this section shows that the increasing of the irrigation area is more reliable in term of water use for agriculture in the future. The result also shows that the framework is able to apply for the assessment of the future management scenario considering climate change.

4. Conclusion

This study uses the SDSM and WEAP model in order to assess the impact of climate change on the hydrology in the Citarum river basin. The basin was modelled quite well by the WEAP application. The model was able to simulate the stream flow during the analysis periods.

The SDSM was used to downscale the coarse scale of GCM output to the finer resolution of basin scale as it is required by the hydrological model. The SDSM performed well to downscale the temperature and precipitation in the study area. The temperature projection indicates the increasing of temperature for the future compared with the baseline period. The daily precipitation is projected to increase in the future compared with the baseline periods.

The water availability was simulated using the combination of SDSM and WEAP hydrological model. The water availability was projected to increase in the future under two emission scenario A2 and B2. The study intended to cater the research in the developing country using the low-cost assessment tool to increase the awareness of water resources change due to future climate change. By comparing the irrigation water demand and water allocate to agriculture, the reliability of agriculture water use is expected to change in the period of 2050s and 2080s while the reliability will not change in the period of 2020s. The reliability under A2 scenario is expected to be higher than B2 scenario.

This study provide a useful tool to assess the impact of climate change on future water availability as the input for future water planning and assess the reliability of agriculture water use in the future. The result provides the input for the decision maker to take decision on how to manage water resources under future climate change.

5. References

[1] Wilby R.L., Dawson C.W., Barrow E.M. 2000. Sdsm — a decision support tool for the assessment of regional climate change impacts. Environmental Modelling & Software 17 (2002) 147–159. UK

[2] IPCC, 2007: R.K. Pachauri, A. Reisinger (Eds.), Climate Change: Summary for Policymaker of Synthesis Report. Core Writing Team, Cambridge University Press, Cambridge

[3] Toth, B., Pietroniro, A., Conly, M.F., Kouwen, N., 2006. Modelling climate change impacts in the Peace and Athabasca catchment and delta: I-hydrological model application. Hydrol. Processes 20, 4197–4214.

[4] Nurmohamed, R., Naipal, S., De Smedt, F., 2007. Modeling hydrological response of the Upper Suiname river basin to climate change. J. Spatial Hydrol. 7 (1), 1–22..

[5] Forbes, K.A., Kienzle, S.W., Coburn, C.A., Byrne, J.M., Rasmussen, J., 2011. Modelling the impacts of selected GCM derived climate scenarios on the future hydrology of a hybrid watershed in the Oldman River watershed, Alberta, Canada. Clim. Change

[6] Wilby RL, Wigley TML. Downscaling general circulation model output a review of methods and limitations. Prog Phys Geogr 1997;21:530–48.

[7] Yates, D., J. Sieber, D. Purkey, and A. Huber-Lee, 2005: WEAP21—A Demand-, Priority-, and Preference-Driven Water Planning Model -- Part 1: Model Characteristics. Water International, 30, 487-500.

[8] Samuels, R., Rimmer, A., Hartmann, A., Alpert, S., 2010. Climate change impacts on Jordan River flow: downscaling application from a regional climate model. Journal of Hydrometeorology 11, 860 e 879.

[9] Joyce, B., V. Mehta, D. Purkey, L. Dale, and M. Hanemann, 2011: Modifying agricultural water management to adapt to climate change in California's central valley. Climatic Change, 109, 299-316

[10] Purkey, D., A. Huber-Lee, D. Yates, M. Hanemann, and S. Herrod-Julius, 2007: Integrating a climate change assessment tool into stakeholder-driven water management decision-making processes in California. Integrated Assessment of Water Resources and Global Change, E. Craswell, M. Bonnell, D. Bossio, S. Demuth, and N. Giesen, Eds., Springer Netherlands, 315-329.

[11] I Putu Santikayasa, Mukand S. Babel, Sangam Shrestha, Damien Jourdain and Roberto S. Clemente, "Evaluation of Water Use Sustainability under Future Climate and Irrigation Management Scenarios in Citarum River Basin, Indonesia," International Journal of Sustainable Development & World Ecology, Vol. 21, No. 2, pp. 181-194, 2014. doi:10.1080/13504509.2014.884023

IOP Conf. Series: Earth and Environmental Science **31** (2016) 012043 doi:10.1088/1755-1315/31/1/012043

Interannual Variability of Rainfall over Indonesia: Impacts of ENSO and IOD and Their Predictability

R Hidayat[1], K Ando[2], Y Masumoto[2], J J Luo[2,3]

[1] Department of Geophysics and Meteorology, Bogor Agricultural University, INDONESIA.
[2] Research Institute for Global Change (RIGC), Japan Agency for Marine-Earth Science and Technology, JAPAN
[3] Center for Australian Weather and Climate Research, Bureau of Meteorology, AUSTRALIA.

E-mail: rahmath@apps.ipb.ac.id

Abstract. Northwestern Java and Makassar rainfall variations are examined in relation to the tropical air-sea coupled phenomena referred to as the Indian Ocean Dipole (IOD) and El Niño-Southern Oscillation (ENSO) using retrospective forecasts known as SINTEX-F. The model predicts the rainfall and its variability realistically in September-November (SON) and in general successfully predicts the interannual variation of northwestern Java rainfall up to 3 months ahead and Makassar rainfall up to 6 months ahead. This indicates potential societal benefits of rainfall predictability in Indonesia. Simple correlation analysis based on observations and model predictions reveals significant influence of IOD and ENSO on the northwestern Java and Makassar. However, the correlation between ENSO and the northwestern Java rainfall becomes insignificant if the IOD influence is excluded. Similarly, the correlation between IOD and Makassar rainfall becomes insignificant if excluding the ENSO influence. These results confirm that a major portion of interannual variation of the northwestern Java (Makassar) rainfall is related to IOD (ENSO) event.

1. Introduction

Regional impacts of El Niño-Southern Oscillation (ENSO) and Indian Ocean Dipole (IOD) on the climate variations in the Indo-Pacific rim countries including Indonesia have attracted much attention in past decades. The climate of Indonesia has been considered to be closely linked to ENSO in the Pacific Ocean. For example, the boreal summer and fall precipitation anomalies in Indonesian region were found to be associated with a drier condition over most part of the country during El Niño or the warm phase of ENSO [1, 2].

After the IOD was identified by Saji et al [3], a number of studies have pointed out the impact of the IOD on anomalies pattern in precipitation over the tropical and mid-latitude regions through its atmospheric teleconnection pattern. A positive IOD event is characterized by cooler-than-normal water in the tropical eastern Indian Ocean and warmer-than-normal water in the tropical western Indian Ocean. This is associated with the condition that the normal convection situated over the eastern Indian Ocean warm pool shifts to the west and brings heavy rainfall over the east Africa and Srilanka and severe droughts over Indonesia and Australia [4, 5, 6]. In addition, Sahu et al [7] pointed out that the extreme high streamflows of the Citarum River in northwestern Java are related to La Niña and negative phase IOD of the Indian Ocean. Meanwhile, the extreme low streamflows of the river are found to be associated with the positive IOD events. Moreover, recent study of Hamada et al [8] described that the IOD events clearly influence rainfall variation in the dry season in northwestern Java.

As the Java Island has a large population and is an important economic region, rainfall variability has strong influence on economic growth and daily activity of millions of people. Extreme dry events lead to devastating drought with losses in crop production over the region, while persistent heavy rains have led to flooding and landslides. Moreover, this is an Island where much of the nation's rice is grown that vulnerable to climate extremes and water shortages. [9] have pointed out that prediction of the Java rainfall is very important for the regional society in order to mitigate drought and flood disaster and agricultural decision making for planting season. In addition, water-related issues are of critical interest to be considered in sustainability and climate risk management efforts in drought and flood of the regions. Thus further study on predictability of rainfall variation resulted from a coupled model would be important as both a scientific motivation and in terms of disaster mitigation. In light of this, the purpose of this work is to investigate modulation of interannual rainfall variability over Indonesia by ENSO and the IOD during boreal fall season by analyzing retrospective forecast of SINTEX-F coupled model, which simulates ENSO and IOD events realistically [10, 11].

2. Data

Nine-member ensemble mean retrospective forecast of total precipitation and sea surface temperature (SST) during 1983-2009 from the ocean-atmosphere-land coupled general circulation model – the Scale Interaction Experiment-Frontier Research Center for Global Change (SINTEX-F) are used in the present study (see [12] for detailed description). The model shows good skill in predicting the IOD up to 2 seasons ahead and ENSO up to 2 years ahead [12, 13]. Observed rainfall anomalies during 1983 to 2007 are collected from nine stations in northwestern Java and one station in Makassar south Sulawesi (Table 1). Quality control of the observed data has been conducted by Hamada et al [8]. We define an index for northwestern Java rainfall by simply averaging the observed rainfall at nine stations. In order to examine spatial pattern of rainfall anomalies Global Precipitation Climatology Project (GPCP) [14] and the Japanese 25-year Reanalysis (JRA-25) [15] with same period are used.

SST anomalies are computed from Extended Reconstructed Sea Surface Temperature (ERSST) v.3 dataset [16] from 1979 to 2008. The dipole mode index (DMI) is adopted from Saji et al [3] which is derived from SST anomalies by taking difference between averaged regions of the western (50°E– 70°E, 10°S–10°N) and eastern (90°E–110°E, 10°S–Eq) tropical Indian Ocean. Influence of Pacific Ocean is represented by Niño3.4 index which is derived by taking area average of SST anomalies over central/eastern equatorial Pacific (5°S–5°N, 170°W–120°W).

Motivated by previous studies that the relationships between the tropical climate indices of ENSO/IOD and Indonesian rainfall are strongest in boreal fall season when there is the greatest coherency between Indonesian climate and large-scale Indo-Pacific Ocean variability [17], we focused our analysis only on the SON season that includes the end of the dry season and onset of the wet season in Indonesia.

In order to separate the influence of IOD and ENSO on northwestern Java and Makassar rainfall, a partial correlation is employed to show a partial relationship between two variables while excluding influences arising from another independent variable. In this way, the partial correlation between DMI and rainfall anomalies excludes the influence due to the correlation between Niño3.4 and the rainfall anomalies. It is written as:

$$r_{13,2} = \frac{(r_{13} - r_{12}.r_{23})}{\sqrt{(1-r_{12}^2)}\sqrt{(1-r_{23}^2)}}$$

(1)

where r_{13} refer to the correlation between DMI and rainfall anomalies, r_{12} denotes the correlation between DMI and Niño3.4 index and r_{23} is the correlation between Niño3.4 and rainfall anomalies. Statistical significance of the correlation coefficients is estimated based on a 2-tailed Student t-test.

3. Annual variations of northwestern Java and southern Sulawesi rainfall

Figure 1a shows annual cycle of the northwestern Java rainfall index which was averaged from nine observation stations. The seasonal variation of rainfall over the region show a distinct annual cycle which the maximum in the months from December to February and the minimum from June to September. The rainfall is at the lowest level in July-August (about 3 mm/day) and peaking up in January - February (about 12 mm/day). A similar annual rainfall pattern is seen over Makassar southern Sulawesi (Figure 1c). In these regions, the peak rainy season corresponds to the northwest monsoon period in boreal winter across the Austral-Indonesian and dry season coincides with the southeast monsoon period in boreal summer. It is noted that the standard deviations of rainfall are relatively high during June to October, revealing most variability is along boreal summer-fall seasons.

Overall features of the observed annual cycle of rainfall (Figures 1a and 1c) are captured by 1-month lead model prediction (Figures 1b and 1d): it shows a wet season that peaks in January and a dry season that peaks in August. This suggests that the pattern of annual variation is predicted well by the SINTEX-F model despite that rainfall amplitude during rainy season (December to April) is weaker in the model.

4. Interannual variations of northwestern Java and southern Sulawesi rainfall

Figures 2a and 2b show interannual variations of the SON model rainfall anomalies of northwestern Java and Makassar superimposed with the associated observed rainfall, DMI and Niño3.4 index, respectively. We find that DMI and Niño3.4 are significantly correlated with temporal rainfall variations for both observed and 9-member ensemble mean prediction at 1-month lead time (predictions initiated from 1 August, 1 September and 1 October for September, October, and November respectively). Years of negative rainfall anomalies are seen to be linked to El Niño and positive IOD years when the eastern pole of IOD is colder than normal. In contrast, years of positive rainfall anomalies are found during La Nina and negative IOD events. Several exceptions are noted for the observed rainfall (Figure 2a) in which positive rainfall anomaly in northwestern Java is seen in 1986 (El Niño year) and negative rainfall anomalies over the two regions in 1989 and 1990 (negative IOD year). Despite some discrepancies in magnitude for certain years, the rainfall variations over northwestern Java and Makassar can be correctly predicted with correlation skill of 0.72 and 0.80, respectively. Besides, the 9-member mean prediction would reduce noise signals and hence better represent the cause-effect relationship between the rainfall and IOD/ENSO forcing.

The significant relationship between ENSO and rainfall variation over Indonesia during SON seasons is consistent with the previous studies based on rain-gauge rainfall data [1]. In El Niño years, the rising branch of the Walker circulation that corresponds to the maximum rainy region shifts toward the east, and the anomalous descending region is located slightly east of Indonesia. Thus, the low-level easterly anomalies over Indonesia reinforce the mean monsoonal winds (i.e. southeasterly) during SON. As a result, this favors to reduce local rainfall. The relation among ENSO and IOD and Indonesian rainfall anomaly is expected since nearly one third of positive IOD events have co-occurred with El Niño events over the past 2-3 decades [18].

In order to investigate the unique relation of rainfall anomalies with IOD from that with ENSO, we employ a partial correlation technique. The strong relationship between IOD/ENSO and Indonesian rainfall anomalies is supported by significant correlation between DMI, Niño3.4 and temporal variation of northwestern Java and Makassar rainfall (Table 2). For comparison, we also examine the correlation for the model predictions in both locations. The model northwestern Java index is negatively correlated with the DMI with a significant correlation of -0.87 (exceeding the 0.01 level of significance), higher than the observed correlation of -0.58. This is not surprising since the model ensemble mean has reduced the noises, which make the correlation coefficients in observed values relatively low. These correlation coefficients remain significant even when ENSO influence is excluded (-0.80 in model and -0.41 in observations). The simultaneous correlation between model northwestern Java rainfall and Niño3.4 is also significant with the value of -0.56, which is slightly higher than -0.47 in the observation. However, the correlation coefficients become insignificant when the IOD influence is excluded, i.e. -0.10 (model) and -0.15 (observed). We also noted that the Niño3.4 index has strong correlation with both model predicted and observed Makassar rainfall anomaly (i.e. -0.76 and -0.74). These correlations remain significant even if we exclude the IOD influence. In contrast, IOD has an insignificant correlation after removing the ENSO influence.

The significant correlation between IOD and rainfall over Indonesia was partly described by Ashok et al 2013 [19]. They pointed out the SST dipole is significantly correlated with the rainfall variability over the eastern and western part of Indian Ocean. As a consequence the East African (Indonesia) region receives above- (below-) normal precipitation during a positive IOD event. In addition, Saji and Yamagata (2003) described that the baroclinic structure of the zonal wind anomaly is suggestive of a zonal overturning cell in the equatorial Indian Ocean. There are two localized anomalous Walker cells: one is over the equatorial Indian Ocean associated with IOD, the other is from 120^0E to the entire equatorial Pacific Ocean in association with ENSO. In association with these anomalous cells, zonal contrast of correlation pattern is reflected in both observation and model prediction (Figure 3).

Figures 3e and 3f show partial correlation between model rainfall anomalies with model DMI and Niño3.4 index, respectively. It is seen that the correlations with IOD in the eastern Indian Ocean are significantly negative, but they become insignificant in the central to eastern part of Indonesian region (Figure 3e). Interestingly, the partial correlation with ENSO does not show a significant dipole correlation pattern in the Indian Ocean. Rather the negative correlation coefficients confined to the central/eastern Indonesia and are accompanied by positive correlation over the tropical Pacific (Figure 3f). These spatial patterns of the correlation coefficient are in general similar to those obtained from GPCP observation and JRA-25 reanalysis data (Figures 3a - 3d), though the negative partial correlations with IOD appear to be confined in eastern Indian Ocean in the GPCP and JRA-25 results (Figures 3a and 3c). The partial correlation with ENSO is centered in central/eastern Indonesia in all the results (Figures 3b, 3d, and 3f). These results suggest that the IOD is dominantly responsible for rainfall variation in eastern Indian Ocean/western Indonesia while most of rainfall variation in central/eastern Indonesia is associated with ENSO events.

5. Rainfall predictability

The interannual variability of rainfall predictions at 1-month lead time for the period 1983-2009 (Figure 2) have shown that the model produces useful skill of rainfall predictions in northwestern Java and Makassar. The predicted rainfall variation over these regions, however, gradually decreases as the lead time increases. Table 3 shows the correlation skill of rainfall predictions gradually decrease to about 0.52 (0.60) at the 3-month (6-month) lead times in northwestern Java (Makassar). The model predictions show significant predictabilities in Makassar region, where the rainfall variation are largely associated with ENSO (Table 3). This is because signals associated with ENSO events are able to be predicted at long lead times [13]. In northwestern Java, on the other hand, where the rainfall variations are mostly affected by IOD, the rainfall predictability is seen to be shorter (i.e. up to 3-month lead time). This might be partly associated with lower predictability of IOD (1-2 seasons ahead) in presence of strong intraseasonal disturbances in the Indian Ocean [12].

6. Concluding remarks

We have documented significant impact of the IOD and ENSO event on Indonesian rainfall variability and their predictability by analyzing retrospective forecast of SINTEX-F coupled model. The model predicts northwestern Java and the Makassar rainfall and its variability realistically with a wet season that peaks in January and a dry season that peaks in August. It is noted that the model predicted rainfall amplitude during December to April is weaker than the observed due partly to model ensemble mean. We have examined relationship among northwestern Java and Makassar rainfalls,

IOD and ENSO. Partial correlation coefficients with IOD are found significant in the western Indonesian region including the northwestern Java, while significant partial correlation coefficients with ENSO are found over the central/eastern part of Indonesian region. It is suggested that the interannual rainfall variation over northwestern Java is essentially related to IOD, in consistent with the IOD-induced changes in atmospheric and oceanic conditions [10]. It is encouraging that rainfall variation in northwestern Java and Makassar can be predicted up to about 3 to 6 months ahead, respectively, with significantly correlation skill (Table 3), according to 9-member mean retrospective forecast experiments for the period 1983–2009.

The skilful prediction of model rainfall during boreal fall season would be beneficial for assessing the potential predictability of Indonesian rainy season onset which is occurred around October-November. This has particular implication for rice production areas over the highly populated island of Java and southern Sulawesi. In this context the onset of the rainy season is particularly critical for the agriculture sector in Indonesia.

Acknowledgments

This study is supported by the Climate Variability Study and Societal Application through Indonesia-Japan "Maritime Continent COE" – Radar-Bouy Network Optimization for Rainfall Prediction Project of Science and Technology.

7. References

[1] Hendon H H 2003 *J. Clim* **16** 1775
[2] Aldrian E and Susanto D 2003 *Int. J. Clim.* **23** 1435
[3] Saji N H, B N Goswami, P N Vinayachandran, and T Yamagata 1999 *Nature* **401** 360
[4] Saji N H and T Yamagata 2003 *Climate Res* **25** 151
[5] Behera S K, Luo J.J, Masson S, Delecluse P, Gualdi S, Navarra A and Yamagata T 2005 *J. Clim* **18** 4514
[6] Ummenhofer C C et al. 2009 *Geoph. Res. Lett.* **36**
[7] Sahu N, S K Behera, Y Yamashiki, K Takara, and T Yamagata, 2011 *Clim. Dyn.* **33**
[8] Hamada, J I et al. 2012 *SOLA* **8** 69
[9] D'Arrigo R and R. Wilson 2008 *Int. J. Clim.* **28** 611
[10] Yamagata T, S K Behera, J J Luo, S Masson, M Jury and S A Rao 2004 *Geophys. Monogr. Ser.* **147** 189
[11] Luo J J, S Masson, E Roeckner, G Madec and T. Yamagata 2005 *J. Climate* **18** 2344
[12] Luo J J, S Masson, S Behera and T Yamagata 2007 *J. Climate* **20** 2178
[13] Luo J J, S Masson, S Behera and T Yamagata 2008 *J. Climate* **21** 84
[14] Adler R F et al. 2008 *J. Hydrometeor.* **4** 1147
[15] Onogi K et al 2007 *J. Meteor. Soc. Japan* **85** 369
[16] Smith T M, R W Reynolds, T C Peterson and J Lawrimore, 2008 *J. Clim.* **21** 2283
[17] Haylock M and J McBride 2001 *J. Clim.* **14** 3882
[18] Rao S A, S K Behera, Y Masumoto and T Yamagata 2002 *Deep-Sea Res.* **49B**, 1549
[19] Ashok K, Guan Z and Yamagata T 2003 *Geoph. Res. Lett.* **30** 1821

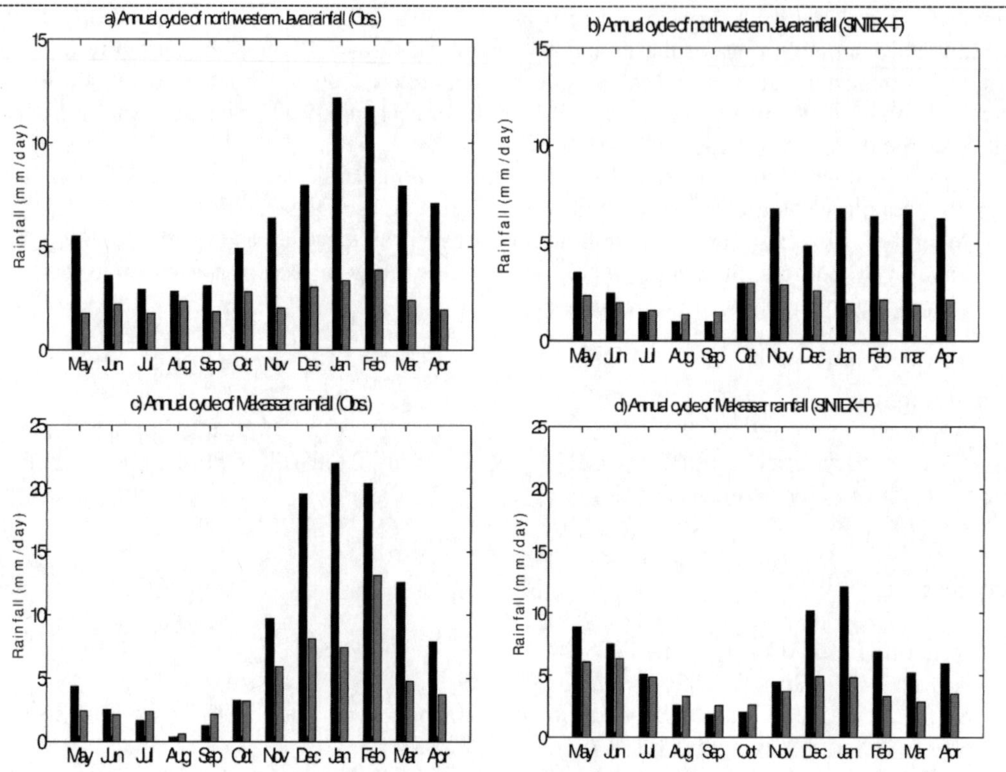

Figure 1. Annual mean (black) and standard deviation (grey) of northwestern Java rainfall derived from (a) observed rain gauges and (b) 1-month lead forecast of SINTEX-F. (c) and (d) As in (a) and (b), but for Makassar/southern Sulawesi. Note that the annual cycle could be weakened in the model due to the 9-member ensemble mean.

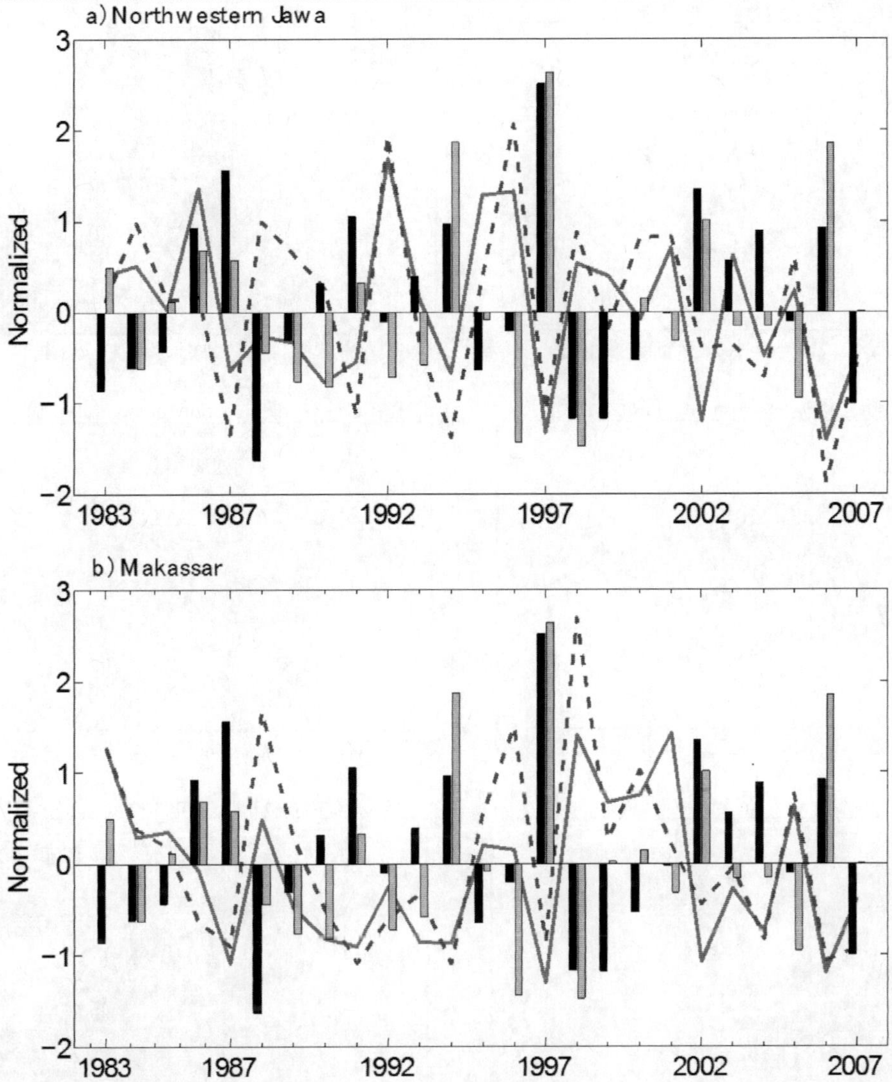

Figure 2. (a) Normalized time series of SON model (observed) northwestern Java rainfall is depicted by dashed blue line (solid red line). Black (grey) bar is observed Niño3.4 (DMI). (b) As in Figure 2a, but for Makassar. The model results are based on 9-member mean prediction at 1-month lead.

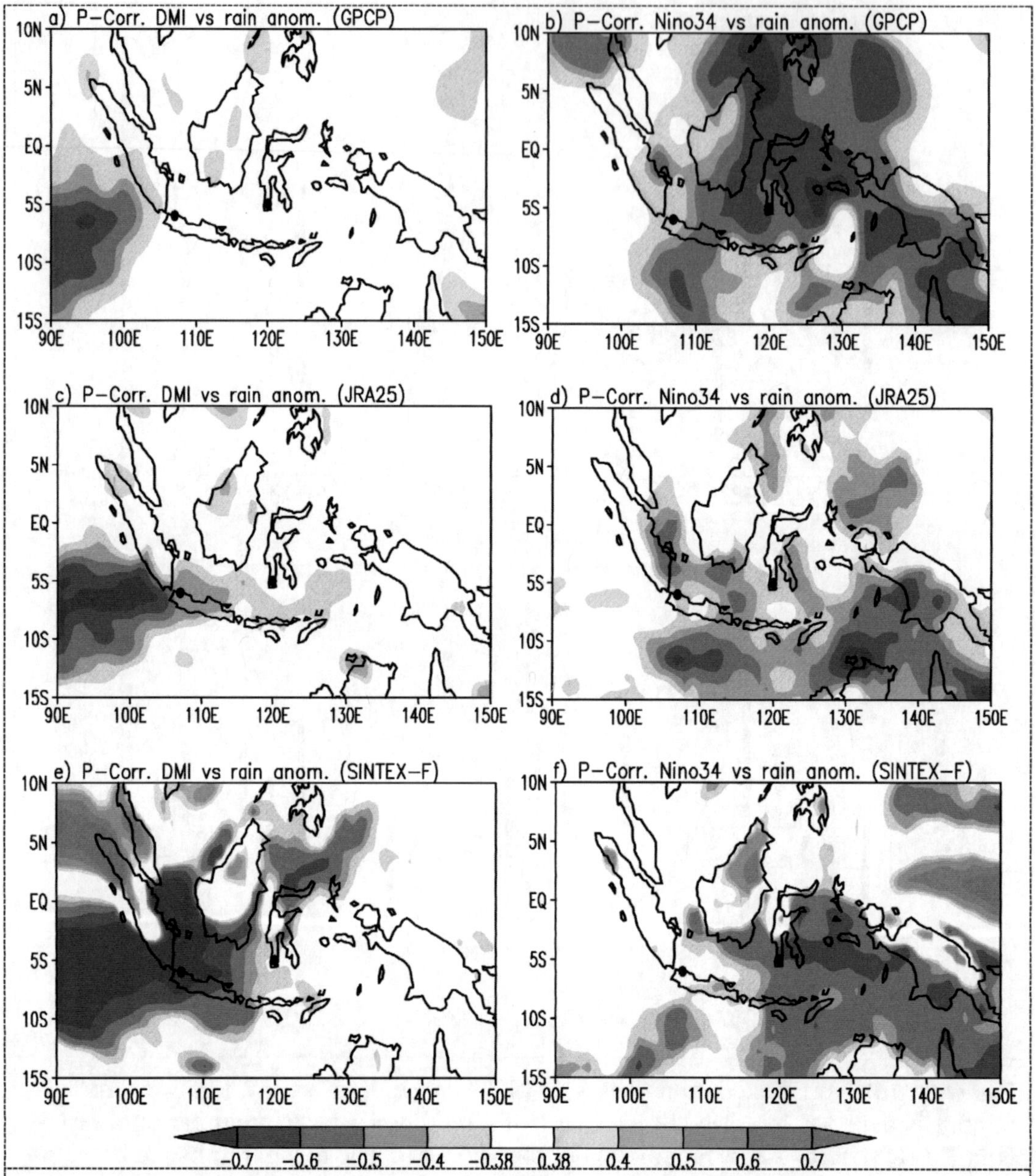

Figure 3. (a) Partial correlation between GPCP rainfall anomalies and observed DMI after removing the effects of ENSO. (b) As in (a), but for the correlation between GPCP rainfall anomalies and Niño3.4 index. (c-f) As in (a-b), but for the results based on JRA25 and 1-month lead prediction of SINTEX-F, respectively. Color shading indicates a significant correlation exceeding 95% confidence level. Black circles indicate northwestern Java and Makassar stations.

Table 1. List of rainfall observation stations observed by Indonesian Agency for Meteorology, Climatology, and Geophysics of Indonesia (BMKG).

WMO No.	Station Name	Lat.	Lon.	Alt.
96733	Pondok Betung	6.25^0S	106.61^0E	-
96737	Serang	6.21^0S	106.10^0E	40 m
96739	Curug	6.23^0S	106.65^0E	46 m
96741	Tanjung Priok	6.10^0S	106.87^0E	2 m
96745	Jakarta Headquarter	6.17^0S	106.82^0E	7 m
96747	Halim	6.27^0S	106.88^0E	26 m
96749	Cengkareng	6.11^0S	106.65^0E	-
96751	Citeko	6.70^0S	106.93^0E	-
96753	Bogor	6.50^0S	106.75^0E	250 m
97180	Makassar	5.04^0S	119.33^0E	14 m

Table 2. The seasonal correlations between SON rainfall anomalies at northwestern Java and Makassar and the DMI and Niño3.4 index based on observations and SINTEX-F model predictions at 1-month lead. pN (pD) is the partial correlation of rainfall with Niño3.4 (DMI) after removing the effects of DMI (Niño3.4). Value exceeding 95% (99%) confidence level is shown with * (**).

	Northwestern Java				Makassar			
	Niño3.4	DMI	pN	pD	Niño3.4	DMI	pN	pD
Obs.	-0.47*	-0.58**	-0.15	-0.41*	-0.76**	-0.51**	-0.65**	-0.04
Model	-0.56**	-0.87**	-0.10	-0.80**	-0.74**	-0.64**	-0.57**	-0.34

Table 3. The seasonal correlations between SON observed rainfalls and 1-month to 6-month lead of SINTEX-F predictions. Value exceeding 99% confidence level is shown with **.

	Northwestern Java			Makassar/Southern Sulawesi					
	1-mo	2-mo	3-mo	1-mo	2-mo	3-mo	4-mo	5-mo	6-mo
Obs	0.72**	0.58**	0.52**	0.80**	0.77**	0.70**	0.63**	0.60**	0.60**

IOP Conf. Series: Earth and Environmental Science **31** (2016) 012044 doi:10.1088/1755-1315/31/1/012044

Utilization of Rice Husk as Pb Adsorbent in Blood Cockles

Eti Rohaeti*[1], Wenny Permata Sari[1] and Irmanida Batubara[1]

[1] Department of Chemistry, Bogor Agriculture University, INDONESIA.

*E-mail: er_a444@hotmail.com

Abstract. Water pollution by lead affects blood cockles, a potential source of food. The aim of this research is to compare rice husk (RH) and rice husk carbon (RHC) in reducing the concentration of lead in blood cockles. RH and RHC were activated with NaOH 1 M, and then the optimal conditions and maximum capacity were determined. This research showed that RH and RHC had maximum adsorbancy capacities of 28.7326 mg/g and 51.5464 mg/g at optimal condition. The optimal adsorption condition for RH in 100 ml Pb solution is 0.32 gram, pH 5, for 4 hours. The optimal adsorption condition for RHC in 100 ml Pb solution is 0.20 gram, pH 5, for 2 hours. Lead content in blood cockles from the north waters of Jakarta (1.9658 mg/kg) is beyond the threshold limit. Lead adsorption by RH and RHC could reduce lead content in blood cockles by about 40% and 31%, respectively.

1. Introduction

The advancement of technology and industries play a role in environmental pollution, especially water pollution. The contaminants in the water may include heavy metals like arsenic (As), lead (Pb), mercury (Hg), and cadmium (Cd). Pb is widely used in metal production, such as in batteries, dyes, and ammunition industries [1], which may greatly contribute to the water pollution. Water pollution by lead affects potential foodstuff that live in water, such as blood cockles. Many studies demonstrate that the level of Pb in blood cockles is higher than the threshold limit allowed by WHO. The level of Pb reaches 1.750 mg/Kg, 2.33 mg/kg, 1.956 mg/kg. 1.37 mg/kg [2-5]. Therefore, efforts are needed to reduce the level of Pb in blood cockles.

Adsorption is a well-known method to remove heavy metals from aqueous solution. This method is simple and relatively cost-effective, thus it is widely used. Rice husk and rice husk carbon are potential adsorbent sources from agricultural waste. Rice husk is a waste (about 20%) from rice production process which availability is quite abundance. In 2013, Statistics Indonesia (BPS) reported that rice production in Indonesia was about 71 million ton. Rice husk capability to adsorb metallic ions is on the active groups of CO–, OH–, Si – H, Si– O– Si, Si–OH [6]. Effecient adsorption of rice husk toward Pb has been reported, among others, by Surchi (2011), in which a decrease of 85% in Pb concentration in the test solution was achieved [7]. Hengpeng, *et al.* (2010) reported that rice husk's adsorption capacity of 70.8 mg/g was obtained after soaking in phosphoric acid [6]. Elham, *et al.* (2010) was also reported that rice husk was able to reduce Pb level in dairy industries wastewater up to 98% [8]. Rice husk has also been reported as an alternative source of raw material for the preparation of actived carbon widely used as adsorbent.

The aim of this research is to compare rice husk and rice husk carbon in reducing lead concentration in blood cockles. The rice husk and rice husk carbon were activated by NaOH 1 M, then their adsorption capacities toward lead solution and lead in blood cockles were determined.

2. EXPERIMENT

Adsorbent preparation:
Rice husk-based adsorbent was prepared with two treatments in order to obtained rice husk powder and rice husk carbon, and then both were activated by using NaOH. The rice husk was prepared with the first treatment: rice husk was washed and dried at 60°C for 24-48 hours. The dried rice husk was grounded to a size of 100 mesh. The rice husk carbon was prepared with the second treatment: rice husk that had already been washed was burned in a special-designed furnace. The rice husk was placed around a cylinder with a hole with fire in it. The rice husk near the cylinder will burned first and then the flame will spread to the farthest rice husk. The rice husk and the rice husk carbon were then activated by soaking in NaOH 1 M with the proportion of the adsorbent to NaOH 3:10. Soaking was continued for 24 hours. Subsequently, the adsorbent was washed with aquadest until NaOH-free adsorbent was obtained, and then dried at 100ºC.

Adsorbent Characterization
Activated rice husk and rice husk carbon adsorbents were characterized by determining the moisture content and ash content, and analyzing the SEM and FTIR spectra. Furthermore, the adsorbent properties toward Pb^{2+} ion were tested.

The moisture content and ash content were determined according to gravimetric method [9]. The surface structure of rice husk carbon adsorbent before and after activation was analyzed using SEM. Functional groups observation was conducted by using IR spectrophotometer to find out the functional groups in the rice husk and rice husk carbon with Pb adsorption capacity.

The adsorbent's adsorption properties toward Pb^{2+} ion were determined by measuring the adsorption capacity of adsorbent using batch method. First, it was determined the optimum conditions for adsorption by measuring the capacity in various pH, adsorbent weight, and contact time. Subsequently, the maximum adsorption capacity and the isotherm pattern of adsorption were determined by measuring the capacity in various adsorbate concentrations under optimum conditions. Pb^{2+} ion level before and after the adsorption were measured by AAS.

Determination of adsorption optimum condition
Rice husk and rice husk carbon 0.20, 0.35, and 0.50 gram were placed into a 100 ml Pb solution 100 ppm that was previously added with HNO_3 to obtained solutions with pH of 2.0, 3.5, and 5.0. The solutions were stirred for 2, 3, and 4 hours and then filtered. The adsorption capacity of each solution was determined according to the following formula:

$$q=\frac{v(C_0-C)}{m}$$

wherein:
q = adsorption capacity
v = solution volume
C_0 = initial adsorbate concentration
C = final adsorbate concentration
m = adsorbent mass

Optimum condition was determined through optimization by using 3 factors box-Behnken experimental design.

Table 1. Pb adsorption capacity optimization variables

Variable		Level (code)		
		1	0	-1
Adsorbent weight	A	0.2	0.35	0.5
pH	B	2	3.5	5
Contact time	C	2	3	4

Determination of Maximum Adsorption Capacity

Maximum adsorption capacity was determined according to the following: 100 ml of $Pb(NO_3)_2$ solution in concentration of 12.75, 35.50, 75.00, 150.00 and 300.00 ppm were each placed into polyethylene bottles. To these solutions, adsorbent (rice husk or rice husk carbon) was added under optimum conditions that had been determined in previous step. Subsequently, the solutions were filtered to separate the adsorbent. Pb level in the solution was analyzed with AAS and the adsorption capacity was calculated. Maximum adsorption capacity, q_{max}, was obtained using adsorption isotherm.

Determination of adsorbent's adsorption capacity toward Pb in blood cockles

Blood cockles sample was obtained from a traditional market, originated from Jakarta Bay. The blood cockles were then cooked (boiled) for 30 minutes and the meat was separated from the shell. To 30 gram of meat, it was added 1,5 gram adsorbent (rice husk or rice husk carbon) and 50 ml water. Adsorption was performed for 30 minutes. Subsequently, the adsorbent was separated. Pb level in the blood cockles was determined before and after cooking, and after the adsorpton. Pb in the blood cockles's meat was determined after destruction in a mixture of 30 ml HNO_3 and 5 ml of concentrated H_2SO_4.

3. RESULTS AND DISCUSSION

Adsorbent Characterization

Fourier Transformation Infra Red Spectra

FTIR spectra was used to find out the functional groups in the samples. Results of analysis (Figure 1) showed that rice husk contained alcohol groups that are seen in a wide stretch in wave number of 3406 cm^{-1}, which is the absorption peak of –OH bond, intensified with the absorption peak in 1080 cm^{-1}, which is the adsorption peak of–C–O bond [10]. Absorption peak in 1648 cm^{-1} showed conjugatable –C=O groups. According to Hengpeng, *et al.* (2010), –C=O and –OH groups are effective in adsorbing Pb heavy metal. There is a decrease in absorption intensity around the wave number of 2940 cm^{-1} in activated rice husk carbon, which is a bonding group of -OH and a methyl radical [6]. This decrease may happen as a result of decomposition of cellulose, hemisellulose, and lignin due to heating to 500°C in the process of carbonization [11].

IOP Conf. Series: Earth and Environmental Science **31** (2016) 012044 doi:10.1088/1755-1315/31/1/012044

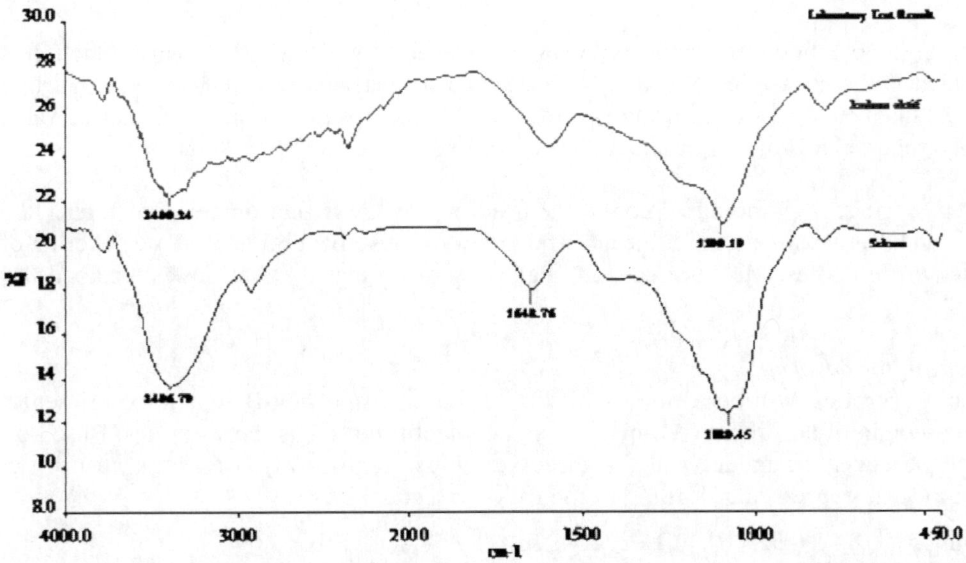

Figure 1. IR spectrum of rice husk and rice husk carbon

Surface morphology of rice husk carbon

Rice husk and rice husk carbon were activated by soaking in NaOH 1M to remove silica. Silica is a mineral found in a large amount in rice husk, which level reaches 20% [12]. Treatment with NaOH will dissolved the silica covering the surface of the pores. The structure and morphology of rice husk carbon before and after NaOH activation are shown in Fig 2. It can be seen that after activation, the pore is clear. The pores formed in various sizes, the pore size shown in the picture is 12.06 x 10.72 μm.

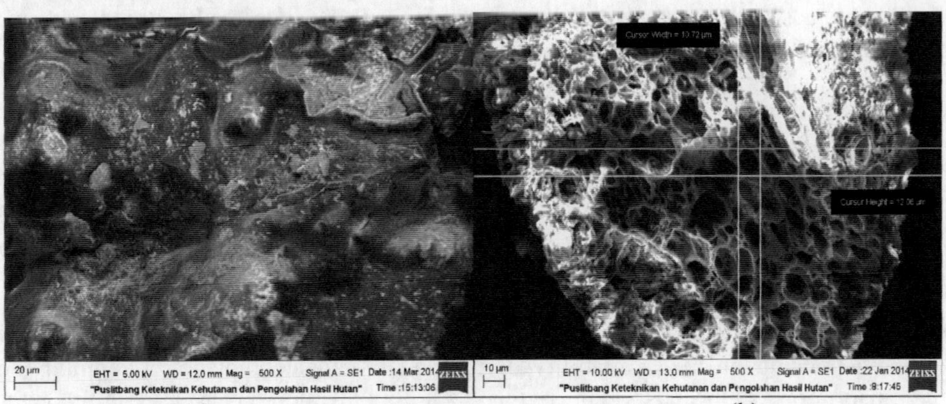

(a) (b)

Figure 2. Morphology of rice husk carbon by SEM before (a) and after (b) activation with NaOH

Moisture content and ash content

The ash content indicates the amount of minerals contained in the sample. The ash content of the husk in this study is 22%. This result is slightly higher than that reported [6, 11, 13]. The difference may be due to the place of origin of the husk and the condition to grow the rice, such as soil condition, irrigation, and fertilizers used.

The ash content in the rice husk pre-treated with NaOH reduced as much as 20%. This reduction is due to SiO, which is the most mineral found in rice husk ash, reacts with NaOH to form natrium silicate (Na_2SiO_3) that could be washed away with water [12]. This result is in line with that of Taha, *et al.* (2011) who reported that there was about 20% reduction of silica weight in the husk.

Moisture content correlates to the adsorptivity of an adsorbent toward adsorbate. The higher the water content in an adsorbent, the smaller the adsorptivity of the adsorbent because of the covering of water molecules on the pores. Moisture content analysis showed that the rice husk contained 8% water.

Optimum condition for adsorption

Optimization of adsorption was performed with Pb^{2+} solution using box-Behnken experimental design by using weight of adsorbent (A), pH (B), and contact time (C) as the variables. Based on the analysis of variance, the model and the three variables significantly affect the adsorption capacity. The model is expressed according to the following equation

$$Q_{\text{activated rice husk}}(\text{mg/g}) = 18.41-0.52A+0.40B+1.74C-1.55AC-2.58A^2+1.17B^2+1.69C^2$$

$$Q_{\text{activated rice husk charcoal}}(\text{mg/g}) = 25.04-8.44A+6.04B+0.20C-7.23AB+2.80A^2-2.94B^2$$

Based on the equation, it can be determined the optimum condition for adsorption, the condition resulting in a maximum adsorption capacity value (Q). The optimum condition for rice husk is 0.32 gram, pH 5, and contact time of 4 hours providing an adsorption capacity prediction of 23.4713 mg/g. This value is higher than that reported by Zulkali *et al.* (2011), which was only 8.6 mg/g. This difference is presumed as the result of increase in the surface area due to silica dissolution in the activation process [14]. Kumar and Bandyopadhyay (2006) reported that rice husk pre-treatment by soaking in NaOH would increase the adsorption capacity from 8 mg/g, initially, to 20 mg/g [15].

The optimum condition for Pb ion adsorption of rice husk carbon is 0.20 gram, pH 5, and contact time of 2 hours. The adsorption capacity under the condition is predicted to be 46.5004 mg/g. Based on the result, Pb adsorption capacity of rice husk carbon is higher than the rice husk.

Adsorption isotherm

The most common way to see the effect of adsorbate concentration to the amount adsorbed is through adsorption isotherm, which is a curve correlating the amount adsorbed to the adsorbate concentration in equilibrium at a specific temperature. The process of adsorption is explained by two types of isotherm, Langmuir and Freundlich.

Langmuir isotherm assumes that every site of adsorption is equivalent and the capacity to adsorp particles is not depending on the availability of particles in adjacent places. Langmuir isotherm assumes that adsorption on the adsorbent surface will form a monolayer. Langmuir isotherm is expressed according to the following formula:

$$\frac{C_e}{q_e} = \frac{1}{Kq_{max}} + \frac{C_e}{q_{max}}$$

Ce is adsorbate concentration in a solution in equilibrium (weight/volume), qe is adsorption capacity expressed in adsorbate weight unit per adsorbent weight, K is Langmuir adsorption constant, q_{mak} is monolayer adsorption capacity.

IOP Conf. Series: Earth and Environmental Science **31** (2016) 012044 doi:10.1088/1755-1315/31/1/012044

Freundlich isotherm proposes heterogeneous energetic distribution of active sites, combined with interaction among the adsorbent molecules. The Freundlich isotherm is expressed by the following equation:

$$\log q_e = \log K + \frac{1}{n}\log C_e$$

Wherein, K is a constant related to the adsorption capacity.

The result of this research showed that Langmuir isotherm is more suitable with the adsorption isotherm patterns of both rice husk and rice husk carbon for the R^2 value in the Langmuir isotherm is higher than the Freundlich isotherm, 0.9586 and 0.9450, respectively. The slope represents the value of 1/max. capacity and the intercept represents the value of 1/K. K is the value of sorption energy constant (Table 2).

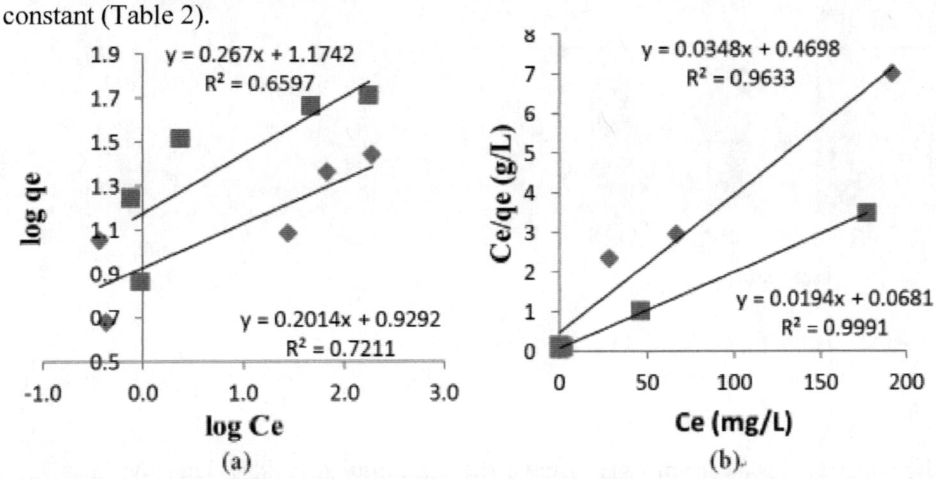

Figure 3. Freundlich Isotherm (a) and Langmuir Isotherm (b) of activated rice husk ■ and rice husk carbon ■

Table 2. Langmuir isotherm for rice husk and rice husk carbon

Sample	Qmax	K	R^2
RH	28.7356	0.0741	0.9633
RHC	51.5464	0.2849	0.9991

Adsorption of Pb^{2+} *in Blood Cockles*

Blood cockles were obtained from a fish-seller in Bekasi, originated from Cilincing, Jakarta, with size of between 1.5-3.5 cm. Amriani *et al.* (2011) reported that the size of blood cockles effecting the Pb accumulated in them [2]. The size of the blood cockles relates to the age of the cockles. The longer the blood cockles live in a Pb polluted water, the more Pb accumulated in them.

The result of this research demonstrated that the uncooked meat of the blood cockles (TR) had been polluted with lead. The lead level was 1.9628 mg/kg, while the threshold limit for Pb in shellfish according to the Head of National Agency of Food and Drug Control Regulation No. HK.00.06.1.52.401, is 1.5 mg/kg.

The lead level in the blood cockles's meat was slightly decrease after cooking (DR), 1.79 mg/kg (Figure 4). However, cooking did not decrease Pb level adequately to be safely consumed [16].

Rice husk and rice husk carbon adsorptions are able to reduce Pb level by 40.15% and 31.50%, respectively, that the final level are under the lower limit of safe consumption. Based on the data, adsorption of Pb in blood cockles by rice husk (DRS) is higher than rice husk carbon (DRA). This may be due to more hydroxyl and carbonyl groups are found in rice husk than in rice husk carbon. Those groups are able to function as ligands and form complexes with Pb. Process of adsorption can also occur in 3 steps: Pb is adsorbed in the surface of the adsorbent, Pb is then moved to the carbon pore, and finally adsorbed in the inner wall of the pore [4].

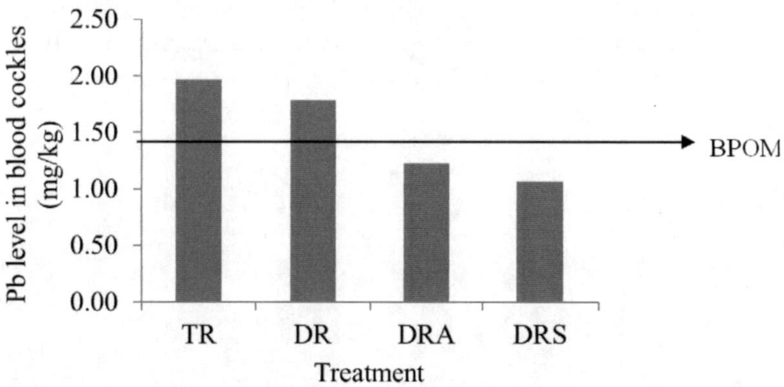

Figure 4. Effect of various treatments to lead level in blood cockles's meat

4. Conclusion:

Both rice husk (RH) and rice husk carbon (RHC) have the capability to adsorp lead ions from a solution. Maximum adsorption capacities of RH and RHC are 28.7326 mg/g and 51.5464 mg/g respectively at optimal condition. The optimal adsorption condition for RH in 100 ml Pb solution is 0.32 gram, pH 5, for 4 hours. The optimal adsorption condition for RHC in 100 ml Pb solution is 0.20 gram, pH
5, for 2 hours. Lead content in blood cockles from the north waters of Jakarta (1.9658 mg/kg) is beyond the threshold limit. Lead adsorption by RH and RHC could reduce lead content in blood cockles by about 40% and 31%, respectively.

References

[1] Manahan S E. 2003. *Toxicological Chemistry and Biochemistry* 3rd Ed. New York (US): CRC Pres.

[2] Amriani, Hendrarto B,Hadiyarto A. 2011. Bioakumulasi logam berat timbal(Pb) dan seng (Zn) pada kerang darah *Anadara granosa* L dan kerang bakau *polymesoda bengalensis l* di perairan teluk kendari. *J Ilmu Lingkungan*. 9(2): 45-50

[3] Suhaimi, A.N, 2013. Heavy Metals and Proximate Composition of Blood Cockle (*Anadara granosa*) from Different Locations of Peninsular Malaysia. Universiti Teknologi MARA.

[4] Rachmawati, R, M'aruf W.F, Anggo A.D. 2013. Pengaruh lama perebusan kerang darah (anadara granosa) dengan arang aktif terhadap pengurangan kadar logam kadmium dan kadar logam timbal. *Jurnal Pengolahan dan Bioteknologi Hasil Perikanan*.2(3):41-50.

[5] Zahir, M.S.M., B.Y. Kamaruzzaman, S.M.Al-Barwani, B.A John, KC.A Jalal, S.Shahbudin, J.C. Goddard. 2011. Bioacumulation of Selected Metals in the Blood Cockle (anadara granosa) from Langkawi Island Malaysia. Oriental Journal of Chemistry. 27-3

[6] Hengpeng Y, Qing Z, Dongyun D. 2010. Adsorptive removal of Cd(II) from aqueous solution using natural and modified rice husk. *Bioresource Technology.* 101:5175–5179. doi:10.1016/j.biortech.2010.02.027.
[7] Surchi, K. M. S. 2011. Agricultural Wastes as Low Cost Adsorbents for Pb Removal:Kinetics, Equilibrium and Thermodynamics. International Journal of Chemistry, 3(3), 103-112. http://dx.doi.org/10.5539/ijc.v3n3p103
[8] Elham A, Hossein T, Mahnoosh H, 2010. Removal of Zn(II) and Pb(II) Ions using Rice Husk in Food Industrial Wastewater. J. Appl. Sci. Environ. Manage. 14(4):159-162.
[9] [AOAC] Association of Official Analytical Chemist. 2005. Official Method of Analysis of The Association of Official Analytical of Chemist. Arlington: The Association of Official Analytical Chemist, Inc
[10] Pavia D L, Lampman G M, Kriz G S. 2001. *Introduction to Spectroscopy* 3rd Ed. Washington(US): Thompson Learning Inc.
[11] Tarley C R T, Arruda M A Z. 2004. Biosorption of heavy metals using rice milling by-products. Charactisation and application for removal of metals from aqueous effluent. Chemosphere. 54:987-995.doi:10.1016/j.chemosphere.2003.09.001
[12] Taha M , Kiat C F, Shaharun M S, Ramli A. 2011. Removal of Ni(II), Zn(II) and Pb(II) ions from single metal aqueous Solution using activated carbon prepared from rice husk. World Academy of Science, Engineering and Technology. 60:291-296.
[13] Krishnarao R V. Subrahmanyam J. Jagadish Kumar T. 2001. Studies on the formation of black particles in rice husk silica ash. *Journal of the European Ceramic Society.* 21:99-104
[14] Zulkali M D D, Ahmad A L, Norulakmal N H. 2006. *Oryza sativa L.* husk as heavy metal adsorbent: Optimization with lead as model solution. Bioresource Technology. 97:21–25.doi:10.1016/j.biortech.2005.02.007.
[15] Kumar U, Bandyopadhyay. 2006. Sorption of cadmium from aqueos solution using pretreated rice husk. Biosource Technology. 97:104-109.doi:10.1016/j.biortech.2005.02.027
[16] [BPOM] Badan Pengawas Obat dan Makanan. 2009. Peraturan BPOM RI Nomor HK.00.06.1.52.4011 tentang penetapan batas maksimum cemaran mikroba dan kimia dalam makanan. Jakarta: Badan Pengawas Obat dan Makanan

Institute of Physics
Dirac House, Temple Back
Bristol BS1 6BE UK

ISSN: 1755-1307
ISBN 978-1-5108-2226-9

CURRAN ASSOCIATES INC.
proceedings.com